MOLECULAR CAPTURE

I0869244

posthumanities

CARY WOLFE Series Editor

63 *Molecular Capture: The Animation of Biology*
Adam Nocek

62 *Clang*
Jacques Derrida

61 *Radioactive Ghosts*
Gabriele Schwab

60 *Gaian Systems: Lynn Margulis, Neocybernetics, and the End of the Anthropocene*
Bruce Clarke

59 *The Probiotic Planet: Using Life to Manage Life*
Jamie Lorimer

58 *Individuation in Light of Notions of Form and Information.* Volume II: *Supplemental Texts*
Gilbert Simondon

57 *Individuation in Light of Notions of Form and Information*
Gilbert Simondon

56 *Thinking Plant Animal Human: Encounters with Communities of Difference*
David Wood

55 *The Elements of Foucault*
Gregg Lambert

54 *Postcinematic Vision: The Coevolution of Moving-Image Media and the Spectator*
Roger F. Cook

53 *Bleak Joys: Aesthetics of Ecology and Impossibility*
Matthew Fuller and Olga Goriunova

52 *Variations on Media Thinking*
Siegfried Zielinski

51 *Aesthesis and Perceptronium: On the Entanglement of Sensation, Cognition, and Matter*
Alexander Wilson

(continued on page 425)

MOLECULAR CAPTURE

The Animation of Biology

ADAM NOCEK

posthumanities 63

University of Minnesota Press | Minneapolis | London

A version of the Postscript was originally published as "Designing Propositions," in *Propositions in the Making: Experiments in a Whiteheadian Laboratory*, ed. Roland Faber, Michael Halewood, and Andrew M. Davis (Rowman and Littlefield, 2019). All rights reserved.

Copyright 2021 by the Regents of the University of Minnesota

All rights reserved. No part of this publication may be reproduced, stored in a retrieval system, or transmitted, in any form or by any means, electronic, mechanical, photocopying, recording, or otherwise, without the prior written permission of the publisher.

Published by the University of Minnesota Press
111 Third Avenue South, Suite 290
Minneapolis, MN 55401–2520
http://www.upress.umn.edu

ISBN 978-1-5179-1033-4 (hc)
ISBN 978-1-5179-1034-1 (pb)

Library of Congress record available at https://lccn.loc.gov/2020023732

Printed in the United States of America on acid-free paper

The University of Minnesota is an equal-opportunity educator and employer.

UMP BmB 2021

CONTENTS

INTRODUCTION
ON SPECULATIVE MEDIA PHILOSOPHY

THE MEDIUM

Molecular animations are strange media objects. Indeed, their simplicity masks vast complexity. *Molecular Capture* is a book that gives voice to this complexity; it is a book that also problematizes it, conceptually transforms it, and then tests whether our experience and understanding of animation are altered through our conceptualization of it. This book does not seek an exhaustive description of molecular animation. Rather, *Molecular Capture* is a book of speculative media philosophy that derives its method from the work of Alfred North Whitehead. What exactly this means will be borne out over the course of seven chapters and a Postscript, but generally, it consists in adhering to Whitehead's method of fabricating speculative propositions that respond to a problem in our experience of the modern world. The problem that molecular animation forces us to confront has to do with the emergence of a media practice that is a site of convergence for vastly different histories, material cultures, and systems of meaning and value.

Put simply, molecular animations are computer simulations of molecular and cellular processes that immerse viewers in the temporal unfolding of molecular worlds. Animators combine vast quantities of molecular and cellular data acquired from online databanks with commercial and custom animation software to achieve cinematic-quality 3D animations of molecular processes that are far too small or complex to

Figure 1. *Protein Packing*. Directed and produced by XVIVO Scientific Animation, 2013.

see otherwise. And while these animations sit comfortably alongside of other data visualization practices and are also at home in the visual culture of science generally, they are also objects of entertainment. They are media that showcase Hollywood-quality graphics, sound, and editing and are often distributed and consumed on popular social media platforms (Facebook, Twitter, YouTube, etc.). As the animator and scientist Gaël McGill writes, "The very same tools that were invented to animate a character like Shrek or Nemo are now being applied to set in motion protein domains and cellular processes" (2008, 1127). Molecular animation is where science and Hollywood collide.

But in order to understand how molecular animation came to be, how it took shape over time and across numerous domains of practice and expertise, and what is significant about it more generally, *Molecular Capture* cannot be content to rehearse narrow histories of the individuals, knowledge practices, institutions, and funding schemes involved in its emergence in the late twentieth and early twenty-first centuries. Although we will encounter all of these elements along the way, one of this book's main contentions is that molecular animation is about so much more than animating molecules for scientific observers and popular audiences. Molecular animation is about the collision of vastly different regimes of seeing and knowing; it is also about the slow and unevenly distributed histories of entertainment and scientific knowledge production converging in a media practice; and finally, molecular animation is about the design of new modes of perceptual subjectivity at the intersection of science and popular culture.

It is in order to bring these various and often conflicting dimensions of molecular animation into full view that the book begins with the practices of molecular animators. It draws on numerous examples of scientists using animation software to visualize complex molecular processes, such as leukocyte extravasation or the movement of the motor protein dynein. It also explores the incredible power of the technology and how it promises to fill in certain gaps or blind spots within the visual field of biological knowledge production. In this context, the book also draws on case studies of scientists in both the experimental life sciences and computational biology who use animation to answer important epistemic questions in their respective fields. And on the other end the spectrum, the book examines how the technology shares important aesthetic features with popular screen media; it demonstrates how molecular animation is firmly situated in a history of digital entertainment media, and for this reason may even undermine core assumptions about knowledge practices within the life sciences. *Molecular Capture* therefore inquires into the conditions for this media practice to emerge at the intersection of two different visual economies: the life sciences on the one hand and popular media culture on the other. The book is interested in uncovering a much longer and nonlinear history of the epistemic practices of biologists, how they are intimately connected to various technologies of visual perception, and how these technologies of perception collide with the visual economies of digital entertainment. And while this collision of cultures, values, and histories lies at the heart of molecular animation, the book examines how there is, as of yet, no adequate explanation for how and why this came to be, and what the consequences of this convergence are. It is in order to respond to this deficit that *Molecular Capture* develops a theoretical account of how heterogeneous histories, values, and material cultures converge in this media practice.

The book therefore situates molecular animation within an evolving field of historical, epistemological, and ultimately political negotiations in order to expose the technology's relevance for the shifting terrain of biological perception and knowledge production. What becomes apparent from these wide-ranging conversations is that molecular animation not only represents a key transformation in the visual knowledge practices of the biosciences but also brings fundamental mutations in political power within neoliberal capitalism into sharp focus. In particular, the convergence of the visual economies of science and entertainment in molecular animation extends neoliberal modes of governance to the perceptual practices of scientific subjects.

THE AUDIENCE

The histories and theories presented in this book are diverse: the history of science, media theory, theoretical and experimental biology, film studies, science and technology studies (STS), philosophy, and design theory are woven together. What motivates such a collision of discourses and perspectives is the need to address a problem that lies at the heart of molecular animation: What are the conditions for computer animation to function as a site of convergence for knowledge production in the life sciences and the practices and values of popular media consumers?

This coincidence of science and entertainment is approached using the analytical methods of media philosophy and media theory.[1] This is not to say that the arguments presented in *Molecular Capture* will be of no interest to scholars and researchers working in the many other fields covered in this book. One of the guiding assumptions is that molecular animation cannot be framed without careful attention paid to the preoccupations of biologists, historians, programmers, filmmakers, and designers; and for this reason, the book will be of interest to these and possibly other audiences. Indeed, there has been significant scholarship in the last few decades demonstrating the close alignment of science and entertainment throughout the twentieth and twenty-first centuries. The convergence of science and popular media has been well documented in media and science studies, and in many ways their overlap is fundamental to our current episteme (Thurtle 2018; Gaycken 2015; Curtis 2015; Ostherr 2013; Landecker 2005; van Dijk 2005). Nevertheless, the way in which this collision of cultures occurs in molecular animation is a problematic worthy of consideration by media philosophers.

Before moving on to address how and why the aforementioned is problematic, I want to head off a possible criticism. One could easily challenge what seems to be a reductive separation of entertainment and scientific cultures in this book. To many, this may seem like a historically inaccurate and theoretically untenable claim. On the one hand, there is wealth of scholarship in cinema and media studies, the history of science, and STS demonstrating that science and entertainment have interwoven histories and materialities. For well over a decade, Hannah Landecker (2005) has shown the cinematic dimensions of biological knowledge production throughout the twentieth and twenty-first centuries. And others, including Kristin Ostherr (2013), have paid special attention to how education becomes an important site for mixing regimes of seeing and knowing as well as for disciplining the medical gaze (for the latter, see also Cartwright 1995). Likewise, Janina Wellmann (2017a) showcases the cinematographic character of embryological knowledge, while ethnographers of computational and structural biology celebrate the affective and embodied landscape of molecular renderings (Myers 2015).

And then on the other hand, one could challenge the epistemological and ontological separation of "science" and "culture," or indeed, of any position that divides social and scientific knowledge. Since at least the 1970s, the field of literature and science has been hard at work dispelling the myth that scientific and cultural modes of production are distinct and that C. P. Snow's two-cultures paradigm has ontological merit (Meyer 2018). This work falls in line with decades of research in the sociology of scientific knowledge and feminist epistemologies of science advanced by Bruno Latour, Donna Haraway, Gaston Bachlelard, Isabelle Stengers, Karen Barad, and others; it also parallels the more strictly philosophical work of Alfred North Whitehead, Henri Bergson, William James, Susanne Langer, Gilles Deleuze, and Félix Guattari, as well as research issuing from the various new materialisms, speculative realisms, eco-criticisms, and so on. This scholarship has had no small effect on media theory and philosophy as well, especially evidenced in the work of Jussi Parikka, Luciana Parisi, Thomas LaMarre, Matthew Fuller, McKenzie Wark, Alexander Galloway, and others (many of whom will be discussed in chapter 4).

This brings me to my point: If the separation of cultures is no longer historically and theoretically viable, then why emphasize their difference? As will become clear throughout the book, my claim is not that significant exchanges between the media cultures of science and entertainment do not exist; indeed, the historical production of regimes of seeing and knowing is inherently mixed, a promiscuous movement between science and popular culture. For this reason, *Molecular Capture* is in full agreement with scholars who insist on their entanglement. However, one of the fundamental assumptions of this book is that we need to bring the complex and asymmetrical exchanges between styles of knowing and seeing into full view. If our current episteme lies at the intersection of popular and scientific epistemologies, then it is essential to understand how this episteme took shape. In other words, not only do we need to make the contaminations explicit but we also need to chart the frictions and tensions, or in other words, the distinctions that scientists continue to make between their work and the consumable products of popular media culture, even if the boundary separating the two is always leaky. To overlook these differences, or to explain them away in a grand metanarrative that synthesizes them is just as reductive as maintaining their separation. There is a reason that scholars continue to narrativize their conjunction: it is not a given for scientists that their work is subject to the same regimes of seeing and knowing as those deployed in popular media culture. And the reverse is also true.

This brings me to a more philosophical point. The research presented in this book is indebted to the philosophies of immanence and process that resist the ontological distinction between scientific and cultural ways of knowing. This is one of the reasons that Whitehead's work features prominently in *Molecular Capture.* At the core of his metaphysics is resisting the "bifurcation of nature": the modern habit of thought that insists on privileging certain abstractions (scientific) over other ones (cultural, subjective, etc.). What this does not mean, however, is that there are not significant differences between knowledge practices. A physicist and a poet do not look at the sunset in the same way; and a biologist and musician do not see the rhythmical patterns in the biological world in the same way either. What counts as valuable knowledge in one domain (biology) does not necessarily count as valuable knowledge in another (popular culture). These differences matter. And for Whitehead, resisting the bifurcation of nature does not mean explaining these differences away or showing how the divergences are all, at bottom, really instances of the same thing. There are real heterogeneities in the same unruly world, and we cannot marshal one to explain all the others. The world is at once convergent and divergent, deeply interpenetrating but also heterogeneous through and through.

What emerges over the course of Part I are heterogeneous visual cultures that have unique, if overlapping, material histories, epistemic practices, and normative frameworks. The visual cultures of biology and entertainment do not possess rigid identities, however: they are themselves heterogeneous, composed of a multiplicity of material and symbolic systems. But they do share consistencies that need to be traced out over the course of the nineteenth, twentieth, and twenty-first centuries. It is in many ways the job of Part I to trace these consistencies and to see what visual economies emerge from them. What becomes apparent is that these heterogeneous cultures converge in a highly specific way in the late twentieth and early twenty-first centuries (they are both divergent and convergent). And while their cross-pollination is nothing new and has been well documented in the scholarship, the collision that this book is interested in has unique features, and molecular animation brings these features into sharp focus.

But if this book shows how molecular animation is an outgrowth of interwoven visual cultures, then here is what is exceptional about this situation: molecular animation traverses both domains, because biological and entertainment media cultures share a visual logic in the twenty-first century. This media form is legible and valuable to each domain for the very same reason: a single logic of the visual ranges over heterogeneous domains of seeing and knowing. This is what makes molecular animation invaluable for investigating biological knowledge production and popular culture consumption: it is a site of convergence. The visual logic that makes animation "scientific" is also the one that makes it "entertainment." But if science and entertainment have always produced hybrid media forms, then the one molecular animation brings into existence raises a problem for philosophical thought: How do many things converge into one thing without the many being reduced to the one? This is a problem that has important ontological and political consequences for our current media culture.

THE METHOD

One of this book's central challenges is to uncover how the collision of visual cultures came to be in molecular animation. I argue that this question raises a problem best addressed to media philosophers who inquire into the ontology of media and their capacity to draw heterogeneities together, to bring divergent lines of history, matter, and meaning into forms of consistency without reducing their divergences to a single frame of reference. But if this presents a problem worthy of philosophical reflection, then it's because it creates a space for a genuine problematic in Deleuze's and Whitehead's sense of the term. When Deleuze discusses the nature of the problem in *Difference and Repetition,* he aims to distinguish it from any conception that would prescribe "the possibility of [its] solution" (1994, 165). A problem is a field of tensions, of strains and pressures that does not get resolved with the appropriate solution. Solutions emerge from this field of tensions, but they do not exhaust it; they only serve to change the nature of the problem, as well as the possibility for future responses.[2] As such, responding to problems, writes Deleuze, is a question of learning from them instead of "knowing solutions" (164). In our case, this would mean that responding to the problem molecular animation sets out for us does not mean offering a final solution. It means learning from the contours and textures of the problem, experimenting with possible responses, only to see what new tensions and questions emerge.

The solutions proposed do not therefore exist in advance; they do not hang in the air waiting for problems to emerge that they can solve. Rather, solutions emerge from the peculiarity of the problem posed by molecular animation and only make sense in relation to it. This is precisely why Deleuze and Félix Guattari write in *What Is Philosophy?* that "a solution has no meaning independently of a problem to be determined in its conditions and unknowns; but these conditions and unknowns," they add, "have no meaning independently of solutions determinable as concepts" (1994, 81). What this means is that just as the solution depends upon the formation of a problem, the problem depends upon the solution to give it sense and meaning. Likewise, the conceptual frameworks that emerge in this book are as much a response to the problem posed by molecular animation as they are a clarification of that problem (this will become especially clear in chapters 6 and 7 as the problem itself gets reworked).

Aside from Deleuze and Guattari, there are few other thinkers who understand the relation between problems and solutions better and more thoroughly than Whitehead. And there is perhaps no other thinker who helps devise a method for responding to the problem we have posed for ourselves in this book quite like Whitehead. As Isabelle Stengers writes in *Thinking with Whitehead*, we should never forget that Whitehead was a mathematician. And mathematicians, she reminds us, "are they who do not bow down before contradictions but transform them into an ingredient of the problem" (2011b, 15). What this means is that oppositions—between the regimes of visuality, for instance—should not be overlooked or explained away. There are real differences between theoretical biologists working with time-based media and entertainment producers providing content for online streaming. Biologists using computer animation do not think about their work as mere entertainment (even if many of their technologies and values are imported from the entertainment industry); and popular content producers do not claim their work meets the rigorous demands of scientists making claims about the molecular world (even if they animate molecules). These are differences and asymmetries that matter; they express power dynamics within and between knowledge practices and should not be overlooked in the service of tidy histories of influence. And yet somehow a convergence between their image logics takes place—heterogeneous regimes of seeing and knowing are held together in molecular animation, and this book endeavors to understand how this came to be. But for Whitehead, as for Deleuze, the contrasts and tensions have to remain at the very heart of the problem and solutions we devise.

What's crucial about mathematicians, as well as Whitehead's approach to philosophy, is that they are also "the ones who dare to 'trust' in the possibility of a solution that remains to be created. Without this 'trust' in a possible solution," Stengers continues, "mathematics would not exist" (2011b, 15). Solutions do not preexist their problems but are fabricated by trusting mathematicians to meet the rigorous demands of the problems posed.[3] The metaphysical machinery Whitehead builds responds to a problem in modern experience. Actual occasions, eternal objects, prehensions, satisfactions, societies, and so on—these are solutions (concepts) designed for this purpose. They do not represent the world as it is "in itself," nor do they express the essence of nature behind appearances. In fact, this cannot be the case, given the nature of the problem that Whitehead poses for himself. While this problem changes over the course of his philosophical career, broadly it can be framed in terms of a question: How do we give an account of experience that refuses to authorize one set of descriptions or experiences at the expense of another?[4] Whitehead has other ways of formulating this problem, and each formulation calls for a slightly different mode of response. In *The Concept of Nature,* for instance, he conceives of it in terms of overcoming the "bifurcation of nature" into physical and mental stuff, or into the fundamental constituents of reality and mere representations of it. This problem is then recast in *Science and the Modern World* into resisting the "fallacy of misplaced concreteness": a modern habit of mind in which the importance of an item of experience is exaggerated; this is "the accidental error of mistaking the abstract for the concrete" (51).[5]

Later, in *Process and Reality,* Whitehead systematically characterizes his method for responding to this fallacy. This is what he calls speculative philosophy, and it follows the path of an airplane: "The true method of discovery," Whitehead explains, "is like the flight of an aeroplane. It starts from the ground of particular observation; it makes a flight in the thin air of imaginative generalization; and it again lands for renewed observation rendered acute by rational interpretation" (PR, 5). The solutions that Whitehead devises are speculative abstractions (imaginative flights) charged with transforming our experience of the world based on a problem found in our experience of it (in the ground of a particular observation). But the "adequacy" of these solutions can only be determined by whether or not they actually transform our experience in relation to the problem posed (this determines the efficacy of the concepts). Do we now refuse to authorize the legitimacy of privileged cases? Or as Stengers puts it, adequacy "obliges philosophers not to invoke any cause allowing them to eliminate, forget, treat as an exception, or disqualify an element of experience. In particular, the scheme must be able to embrace the very thing that would be invoked in a mode of a challenge, contradiction, or scandal: but what do you do with 'this'?" (2011b, 246).

What is important to note is that upon landing from the speculative flight back in experience, speculative philosophy does not end its work. Once experience is enriched by metaphysical description, which is to say, our experience of the world has been transformed by it, then new problems emerge and new proposals for speculative adventure take shape. In Deleuze's vocabulary, the transcendental field of the problematic has been changed by the solution but does not go away. Whitehead famously writes, "Philosophy begins in wonder. And at the end when philosophic thought has done its best the wonder remains" (MT, 168). Speculative philosophy is an ongoing adventure, and "the merest hint of dogmatic certainty as to finality of statement is an exhibition of folly" (PR, xiv).[6]

It is with Whitehead's speculative "method of discovery" in mind that *Molecular Capture* responds to the problem molecular animation poses for thought. The spirit of this method is captured in the three parts of this book: Part I (chapters 1, 2, and 3) manifests the problematic nature of science and entertainment in molecular animation; Part II (chapters 4 and 5) takes flight into speculative abstraction in order to devise a scheme for understanding animation's unique composition of histories, values, and materials; and then Part III (chapters 6 and 7) lands from philosophical abstraction in order to reassess animation with fresh eyes. Finally, the Postscript takes the processual nature of speculative thought seriously and inquires into what new questions speculation has opened up and what new paths for philosophic inquiry demand our attention. This section will help to demonstrate that molecular animation is indeed worthy of speculative philosophy: it generates problems that do not get resolved—new dimensions of the problematic continue to be produced, which call for the fabrication of new modes of response.

THE CONTEXT

What I'm proposing about computerized molecular animation might seem surprising or even a little unwarranted. This is in part because the media practice has largely escaped critical attention within the humanities and social sciences. For this reason, there may seem to be little precedent for the speculative propositions I construct in this book, and this may seem alarming given the importance I attribute to molecular animation.

There are several important exceptions to the relative silence on the use of computer animation in the biosciences within the humanities and social sciences. Phillip Thurtle's article "Animation and Vitality" (2014) reframes our understanding of the technology in terms of compositing within the art of animation and then shows how molecular animation enables us to comprehend the rates of change between scales (or layers) of biological systems. And in his most recent book, *Biology in the Grid,* Thurtle (2018, 208) draws on the contemporary molecular animations of Drew Berry and Janet Iwasa as well as Ed Lewis's low-tech animations of gene regulatory networks to understand how biologists gather disparate elements in order to build the biological world up. Natasha Myers also addresses molecular animation in *Rendering Life Molecular: Models, Modelers, and Excitable Matter* (2015) and pays special attention to how it fits within her larger ethnographic study of 3D crystallographic modeling. In particular, she demonstrates how animations such as *The Inner Life of the Cell* (2006), which I discuss at length in chapter 1, open up "new spaces for molecular storytelling," especially stories that "destabilize mechanistic theories of life" (Myers 2015, 187). Christopher Kelty and Hannah Landecker have also developed a "theory of animation" in relation to the life sciences and are especially keen to show, not unlike Myers and Thurtle, how moving images (from early time-lapse microcinematography to computer animation) animate perception and knowledge in biology. Likewise, Janina Wellmann underscores the importance of animation for biology. But instead of looking to molecular animation, she grapples with the technologies used for creating "digital embryos," or "*in vivo* images," in developmental biology and argues for a broader conception of "animation" (via Tom Gunning) that captures the cinematic quality of development in these computer images. In the end, these images are thought to *animate* how developmental biologists understand embryogenesis.

These accounts stand out as key moments in the theorization of animation within the life sciences. And while there exist many earlier attempts to "animate" biology using cinematic or other moving-image technologies, these are some of the most important efforts to conceptualize the use of computer animation technologies in the contemporary biosciences. And what's more, each of them emphasizes, and not unlike what I am doing here, the fact that computer animation is a largely undertheorized and potentially revolutionary technology for the study of living organisms. But it is also the nature of the transformation that computer animation promises that unites these pieces as well. For these authors, animation forges new and exciting epistemic pathways in biology and opens up a space for scientific thought and perception to reflect the dynamic and processual systems they study. These arguments are less about representation and more about enabling scientific thought to unmoor itself from mechanistic paradigms of biological knowledge production (e.g., Neo-Darwinism).[7] Thought itself becomes "vital." What's more, these conceptions of media closely parallel claims made about time-lapse microcinematography in the early twentieth century (which I discuss in chapter 7),[8] as well as a certain tendency in media studies to promote the "vitality of media" and their capacity to forge new modes of embodied and affective engagement with the world.[9]

While these works make important contributions to an emerging area of research at the intersection of media and science—and for this reason, I consider myself deeply indebted to them[10]—they are largely silent on the problem that animates this book. I argue that these accounts need to be supplemented with a critical and speculative framework that can lend specificity to how and why media and epistemic cultures converge in this media practice. In other words, we need to account for (1) how heterogeneous regimes of seeing and knowing coincide in molecular animation without losing their heterogeneity, and (2) what mechanisms brought this heterogeneity of elements together. In Whitehead's vocabulary, these are the initial constraints of the solution to the problem that animates our media philosophy. They dictate whether our propositional scheme is adequate and can transform our experience and understanding of animation. It is therefore the nature of the problem posed, as well as our mode of critical and speculative response to it, that sets this book apart from previous scholarship on animation and biology.

It is in order to address the heterogeneity of epistemic, visual, and material cultures assembled in animation (the first constraint) that I draw on Whitehead's notion of "social order." I use his conception of social assembly to build what I call an "ecosocial theory of media" that can account for how diverse elements, with various histories, from different spheres of meaning and value, assemble in a media system and significantly contribute to life science research. In particular, Whitehead's sociology addresses how material and symbolic systems with distinct histories and relevancies can also be interpenetrating and connected; heterogeneities are woven together in his speculative sociology. This is where the book is also a part of a wider conversation in speculative media theory about the formation of today's digital and information systems out of forgotten histories of media (media archeology, Parikka 2012a). The book addresses how these media-archeological perspectives are also closely aligned with ecological and materialist theories of media, especially insofar as media archeologies expose the variety of materials and processes—from the technical to the biological and the geological—folded into the evolving field of relations that make up today's media systems and infrastructures. Matthew Fuller (2005), Jussi Parikka (2015), Siegfried Zielinski (2006), and a host of other contemporary media theorists have made significant contributions to this growing area of research, whose long list of predecessors include Marshall McLuhan, Harold Innis, Walter Ong, Félix Guattari, Lewis Mumford, Vilém Flusser, Gregory Bateson, Whitehead, and others.

The aim of this speculative sociology of media, then, is to frame molecular animation in terms of the diversity of its composition while paying due attention to how this diversity is irreducibly coordinated. On the one hand, this means accounting for the fact that molecular animation is situated in a history of popular cinema and media, and as such emerges right alongside and is virtually indistinguishable from a wide range of digital entertainment media in the twenty-first century. This will take us into cinema and media studies and will call upon us to engage with debates over the ontology of the digital image, the rise of computer animation, and the ubiquity of screens in today's entertainment culture.[11] It will also require serious consideration of molecular animators who make no claim to be doing science and whose aim is to fictionalize, popularize, and ultimately, Hollywoodize biology. But on the other hand, a speculative sociological engagement with molecular animation also requires us to account for its emergence within the history of visual epistemic practices in the life sciences. This is where this book's conception of media also differs from many others: it demands a careful assessment of visual media's relation to the history of knowledge production in biology, as well as an account of how these visual epistemic practices are precisely what make the same molecular animations relevant for many scientists today. This involves, for instance, reading the knowledge practices embedded in the use of light and electron microscopy in the life sciences as a part of the history of molecular animation; it also means framing X-ray crystallographic methods, florescence microscopy, mathematical models, 3D modeling, and microcinematography in terms of this history of scientific media. Thus, in addition to having a foot in cinema and media studies, this book is also primed to engage in debates in the history of science on questions of visual epistemology at the intersection of experimental and theoretical biology. And some of the well-known voices in the history of science—e.g., Hans-Jörg Rheinberger, Evelyn Fox Keller, Lily E. Kay—are among the most important interlocutors in this stretch of the book.

Still, one of the central challenges of *Molecular Capture* is to elaborate a conception of molecular animation that can account not only for how but also for why the histories, practices, and values of the biosciences and popular media converge the way they do in this media form (the second constraint). In other words, what reason do these visual cultures have for assembling the way they do? And what are the larger stakes of this convergence? It is in order to address these questions that the book turns to Michel Foucault's genealogy of governmentality. What Foucault's work, in conversation with Whitehead's, is able to expose are the histories of perceptual governance concealed within this media ecosystem. While computer animation often parades as an expressive technology for scientists (see above), the book demonstrates how a more subtle reading of the technology, one that takes stock of neoliberal governmentality's need to merge cultural and scientific production, reveals its function as an apparatus of capture, or a dispositif, that has taken well over a century to develop. The book argues that governmentality's slow and uneven development has managed to draw the visual economies of science and entertainment together into a molecular animation apparatus. For this reason, and in line with previous work on molecular animation, the book suggests that animation makes a key intervention into the epistemological practices of the biosciences. However, and unlike previous scholarship, *Molecular Capture* contends that this intervention is a result of essential transformations in the distribution of political and economic power, and as such reveals a much deeper set of historical negotiations concerning governmentality's ability to steer scientific knowledge production and the practices of popular media consumption in the same direction.

This political reading of scientific media engages a wide range of critical theories of science, media, and political and economic power. In this context, there are a number of silent interlocutors who paved the way for the study I am conducting here. Eugene Thacker's (2005) early work on the exchange of "biomedia" within wider systems of biopolitical control, as well as Robert Mitchell's (2016) more recent study of biopolitics, media, and population management, are very much in the background of this work. Likewise, Paul Rabinow's (2005) and Nikolas Rose's (2001) discussions of the new forms of biopower (biosociality and ethopolitics) shaping our rapidly commercializing biosciences provide important context. Similarly, Melinda Cooper's (2008) and Kaushik Sunder Rajan's (2006) foundational analyses of neoliberal governance and the rise of biotechnology, as well as Cooper and Catherine Waldby's work on "clinical labor," have all been instrumental in shaping the discursive framework in which this book is situated. Finally, Orit Halpern's (2014) skillful history of the transformation of the perceptual and epistemic spaces of mid-twentieth-century subjects due to a convergence of cybernetics, computer science, design, and governmentality is an important touchstone for my argument about the control of perceptual subjectivity through interface design in chapter 7.

This book also engages a number of critical theories of media that intervene in the emergence of perceptual governance. In this context, Jonathan Beller's work *The Cinematic Mode of Production* (2006) is especially important for us. Despite certain reservations I have about his unflagging endorsement of a Marxist critique of media, there are few theorists who make the economic control of perceptual and affective space by cinema and entertainment media more apparent than Beller. And in particular, his critical assessment of the labor of looking—"too look is to labor" (2006, 2)—will be essential for developing our understanding of the governance of perception in the molecular animation apparatus. This book also addresses how theorists working within the tradition of (post-)autonomist Marxism expose the subjectivizing and desubjectivizing forms of work that have emerged under the reign of computational capitalism, as well as the range of psychopathologies that issue from these new working conditions (Berardi 2009, 2011). We also briefly explore Byung-Chul Han's notions of the "exhibition" and "pornographic" society in the context of his critique of the positivity, immediacy, and visibility of today's media-saturated culture.[12] The relation that these and other critical theories of media have to this project will come into sharp focus in chapters 6 and 7, where we examine the neoliberal governance of perceptual space in the animation apparatus by drawing on Deleuze's conceptions of control and *dividuality* in order to push Foucauldian notions of neoliberal human capital in new directions.

Finally, *Molecular Capture* would not be possible without the scholarship in cinema and media studies, the history of science, and STS that directly engages with the relation between biology and entertainment culture. In this regard, Hannah Landecker's (2007) work on microcinematography and biological knowledge production has proven to be groundbreaking; and so has Oliver Gaycken's (2015) important book that examines popular-science film and its impact on the production and circulation of knowledge in the early twentieth century. Likewise, Scott Curtis's *The Shape of Spectatorship* asks methodological questions about the use of film in early twentieth-century scientific and medical fields that are also germane to this book. And in earlier work, Lisa Cartwright sets the stage for interrogating relations of disciplinary power in relation to visual epistemology in medicine by examining how "the motion picture, in conjunction with more familiar nineteenth-century medical recording and viewing instruments and techniques . . . was a crucial instrument in the emergence of a distinctly modernist mode of representation in Western scientific and public culture" (1995, xi). These works, along with others that examine the close relation among science, visual technologies, and knowledge production, pave the way for *Molecular Capture* to undertake its study of computer animation and entertainment culture in the twenty-first century biosciences.

There are a number of features that set this book apart from the studies just mentioned, however. First, this is not a book written for historians; nor is it a book with ethnographers, empirical researchers, or even film studies scholars primarily in mind. While historical analysis, empirical research, and formal film analysis are everywhere present (especially in Parts I and III), what distinguishes *Molecular Capture* is that it relies on speculative philosophy and critical media theory to anchor its claims about biology and cinema (hence, the sections on philosophy are found in the center, Part II). In other words, this book's proposal that the capture and control of the perceptual economies of scientific subjects is achieved through the design of the animation dispositif is most effectively understood using philosophical modes of inquiry, in close connection with other theories and histories of media.

Second, this book is both speculative and critical. What this means is twofold: On the one hand, it entails the deployment of theoretical abstractions to enhance our understanding of the history and future of media practices. While this is a method commonly used in fields such as media ecology and media philosophy, it is not widely practiced by those who study science and cinema (save occasional references to contemporary theory). And on the other hand, *Molecular Capture* is explicitly critical of the history and future of the science-cinema-entertainment apparatus. In the main, studies of biology and moving-image culture are less overtly critical and more affirmative (although sometimes agnostic) about the entanglement of moving images and biology (with important exceptions: see Cartwright 1995). While this study is "affirmative" in the Foucauldian sense that power is productive and not repressive, it is deeply critical of the current regimes of power and argues that they require redirection.

And third, *Molecular Capture* draws on biology and moving-image culture to advance certain philosophical arguments (and not only the other way around). If Whitehead and Foucault seem to be on different ends of the theoretical spectrum (see chapter 5), then *Molecular Capture* creates a space for them to engage in productive dialogue in order to better interrogate our contemporary media culture. The book shows how speculative metaphysics and the critique of power (which are too often kept separate) are interwoven theoretical frames that permit a genealogical reframing of the formation, design, and reimagination of an especially powerful media apparatus. Media therefore become the raw material for the production of new philosophical concepts.

But in the end, it is in bridging the aforementioned discourses that the book is able to reimagine what it means to visually capture the molecular world: this world is at once "captured" in the moving image and "captured" in a dispositif. The title of this book plays with these two senses of the word "capture." If molecules are visually captured in computer animation and are thus a part of a longer genealogy of capturing molecular structures and processes using visual technologies, then they are also captured in the sense that Foucault and others describe the process by which heterogeneous elements are "made visible" as a part of a larger strategy of governance. *Molecular Capture* is a book about how these two senses of capture converge. Molecular animation is the site where visual and political capture collide.

THE CHAPTERS

This book's wager is that Whitehead's speculative method is valuable for understanding a media practice with political stakes. As we will see, Whitehead's own notions will have to be revised in order to accommodate the technological, social, and political–economic realities of the twenty-first century. However, *Molecular Capture* is less concerned with whether the concepts Whitehead devised to interpret the textures of experience in the 1920s and 1930s hold up today (although some of them certainly do) and more interested in whether his "method of discovery" might be capable of animating our understanding of a contemporary media practice.

To put this method to the test, Part I of *Molecular Capture* "starts from the ground of particular observation" by situating molecular animation as a media practice firmly rooted in extant conversations about digital entertainment media, scientific visualization, and knowledge production. It is in this context that the problematic nature of molecular animation takes shape. Chapter 1 places molecular animation in the context of the "digital turn" in popular cinema and media. The book draws on numerous examples of molecular animation to demonstrate how the medium distinguishes itself from other data visualization techniques in the biosciences by its distinctive use of time-based visual media, and then shows how these practices become legible in terms of a wider set of conversations in cinema and media studies over the computerization of the cinema as well as animation's role within this shift. In this context, we see how the medium expresses a specific "tendency" of the moving image shared by the popular mediasphere more generally (LaMarre 2009), and how this confirms the suspicions of many scientists that computer animation should not be used for "serious" science—it is an entertainment medium.

Chapters 2 and 3 chart a very different conceptual itinerary for molecular animation. Together, they demonstrate how this very same tendency of the moving image is also firmly rooted in the visual culture of the life sciences (*pace* the objectors) and that it helps to resolve crucial differences between the experimental and theoretical cultures of biology. Drawing heavily on the history of science, chapter 2 examines the centrality of visual technologies to the history of epistemology in the experimental biosciences—from cell theory and classical genetics to molecular biology—and then shows how this domain of biological research is nevertheless left with serious "blind spots" in its field of vision due to the fact that overall systems of interaction are not visible.

Chapter 3 then examines the various and complex struggles that mathematical and theoretical biology have had being accepted into mainstream biology (experimental biology), but then shows how it slowly won the favor of experimentalists because it allowed wider systems of interaction to be mathematically modeled—an especially important requirement in the post-omics era of life science research. But while entire systems can now be understood through systems modeling, these models nevertheless lack the experimental and observational verification of experimental biology. Using various case studies, the chapter shows how molecular animation brings these experimental dimensions of biology into systems modeling. More than this, it considers how the very same tendency of the image that ties molecular animation to popular media practices is also what solves key epistemic problems in the life sciences. Entertainment culture and scientific epistemology have overlapping values.

At the end of Part I we are left with the problem of understanding how heterogeneous visual cultures converge in molecular animation through shared values. Part II then "makes a flight in the thin air of imaginative generalization" in an effort to develop a general theory of media capable of responding to this problem. Chapter 4 takes the first step by turning to Whitehead's theory of social order and valuation in the context of a wider set of conversations in media ecology and media materialism. It is here that the book explicitly engages Whitehead's own concepts and examines how his sociology addresses the way in which radically diverse material systems, with various histories and values, converge in social orders to reproduce the same "defining characteristic" or "common element of form" through shared "valuations." This reproduction is itself finite and contingent upon wider and narrower social orders supporting its existence. The chapter demonstrates how social reproduction is "ecosocial" and gives us the building blocks for a theory of media that can account for the irreducible coordination of visual cultures in a media system through shared values.

Chapter 5 examines how this ecosocial theory of (evaluative) media only gets us part of what we need, since Whitehead cannot answer why divergent cultures converge in such a media system, and what wider ecosocial relations would support these shared valuations in any case. Thus, our media theory needs to be complemented by a theoretical account of the mechanisms, devices, and techniques for coordinating heterogeneous systems, especially in human societies. This will bring up a much larger set of issues in philosophy and cultural theory about Whitehead's supposed inability to respond to the specificities of human history—especially political and economic history. But it will also open the door for us to think about new conceptual alliances, and with Foucault in particular. This chapter therefore stages an (unexpected) encounter between Whitehead and Foucault in order to conceive of the convergence of heterogeneities in media in terms of relations of power and governance. This conceptual alliance will equip us to understand the formation of socially organized media systems in terms of governmental reason.

Part III will then do the difficult work of "land[ing] for renewed observation rendered acute by rational interpretation." As will become apparent, this third stage of speculative philosophy does not simply apply theory, but it involves understanding how the speculative tools developed in Part II reframe what the problem means (Part I). In this spirit, chapters 6 and 7 show how the conceptual framework developed in Part II sharpens our perception of how social environments converge the way they do because of the techniques of government. In other words, these two chapters explain how these visual cultures are drawn together in a highly specific way through the very the same modes of neoliberal perceptual governance. Chapter 6 examines how techniques of government have controlled popular screen culture from the very beginning of the cinema—via a careful reading of the history of critical media studies and Foucault on neoliberalism and Deleuze on control—in order to produce specific modes of perceptual subjectivity. The book argues that if molecular animation practices share a history with the practices of digital entertainment media (which they do), then these animation practices are also firmly situated in the history of embodied perceptual governance that entertainment media are (chapter 1 is then supplemented by chapter 6).

Chapter 7 argues that the forms of governance shaping popular media culture are also shaping the visual epistemic practices in the life sciences. This will challenge us to reframe our conception of the history of visual epistemology in chapters 2 and 3 in light of a genealogy of the techniques of government. As a consequence of this, a host of visual practices emerge as especially important attempts to govern the perceptual spaces of the life sciences (e.g., microcinematography), and for this reason become essential ingredients in the genealogy of perceptual governance. The result of this inquiry is that we come to appreciate how molecular animation is a product of governmental reason working through diverse channels throughout the twentieth and twenty-first centuries to draw the visual cultures of science and entertainment together into an apparatus of capture. The molecular animation apparatus is an achievement of governmentality.

The Postscript attempts to pick up on loose ends and leverage them for new speculative adventures. In particular, the book underscores how apparatus construction is not accidental, which is to say, outside of the highly rationalized practices of government, but it is *by design.* The book's provocation in this final section is to conceive of the rationalized experimentation that produces apparatuses as a part of a wider governmental design strategy. One of the consequences of this claim is that the entire book can be reread through the lens of design and furthermore as an attempt to sketch a theory of media design in the age of perceptual governmentality. In this context, one of the central theses will be that design is the bridge between the "art" and "reason" of governmentality. By drawing out governmentality's connection to design, the book opens up a whole series of questions about the possibility for governmentality to function as a distributed design practice in the context of molecular animation and contemporary media societies more generally. This conception of design only comes into view at the end of the book, after the first pass at speculative redescription is complete and new speculative adventures are beginning to take shape. This will set the stage for engaging a wide range of new conceptual tools and practices in design and architecture as well as the possibility of conceiving a critical and speculative intervention into the design of the molecular animation apparatus.

If *Molecular Capture* is a book of speculative media philosophy, then this means the success of its imaginative flight can only be determined after it returns to the ground from which it took off. It is only at this point that we can assess whether our experience and understanding of molecular animation has been transformed by the concepts we have constructed and whether we are reenergized for new adventures of thought. It is in this spirit that the book resists attempts at "sufficient philosophy" (Laruelle 2013) and instead embraces what Stengers (2011b) calls the "pragmatic art of consequences" in Whitehead's thought. *Molecular Capture* dares to trust in the possibility that a pressing problem within molecular animation can be articulated, addressed, and reimagined in the course of seven chapters and a Postscript.

Although this book is primarily intended for media philosophers and theorists, there are likely other audiences who will find portions of it valuable. Because *Molecular Capture* traverses more discourses and methods than is customary in a single monograph these days, I have included several itineraries through the text, even though the full weight of the argument can only be appreciated when the book is taken as a whole. Nevertheless, historians of science and STS scholars may be particularly interested in chapters 2, 3, and 7, since this is where the book makes historical arguments about the intersection of theoretical and experimental methods in the life sciences through visual epistemic practices. Likewise, theoretical biologists are likely to find useful material in chapters 2 and 3 (but especially 3), given that the book argues for blending not only experimental and theoretical methods but also bottom-up and top-down approaches to systems biology. Film and media studies scholars might find chapters 1, 6, and 7 valuable, since these chapters review key arguments in digital visual culture as well as the emergence of digital media apparatuses that shape and govern the sensorium. Continental philosophers will find fresh arguments about the interweaving of Whitehead's speculative metaphysics and Foucault's late theories of governmentality in chapters 4 and 5, and they will also find Foucault's neoliberal governmentality reworked with an eye to the proliferation of digital media apparatuses in chapter 6. Design theorists and philosophers will discover important material in chapters 6 and 7, especially those interested in "ontological design" (where design *designs*), but they will be most at home in the Postscript, which explicitly tackles recent debates in critical design theory, as well as speculative media philosophy as a whole, in design theoretical terms. Finally, media philosophers and theorists who are not compelled to engage in detailed histories of biology, technology, and epistemology, and are more interested in the stakes for speculative media ecology and media materialism more narrowly construed, will find useful overviews of the proposed media philosophy in chapters 4, 5, and 6.

PART I

The true method of discovery . . . starts from the ground of particular observation.

—Alfred North Whitehead,
Process and Reality

1

MOLECULAR ENTERTAINMENT

While I was preparing this manuscript, my wife forwarded me a link to her Facebook page. When I opened it, there was an uploaded molecular animation (see Figure 2) followed by a dozen or so comments—"Fantastic," "Amazing," "Science has come so far," and so on—and links to other animated molecules. To my surprise, my department chair forwarded the same animation clip from his Facebook page the following day, but he also included in the comments section a link to the trailer from *Fantastic Voyage* (1966) and wrote that "scientific visualization (aka 'fantasy') was so much more fun back in 1966."[1]

Figure 2. *Kinesin Motor Protein.* Directed by John Liebler of Art of the Cell: Medical and Scientific 3D Animation, 2014.

My wife and department chair were no doubt aware of how humorous I would find this: they are a part of the social and cultural life of biology. Scientific visualizations have become cultural artifacts circulated and consumed by mass audiences. That molecular animation has made its way onto Facebook should come as no surprise. It's telling that to this day, one of the best-known molecular animations, *The Inner Life of the Cell,* made by David Bolinsky and others at XVIVO Scientific Animation, was first screened at the 2006 computer graphics and arts conference, SIGGRAPH. This seems to confirm what the historians of science Lorraine Daston and Peter Galison noted over a decade ago about scientific images more generally: they are no longer "competing with art or even employing art but positioned as art itself" (2007, 384). Where scientific visualizations may have once circulated only among professionals in scientific atlases, today the boundary separating scientific and artistic image consumption is increasingly blurred.[2] Molecular animation would seem to make this point stunningly clear: molecular imaging is cinematic, and it is quickly becoming a cultural commodity.[3] With this wider net cast, scientific images are now circulating over an ever-increasing number of media channels and seem to be leaving the book form of the scientific atlas to collect dust as its successors are shared instantaneously, updated in real time, and are consumed like any other form of video entertainment.[4]

Molecular animations would appear to exemplify Daston and Galison's point: not only do they have an increasing number of circulation channels—on social media sites such as Facebook, Twitter, and Instagram—but they also seem to be designed for entertainment consumption just as much as they are for scientific modeling and knowledge production. Indeed, they are built out of the same tools and techniques as Pixar and DreamWorks animations. Recall that Gaël McGill, director of Molecular Visualization at Harvard Medical School, put it this way in 2008: "The very same tools that were invented to animate a character like Shrek or Nemo are now being applied to set in motion protein domains and cellular processes" (2008, 1127).

In this chapter I attempt to characterize the relation between molecular animation and the cultural production of scientific images by situating the use of computer animation in the biosciences in the context of the history of cinema and entertainment media. In doing so,

I draw connections between molecular animation and other data visualization techniques in the life sciences but then argue that molecular animation is distinguished by its use of time-based media as well as a wide variety of digital editing techniques. These techniques serve to make certain parallels between molecular animation and popular digital media practices apparent. I demonstrate how these parallels become most explicit in the context of a wider set of conversations about animation and the digital turn in popular cinema. I then use this to illustrate how molecular animation seems to be more legible to the diverse practices of entertainment media and the cinematic arts than to the practices of data visualization within bioscientific research. It is this convergence between molecular animation and popular media that secures molecular animation's place within a genealogy of popular moving images. Along the way, I show how this genealogy could also be used to support or even justify the critics of molecular animation within the scientific community, and I look to the work of Ian Hacking, Lev Manovich, Thomas LaMarre, and others to support this thesis.

MOLECULAR GRAMMAR

In *Rendering Life Molecular: Models, Modelers, and Excitable Matter,* Natasha Myers discusses *The Inner Life of the Cell* at length. She introduces the short film in an exquisite passage that's worth quoting in full:

> *[The Inner Life of the Cell]* warps time, space, and scale to invite viewers on an Alice-in-Wonderland journey to get down inside a cell and take a look around. Once viewers are inside, this seething cell pullulates with life. A musical score lilts and swells as viewers orient themselves in a spacious expanse of morphing bodies. . . . In scenes that recall the sublime and lonely images of space flight and deep-sea exploration, the animators choreograph moments of dramatic tension and sheer elation. (2015, 182)

For Myers, animating molecular worlds seems more akin to directing a Hollywood feature film or choreographing a dance performance than it is to the grueling and thankless labor of molecular modeling.[5] And it is for just this reason that molecular animations are beginning to draw significant attention in data visualization communities—both positive

Figure 3. *The Inner Life of the Cell*. Directed by XVIVO Scientific Animation, 2006.

and negative.[6] Molecular animations (using computer animation software) are relatively new to molecular data visualization and the visual culture of the life sciences more generally (1990s), and so they certainly do not exhaust the possibilities for visualizing massive amounts of cellular and molecular data.

Let's briefly look at the molecular process that *The Inner Life of the Cell* seeks to visualize: leukocyte extravasation. Leukocyte extravasation, or the rolling movement of a white blood cell in the human body out of the circulatory system toward a site of infection or damage in a host, is the immune system's first line of defense. And as such, researchers are keen to use a range of visual technologies—from electron microscopy to computer graphics—to visualize this important process. For this reason, there are numerous options for visualizing, modeling, and simulating leukocyte extravasation, each with different affordances. For instance, the Charles Mackay laboratory researches the mechanisms of immune cell migration, and in 2000 Ulrich H. von Andrian and Charles Mackay published the graphic in Figure 4 in *The New England Journal of Medicine.* The leukocyte has a "cartoonish," irregular shape (so as not to be confused with the red blood cell), and the adhesion molecules are depicted mechanistically, as if they were mechanical parts of a well-oiled machine—a "molecular machine." Aside from the fact the leukocyte is motionless in this graphic and is far less entertaining than its depiction in the animation, the graphic distinguishes itself by illustrating the causal relation among adhesion molecules that regulate the rolling movement of the white blood cell along endothelium cells. In this regard, the graphic also bears some resemblance to Molecular Interaction Maps (MIMs) that systems biologists use to depict cellular and molecular interactions, though it does not hold to their standardized notation, minimalist aesthetic, and exhaustive detail (see Figure 5).[7]

The point is that von Andrian and Mackay's visualization is both highly interpretive (insofar as it attempts to communicate the mechanistic character of unseen molecular interactions through "cartoon-like"

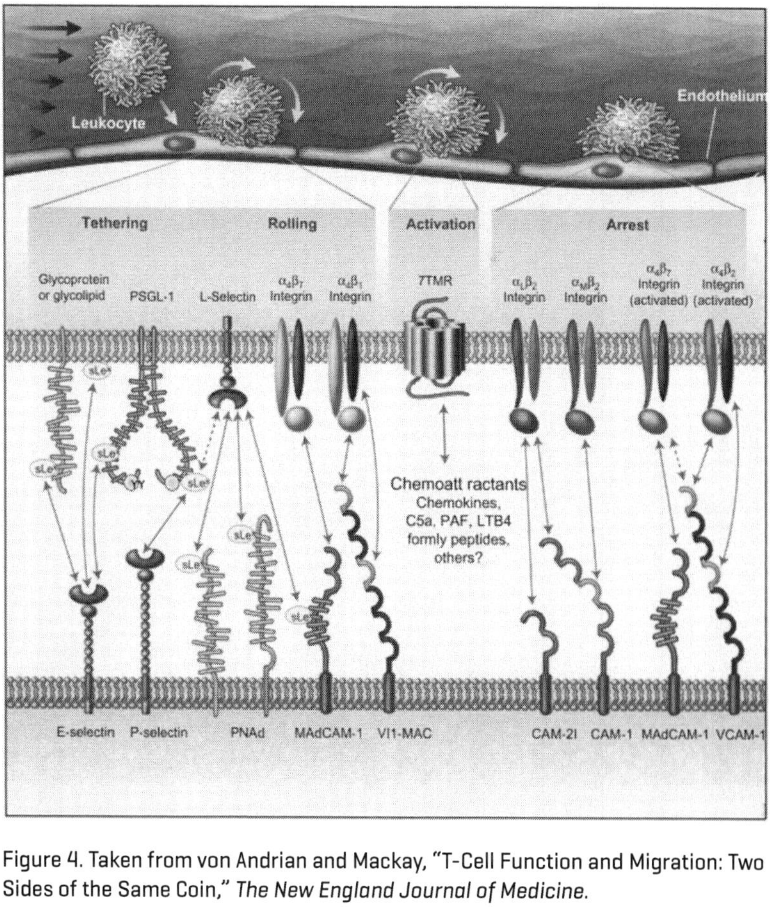

Figure 4. Taken from von Andrian and Mackay, "T-Cell Function and Migration: Two Sides of the Same Coin," *The New England Journal of Medicine.*

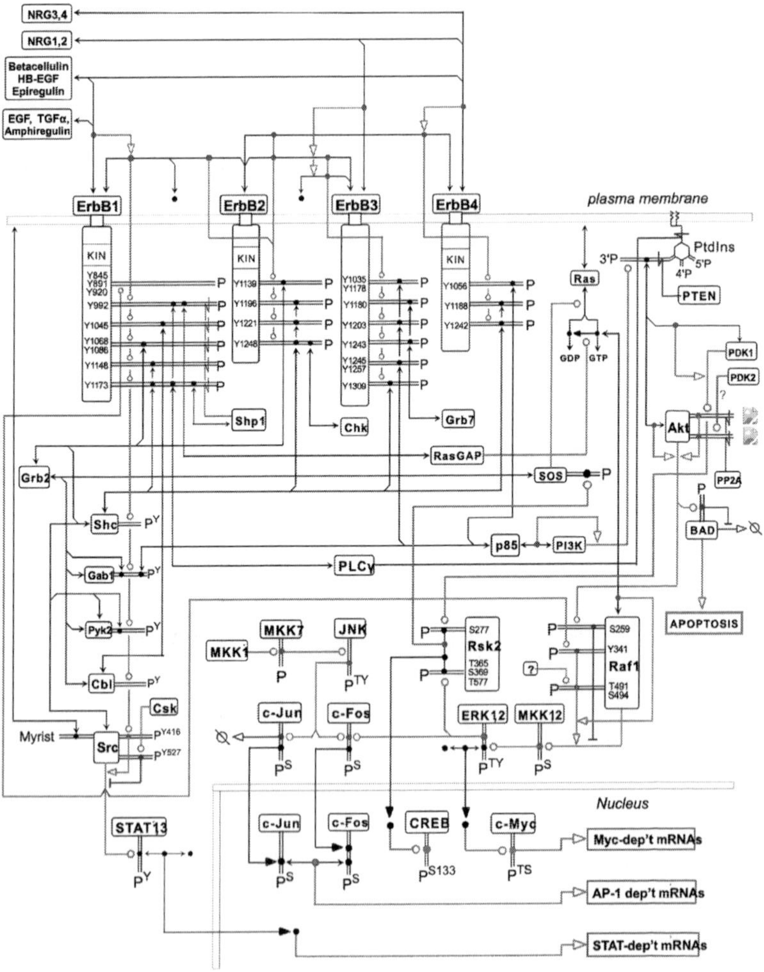

Figure 5. A Molecular Interaction Map (MIM) of signaling from epidermal growth factor (EGF) receptors. Kurt W. Kohn, Mirit I. Aladjem, Sohyoung Kim, John N. Weinstein.

graphics) and informative (insofar as it maps the causal relations among cellular and molecular processes). It goes without saying that *The Inner Life of the Cell* does not provide the same quality of information that Figures 4 and 5 do: it labels neither the functions nor the causal sequence given to those functions. But the animation does not communicate less information than the others do. This is a point that will be nuanced throughout this chapter, but suffice it to say that the molecular animation informs perception and thought differently and not less. Just as the visual semiotics of Figure 4 do not stop at scientific content, so too molecular images inform the human sensorium in ways that exceed the bare communication of "scientific facts." This point is obvious enough if we look again at Figure 4. While spatial organization, directional signage, and text all conspire to signify a routine molecular process, the image also relays a particular feeling for how this process unfolds. Without denoting it in words, the image forcefully conveys that leukocyte migration is mechanical. The graphic is full of figures that bear a striking resemblance to gears, pulleys, and levers that help the leukocyte travel down a conveyer belt. Referring to the figures as "cartoonish" actually underestimates how they function as bearers of meaning: they signify that molecular interaction unfolds with the automation and efficiency of an assembly line in the era of Fordist mass production.

In other words, the sensorium is informed by *how* scientific content is presented. The content is inseparable from *how* it is made available to thought and perception. Marshall McLuhan made a similar point decades ago when he proclaimed that "the medium is the message," but I want to put a much finer point on this by elucidating what this means in the very specific context of the visual culture of science. "Facts" about leukocyte migration are inseparable from the media that deliver them. The visual form that this molecular process acquires is therefore never *mere* form, an aesthetic addition that is inessential; but it is the environment that orients thought and perception. The "fact" that leukocyte migration has a determinate set of operations, which are finite and knowable, makes sense in the context of a mechanical universe of molecular interaction; and this universe is what the graphic delivers.[8]

But what happens when we situate these insights about facts and forms in the context of molecular animation? *The Inner Life of the Cell*

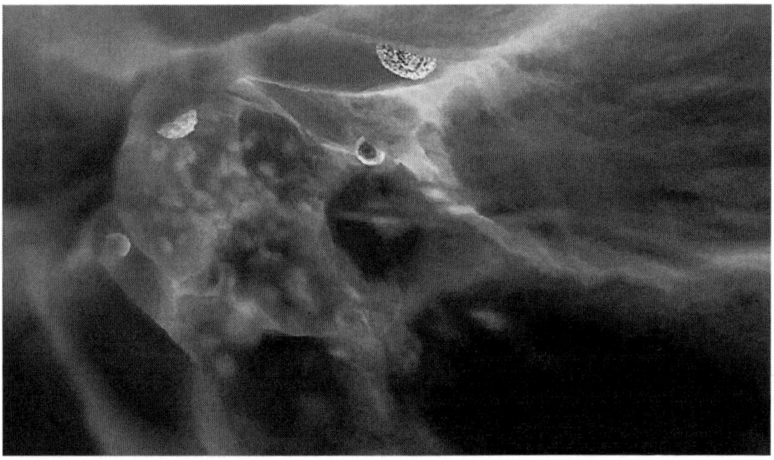

Figure 6. Opening sequence of *The Inner Life of the Cell.*

takes the molecular process depicted in Figure 4, animates it, and relays a very different sense of the same process.

At the outset of the short film, the viewer is immersed in a cellular environment. The leukocyte that we saw in Figure 4, whose movement down the endothelium "conveyer belt" is represented with directional arrows, now rolls down the cellular surface *in time.* The viewer is not at a distance or removed from this molecular process. We are not "modest witnesses" of nature, as we are in Figure 4 and virtually all other molecular visualizations;[9] rather, we are fully embedded in the molecular world. The film opens with a deep-focus shot of white blood cells traveling down the endothelium toward the camera (above). To create the effect that the spectator is participating in the action, the camera's own movements are unsteady and slightly jerky—as if the scene were shot with a handheld camera—as it continues to track the cell's migration in a following shot. The grammar of Hollywood cinematography is unmistakable in this opening sequence, and it enables the viewer to feel as if they were a part of this world and not above or outside of it.

What's intriguing is that camera mobility and agency are quickly established. After the opening sequence, the camera begins to move freely, traveling between interior and exterior cellular environments with speed and efficiency. The camera moves independently of the figures on screen, often zooming in and out of one molecular process only to travel to a new site of interaction with little regard for following and establishing a relationship with its molecular characters. By minute two, the leukocyte is all but forgotten and is only re-found at the end of the film. Furthermore, the frame is the most active agent in the short film: it is constantly in motion and unencumbered by the many spatial and temporal differences among molecular and cellular environments mathematically known to exist. The mobile frame functions as the great equalizer, leveling distinctions and bringing heterogeneous systems within its orbit. If the viewer is immersed in cellular and molecular worlds through computer animation, then she is not submerged in it as another molecule—there is no POV shot to establish this. She is more like an alien in this world, adopting a perspective that no other system can.

What's striking is that leukocyte migration receives an entirely different significance in *The Inner Life of the Cell* than it does in von Andrian and Mackay's graphic. Where the latter depicts the process as linear and mechanical, the former depicts it as anything but causally determined, at least in any straightforward way. Indeed, the space of molecular interaction is now a relational space, as all elements appear to be interconnected. This point is driven home when the camera returns to the leukocyte at the end of the film, suggesting that leukocyte migration is a part of a gigantic network of molecular interactivity that is continually in motion, forming and reforming itself, and never static; all of these relations are unfolding in time. The many timescales of molecular and cellular processes are made visible.[10] Some are incredibly fast—a cycle is completed almost instantly—while others are incredibly slow, too slow for the human to perceive. Computer animation makes these various and complex temporalities perceptible. Where more traditional data visualization and molecular imaging techniques have to forgo including the temporal complexity of molecular systems (see chapter 2), animation brings this elusive dimension of molecular systems into view. The molecular process depicted in Figure 4 all of a sudden looks like a very different process: while many of the same elements are there—leukocyte, endothelium, and so on—their meaning and significance have changed dramatically.

Since the release of *The Inner Life of the Cell* in 2006, the field of computerized molecular animation has exploded: from the spread of online galleries and educational resources—such as MolecularMovies .org, Biovisions at Harvard University, Cell Image Library, and *E. O. Wilson's Life on Earth*[11]—and major conferences on molecular animation[12] to the publication of numerous scientific papers in top-tier journals and funding for major National Science Foundation (NSF) projects.[13] While not all computer animations are created equal—for instance, Janet Iwasa (2010) notes that the goal of Molecular Flipbook is "not to make cinematic-quality animations"—the overwhelming majority of animations not only replicate many of the cinematic techniques used in *The Inner Life of the Cell* but also build on and even surpass them.

Protein Packing (2013) is the third film in *The Inner Life of the Cell* series made by XVIVO Scientific Animation for Harvard University. Much like *The Inner Life of the Cell,* camera mobility is established at the outset of the film. What distinguishes *Protein Packing* is that ominous nondiegetic sound accompanies the spectator's movement into the image toward a neuronal cell. Together these techniques pique the spectator's interest and curiosity as the camera moves into the image. But if the cell is what lies in the immediate field of vision, it is certainly not the camera's destination; it is not what is luring the camera into the depths of the image. Rather, it is what lies behind the cell wall that grabs and focuses attention. Resembling countless Hollywood films where there is something unknown and strange lurking behind the door, the cell wall is the door that conceals another world—a world that is likely to contain very different spatial and temporal magnitudes. On-screen desire is led to what is in excess of it—off-screen space.[14]

But like all good Hollywood features, the door is opened. And in *Protein Packing,* this means that the spectator penetrates the cell wall: the camera literally finds a portal (see Figure 7b) that leads to a world that does not obey the same spatial and temporal logic. The spectator enters a world inhabited by a frenetic ensemble of elements (proteins), whose jostling and crowding (protein crowding) bear little resemblance to the subdued activity in the extracellular world it just left behind. The spectator is overwhelmed by what seems like molecular chaos (see Figure 7c). The camera itself cannot seem to find stability: the image shakes as much as the proteins do. Eventually, the camera finds the coordinated operations of molecules, such as motor proteins transporting synaptic vesicles on microtubules in the cytoplasm. Other molecular interactions (e.g., clathrin molecules assembling into a coat promoting the curvature of the golgi membrane, which it then sheds) appear organized as well, even rhythmic, resembling the choreographed movements of a ballet, a "molecular ballet" (see Figure 7d). But after lingering on these organized movements momentarily, the camera zooms out, leaving the chaos of proteins jostling to fill the image before it fades out.

Much like *The Inner Life of the Cell, Protein Packing* establishes camera autonomy at the outset. But what's noteworthy about this film, and unlike many other molecular features, is that it projects cinematic desire off-screen, to what lies outside of the frame in the opening sequence. And yet *Protein Packing* anchors this off-screen space (the cell interior) firmly within the frame. The cell interior functions as the excess that organizes on-screen desire. Absence is what structures presence: what is outside the frame is actually what is most inside of it. The cell interior does not remain a virtual presence, however, a ghost that haunts on-screen space: the camera penetrates the cell wall, actualizing the desire to seize the kernel of excess, to consume what is off-screen. What it discovers is a world that does not obey the same rules as the one that it just left. It is a world that obeys an altogether different logic. If the camera is not comfortable at first, finding proteins to be overcrowded, noisy, and obstreperous, eventually it discovers organization, something that it recognizes and that can help it to find its bearings. The outside is now consumable.

Throughout his career Deleuze wrote a good deal about the "outside," and in many different contexts—philosophical, political, aesthetic. In his more strictly philosophical works, he refers to the outside as the virtual, after the philosophy of Henri Bergson, and I think this notion can shed important light on the relation between inside and outside space in *Protein Packing.* What is crucial to understand about the virtual, and especially in the context of digital media, is that it is not assimilable to colloquial notions of "virtuality": namely, cyberspace, virtual reality (VR), simulation, and so on.[15] Where in the latter context, the virtual seems less real than actual experience, for Deleuze, the virtual is just as *real* as the actual: it designates the *real conditions* of actual experience. The virtual and actual are just real in different senses (Deleuze 1994, 211).

The requirement that the virtual–actual pair be real in different ways could not be more important for Deleuze. In fact, this is what

a

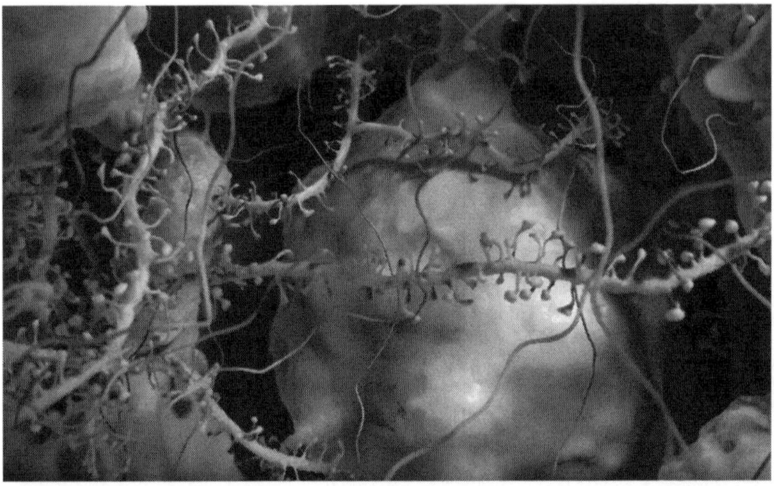

b

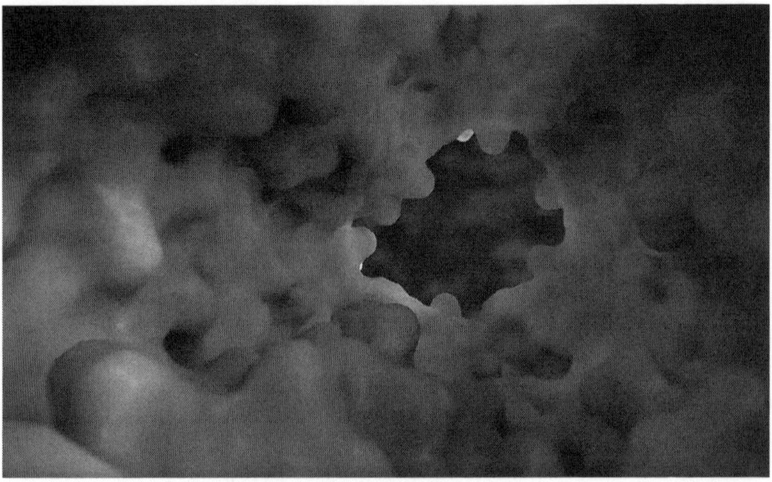

c

d

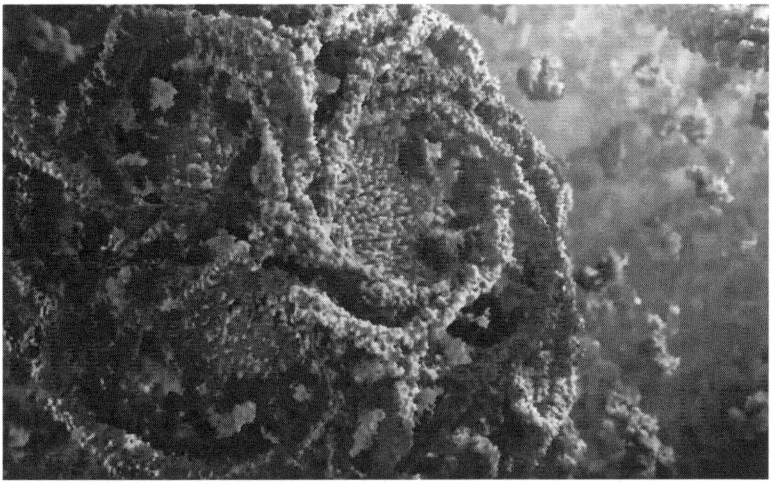

Figure 7. *Protein Packing*. Directed by XVIVO Scientific Animation, 2013. *(a)* The camera approaches a neuronal cell. *(b)* The camera discovers a portal. *(c)* Molecular chaos. *(d)* Shedding of clathrin molecules, which resembles a molecular ballet.

Immanuel Kant did not understand about the transcendental conditions of experience and is what makes his transcendental philosophy largely unsatisfactory. If Kant "is the one who discovers the prodigious domain of the transcendental," then he did not realize what was required of this domain: he "traces the so-called transcendental structures from the empirical acts of psychological consciousness" (Deleuze 1994, 135). What this means is that Kant's transcendental realm does not actually provide the conditions of experience, since it is derived from what it is supposed to condition. If transcendental conditions (the virtual) resemble the conditioned (the actual), as they do in Kant, then the genesis of experience is not explained—the virtual and actual mirror each other in an endless repetition of the same, without having offered an account of how the givens of experience are actually given. This is why Deleuze, after Bergson, claims that Kant gives us the conditions of *possible* experience instead of the conditions of *real* experience—since the possible is just like the actual, it only lacks reality (211–12). For this reason, the virtual *qua* transcendental condition cannot obey the same rules as the actual: where the actual is defined by discrete subjects and objects, the virtual, by contrast, is characterized by continuous multiplicities (Deleuze 1994, chapter 4).

Deleuze's thesis about virtual conditions helps us to conceptualize the opening sequence of *Protein Packing.* If on-screen space is organized by what is off-screen (the cell interior), then this off-screen space functions as the condition for what is present in the image. But this condition is also, and importantly, governed by a logic that is nonassimilable to what it conditions: the logic governing the distribution of elements inside the cell is of an altogether different nature than what determines the organization of elements outside the cell. It is for this reason that the cell interior functions like a *transcendental* condition for the image.[16] There seems to be a very literal enactment of Deleuze's philosophy of the virtual in this opening sequence. What the film communicates as a whole, however, is something altogether different. The spirit of the opening seconds of the film is quickly belied: the camera pierces through the cell wall to get the cellular interior into its grip. While protein crowding is jarring to the camera at first, organized movement is soon rediscovered—purpose and meaningful activity are restored. The virtuality of the cellular interior is rendered comprehensible (and therefore no longer virtual); it is something that can be brought into view instead of remaining a spectral presence haunting the cellular exterior.

Protein Packing demonstrates an important point about molecular animation generally—namely, that the visual technology has the capacity to eliminate the outside of the frame. Of course, *The Inner Life of the Cell* and most other animations in the biomedical sciences are excellent examples of the technology's ability to layer time into the visualization of cellular and molecular processes. These films are therefore a part of an important genealogy of scientific and medical visualization that animates what is often rendered static and lifeless (e.g., Figures 4 and 5). Subjects are no longer mute observers standing over and above processes unfolding "out there"; the mobile camera puts the perceptual subject directly in touch with the biochemical world and empowers it to overcome the spatiotemporal obstacles that would normally limit other perceptual systems. But what *Protein Packing* renders acute is the fact that mobile perception can surmount any exogenous force. Everything is available to visual perception. The camera doesn't just link heterogeneous systems; *it is the link between any and all heterogeneous systems.* I think this is what the film demonstrates so forcefully: even Deleuze's virtual is no match for cine-perception. Thus, what *Protein Packing* makes plain is that there is no outside to molecular perception.

Given this, I want to venture the hypothesis that if there is a dominant tendency or logic of the visual in computerized molecular animation, then it is this: the animate image empowers visual perception to surmount any and all relation to externality.[17] There is an important qualification to be made, however: insofar as the animate image lends vision the capacity to conquer all exogenous forces, then there is an important sense in which there is no possibility of exhausting the molecular and cellular forces that need to be conquered, that need to be made actual for visual perception. This is what keeps the frame mobile, flexible, and even desirous. But it is precisely this tendency or logic of the visual in computerized molecular animation that needs to be conceptually and genealogically situated. This is a tendency that did not emerge ex nihilo and has a rich and complex history linked to various practices, discourses, and technicities of perception. Although we have noted vague similarities between computer animation and other contemporary data visualization techniques (see above), in the remainder of this chapter I contend that there are also clear affinities between the logic of the visual expressed in molecular animation and the one that popular media culture reproduces. It is the nature of this relation that has yet to be characterized in either the scientific literature or in film and media scholarship. And it is just this deficit in the literature that has made molecular animation's relation to popular media and scientific epistemologies difficult to understand; it is also why these connections demand the careful attention of media philosophers.

In the remainder of the chapter, I show how the characteristics of this visual logic fit into a much wider conversation in cinema and media studies about the computerization of the cinematic image and its relation to animation within popular screen media. I argue that the connection between molecular animation and popular media culture becomes readily apparent in debates about the ontology of the cinematic/animatic image in film and media studies. I then supplement the shortcomings of these discussions with LaMarre's assessment of cinematic and animetic *tendencies* of the image. This overview provides a framework for understanding molecular animation in terms of a broader set of visual media logics that have taken hold of the popular mediasphere. It is this genealogy that also gives critical purchase to the suspicion among some scientists that computer animation should not be used for "serious" scientific research. In chapters 2 and 3 I offer a very different way of historically situating this very same logic of the visual.

NEW HISTORIES OF CINEMA AND ANIMATION

Animation is no longer the ugly stepchild of the cinema. In the last few decades, dozens of manuscripts and edited collections have been published on the varieties and potentials of the animated image—in, for example, its CGI, cel, planar, and experimental forms—in Western and non-Western contexts.[18] It's also not uncommon to see essays on animation published right alongside more traditional essays on film and media criticism. What's more, computer animation has clearly been embraced by academic audiences well beyond cinema and media studies and has been integrated into discourses within the life sciences, architecture and design, education, experimental media arts, and other areas. It is not inconsequential that animation has not always been so readily accepted by cinema studies and the wider academic community. And in this context, it is not uncommon to hear cinema and media scholars discuss animation's movement from the margins or peripheries of scholarly interest right into the center (Gunning 2014; Beckman 2014b).

What grounds this spatial metaphor is far more complicated than cinema's warm embrace of an overlooked set of practices. One might argue that these shifts in concern happen against the backdrop of what many scholars have diagnosed as a "crisis" in the cinema.[19] The terms and severity of this so-called crisis have been widely debated in the last couple decades, and there is perhaps still no consensus on the fate of the cinema. Nevertheless, at the heart of these wide-ranging conversations is an abiding concern for the effects of digitization on the cinematic image. For some, "digital cinema" is not cinema at all, while for others it is an opportunity to disrupt narrow definitions of *what cinema is* (Andrew 2010) and to forge more inclusive conceptions of expanded cinematic spaces that include ubiquitous and mobile screens among the many varieties of cinematic experience today.[20] Whatever the case, the challenge that computerization seems to pose for the ontologies and epistemologies of the cinema opened up a space for a lively, if sometimes tired, debate about the role of animation within moving-image culture. And while I cannot provide an exhaustive overview of animation's rise to scholarly prominence, and the implications it may have for the cinema,[21] I want to address certain aspects of this conversation in order to situate the emergence of molecular animation's visual logic within a wider set of media logics. And in doing so, I will also have the occasion to nuance our understanding of what animation is in relation to the cinema.

Now, it may be slightly misleading to begin our analysis of animation's relation to the cinema with the (potential) challenge that computer graphics pose to the hegemony of celluloid within moving-image culture. It covers over the fact that film and media theorists have for some time now shed light on the relation (and even identity) that animation and film share. Sergei Eisenstein, Béla Balázs, Germaine Dulac, and Dziga Vertov have all offered valuable meditations on the animate image.[22] And more recently, film scholars such as Stanley Cavell, Dudley Andrew, Tom Gunning, D. N. Rodowick, Kristin Thompson, Esther Leslie, Alan Cholodenko, and others have sought to expose the overlaps between the technologies, even if their exact relation remains fairly ambiguous in their writings. For example, Cholodenko has argued for decades now (even before the so-called turn to animation) that all cinema is at bottom animation: "Every encounter with film is an encounter with animation—cinema, that is, live action film, included" (2008, 1; 1991). Similarly, in *The Virtual Life of Film,* Rodowick proposes that every celluloid film is an animation: "Succession," he writes, "means that fundamentally every film is an animated film as the automated reconstitution of movement from a succession of still images" (2007, 53). However, in her introduction to *Pervasive Animation,* Suzanne Buchan (2013) situates Rodowick's claim, along with Andrew's and Sean Cubitt's (2004), in terms of a much deeper set of commitments to thinking about cinema as a photo-indexical medium. For even if the cinema is animation, and animation functions as cinema in its "purist form" (Andrew 2010, 30), such contentions are grounded in intellectual projects that uncover the specificity of cinema as photo-indexical. Here, Rodowick corrects certain assumptions about the differences between animation and cinema:

> Regardless of the wonderfully imaginative uses to which they are put, and the spatial plasticity they record, cell animations obviously have a strong indexical quality. Simply speaking, each photographed frame records an event and its result: the succession of hand-drawn images and cells reproduced as artificial movement through the animation of succession. (2007, 121)

For Rodowick, no matter how fantastical the worlds depicted on hand-drawn or painted images are, they are all recorded in succession just like any other live-action films—there is a pro-filmic event in cel animation. The differences between animation and cinema are therefore overblown: if every film is animated then every animation is also a film.

The sidelining of animation in the history of moving images may have less to do with ontological differences than it has to do with the persistence of a realist ideology. Citing Thompson, Buchan underscores the dominance of realist ideology and its effect on animation: "wishing to conceal its own constructedness from audiences" (2013, 4), "the novelty effect in live-action films" was pushed into the background and so animation was marginalized in the history of moving images as a result

(K. Thompson 1980, 106). Some version of this story has been told many times over in cinema and media studies and should not strike us as especially new or surprising (see also Manovich 2001). The most interesting aspect of it concerns how animation is recognized as an important part of the history of moving images only insofar as it is identified with the cinema: animation is cinema freed from the constraints of "ordinary space and time" (Andrew 2010, 30; Buchan 2013).[23]

What is intriguing about this narrative is the way in which computer graphics challenge many of the core assumptions that scholars have about the cinema, as well as animation's role within it. With the return of manually constructed images in the 1990s, there could no longer be the same confidence in the identity of cinema across its many permutations—from classical Hollywood film to the avant-garde and Disney animations—with a photographic ontology. For Manovich, live-action footage turns out to be one among the many other raw materials that can be manipulated on the computer (Manovich 2001, 300–301). This cuts to the very core of what cinema is: "Cinema is the art of the index; it is an attempt to make art out of a footprint" (295). And so for many scholars in the late 1990s and early 2000s, the emergence of digital cinema and the prominence of computer animation within it (Manovich: "Cinematography is subordinated to 3-D computer animation" [305]) reveals the narrow assumptions underwriting the history of theorizing cinema, and especially its commitment to live action (and not just realism). What is also revealing is that the hegemony of live action also determined the conditions under which animation could be legitimated within moving-image culture. Put another way: animation had been colonized by live-action cinema.[24] The emergence of digital cinema and computer animation challenged scholars to reevaluate the epistemologies and ontologies of the moving image.

It is in the context of this search for other narratives and meanings for cinema and animation that we find Manovich's *The Language of New Media* (2001) situated. There, Manovich famously reverses the hierarchy of cinema and animation: "The history of the moving image thus makes full circle. *Born from animation, cinema pushed animation to its periphery, only in the end to become one particular case of animation*" (2001, 302). For more than a decade now, this bold assertion has received no shortage of criticism. Mark B. N. Hansen meticulously picks apart

Manovich's position and argues that it equivocates with the notion of the cinema, and for this reason is unable to appreciate what is truly *new* about new media (Hansen 2004, 32). And even though Manovich's thesis has become a requisite citation for animation scholars charting the medium's relation to the cinema, it is almost universally criticized and charged with a range of technical, historical, and ontological essentialisms, including its lack of attention to the political and social realities of digital media.[25]

Still, in a review of Manovich's book ten years after its publication, Alexander Galloway (2011) reflects on its continued relevance as one of the essential contributions to the first generation of internet culture research. What still seems particularly attractive about the book, despite its many equivocations, essentialisms, and distinctly modernist tone, is that it takes seriously the idea that digital technologies are expressive formal objects. In other words, new media objects are aesthetic objects that can be rigorously classified; and this classification continues to be of interest years after the publication of *The Language of New Media*. This is not to say that Manovich's work has not evolved since 2001; and indeed, it has evolved since Galloway wrote the review of it in 2011. In *Software Takes Command* (2013), Manovich deepens his earlier insights about new media, arguing that what we traditionally think of as medium specificity (film, cinema, video, etc.) is really a function of ubiquitous software. The animation/cinema divide is less important than software's metamediation of all data: "There is no such thing as 'digital media.' There is only software," insists Manovich; "The 'properties' of digital media are defined by the particular software as opposed to solely being contained in the actual content (i.e., inside digital files)" (2013, 152). What this means, and is really at the heart of the book, is that the software interface mediates data fields that are largely inaccessible. Software is the mediating layer that allows phenomenal features, such as digital images, to manifest.[26] And while these observations about software metamediation are not unimportant to *The Language of New Media*—indeed, Manovich calls for the invention of "software theory" to understand the "logic of new media" (2001, 48)[27]—the tone and emphasis has changed somewhat by the time he writes *Software Takes Command*.

What remains consistent is Manovich's commitment to the formal analysis of media undergoing computerization and to the marginalization of questions related to software and political economy, as well as other questions related to the cultural and historical production of software logic (2013, 224). And what remains intriguing and useful about *The Language of New Media* for our discussion of computer animation, even if some of its conceptual architecture is dated, is that the logic of the computerized moving image (and its connection to the history of animation/cinema in particular) is foregrounded in the context of a formal taxonomy of new media—or more precisely, in terms of what Manovich later calls the metamedium of software. And what this taxonomy of the image makes clear is that there is no ontological difference between animated molecules used in scientific settings and other digital media content. While this taxonomy will prove helpful in a number of ways, it will ultimately leave us unsatisfied and searching for criteria that allow us to make more useful distinctions between computer images intended for scientific audiences and those intended for popular audiences. Why this matters will become clear below. In what follows, I return to the molecular animations discussed above and use Manovich's classificatory scheme to unpack their meaning and relevance in the context of the renewed importance of animation in the age of computerized media. (It may be useful to note that these features do not drop away from Manovich's later work, although they do not receive the same emphasis in his analysis of the ubiquity of software mediation.)

THE LANGUAGE OF MOLECULAR MEDIA

According to Manovich, there are five distinguishing principles of new media: numerical representation, modularity, automation, variability, and transcoding (Manovich 2001, 18–48). While these principles represent "general tendencies of a culture undergoing computerization" (27), the language of new media is not all that new: the so-called "formal essence of the technology" is expressed in various ways throughout the history of media practices. Indeed, layering, compositing, recombination, modularity, variability, automation, and even binary code, which are all cornerstones of working with digital images, are present in much earlier media practices, including twentieth-century avant-garde film (including the work of Norm McLaren, Len Lye, Stan Brackage, Dziga Vertov, and others). Especially for the avant-garde, techniques such as montage, collage, painting, scratching, as well as other practices were used to "[revolt] against the indexical identity of the cinema" (306) and in many ways prefigure digital cinema.[28] For this reason, "one general effect of the digital revolution," Manovich proclaims,

> is that avant-garde aesthetic strategies came to be embedded in the commands and interface metaphors of computer software. In short, *the avant-garde became materialized in a computer.* Digital cinema is a case in point. The avant-garde strategy of collage reemerged as the "cut-and-paste" command, the most basic operation one can perform on digital data. The idea of painting on film became embedded in the paint functions of film-editing software. The avant-garde move to combine animation, printed texts, and live action footage is repeated in the convergence of animation, title generation, paint, compositing, and editing systems into all-in-one packages. (306–7)

Despite what may seem like a gross oversimplification of media history, combined with a deeply apolitical reading of the avant-garde and digital media aesthetics, Manovich manages to shore up a prominent place for animation in the history of cinema not by pitting animation against cinema but by finding deeper continuities between the two. And he does so all the while breaking animation out of the straitjacket of photo-indexicality. This is no small feat. The result is that pride of place is given to the manipulation, recombination, compositing, sampling, modularity, and variability of images, or in short, to those features that strip the image of a photographic ontology. Those traces of artifice that were relegated to the margins of cinema as animation have reemerged in digital cinema qua computer animation.

In order to frame the molecular animations we already know (*The Inner Life of the Cell* and *Protein Packing,* along with just about every other computerized molecular animation)[29] in terms of Manovich's taxonomy of computerized media, we would have to understand these visualizations in terms of their mathematical reality—that is to say, as numerically defined objects (first principle). And these numerical objects are subject to various forms of "algorithmic manipulation. For instance, by applying the appropriate algorithms, we can automatically remove 'noise' from a photograph, improve its contrast, locate the edges of the shapes, or change its proportions" (2001, 27). Each of the "objects" in a molecular animation have been algorithmically generated from Maya Autodesk, or from some other animation software, using a variety of programming languages, and are therefore programmed to have the shapes, colors, contrasts, movements, and behaviors they do.

Additionally, these objects are "modular" (second principle) or "fractal," which is to say, each element in an animation can be combined into a larger object without the individual object losing its identity (30). We can see the principle of modularity at work in many different ways. For instance, *The Inner Life of the Cell* appears on many different media platforms, for a variety of users, and it is constantly recombined with other content. In fact, the film itself rarely appears in its full version; it is broken down into shorter videos and stills and then combined with sound, (hyper)text, and other image components. Clips and photographic stills of *The Inner Life of the Cell,* along with countless other molecular animations, appear on the websites of TED and pharmaceutical companies, as well as on university and lab webpages, social media sites, television shows, documentaries, and so on. Molecules are viewed and consumed right alongside other digital media content. Just as John Liebler's animation of the kinesin protein appeared in the feed of thousands of Facebook users (see Figure 2), so have countless other animate molecules. In short, the images retain their identity even as they are absorbed into larger media platforms and are circulated and consumed like any other digital object.

The programing for molecular animation also has a modular structure. Molecular animators frequently work with the animation software Maya Autodesk (although more and more alternatives are available now) and use MEL, Maya's embedded programming language, along with Python, an object-oriented programming language (OOP), in conjunction with various custom plug-ins (such as Molecular Maya [mMaya]). These programming languages allow programmers to work with individual units, called "objects," that are self-contained, but can combine and recombine with other units without them losing their identity. This makes it possible to "automate" (third principle) many of the programming tasks (I will return to this in chapter 3 when I discuss OOP). For example, with mMaya, the geometries of molecular structures are preserved (from the Protein Data Bank files) in Maya, allowing programmers to automate the animation of many structures, including DNA.[30]

Perhaps more interesting still is how molecular animation expresses the fourth principle of new media: "variability." Manovich writes that "[a] new media object is not something fixed once and for all, but something that can exist in different, potentially infinite versions" (2001, 36). Similarly, in *Software Takes Command,* he notes that "variability is played out as seemingly endless color schemes and transparency settings. It does not matter how many times we have already seen the same space, it always can appear in a new way" (2013, 322). And indeed, one of the most compelling features of molecular animation is its capacity to explore seemingly infinite versions of cellular and molecular processes. For instance, each shot in *The Inner Life of the Cell* and *Protein Packing* varies the possibilities for viewing molecular and cellular operations with countless angles layered together with new visual phenomena (in the foreground and background), combined with a wide variety of other elements (sound, text, hypertext, etc.). Time itself becomes another layer of manipulation in these animations, varying the possibilities for perceiving the speeds and slownesses of biochemical worlds.[31] These elements work together in molecular animations to generate cascading possibilities for viewing molecular and cellular activity.

Let's take a closer look at what the principle of variability means in the context of molecular animations. In their 2012 film *X Inactivation and Epigenetics,* which was produced for the Walter and Eliza Hall

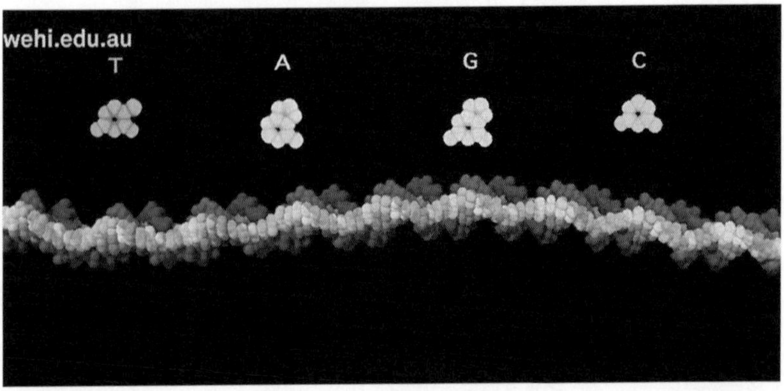

Figure 8. *X Inactivation and Epigenetics.* Directed by Etsuko Uno and Drew Berry. The four bases.

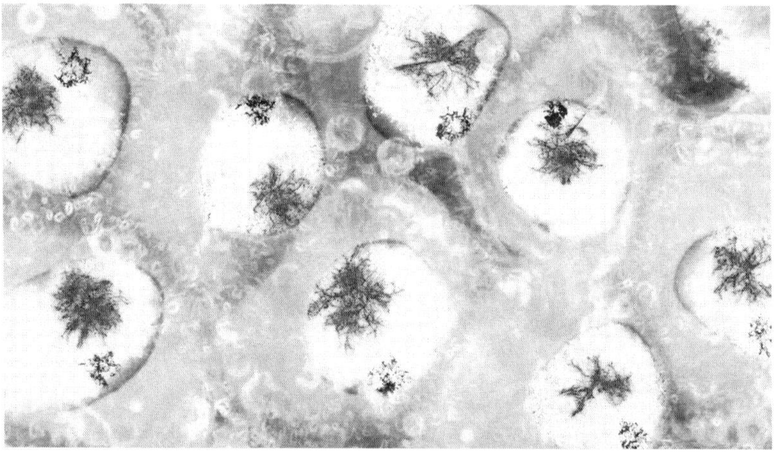

Figure 9. *X Inactivation and Epigenetics.* Female cells with active and inactive X chromosomes.

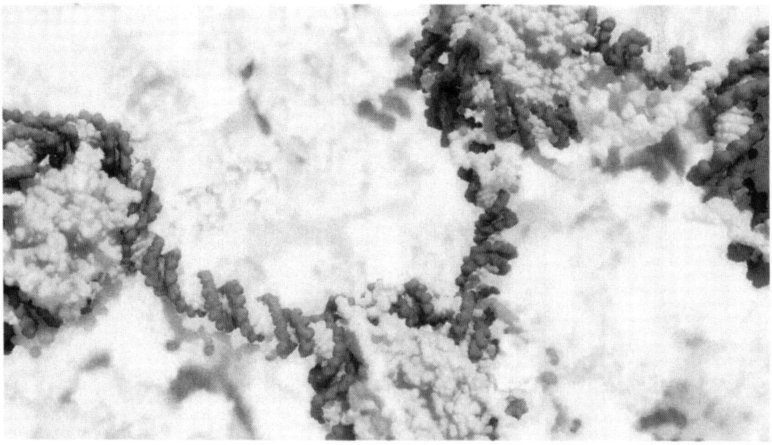

Figure 10. *X Inactivation and Epigenetics.* Active X chromosome.

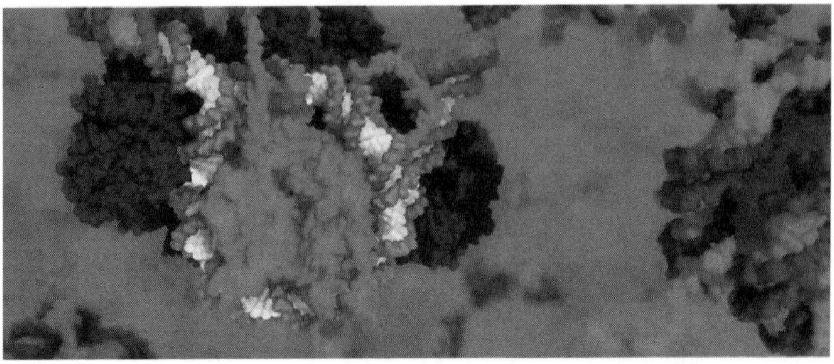

Figure 11. *X Inactivation and Epigenetics.* Nucleosome sliding.

Figure 12. *X Inactivation and Epigenetics.* Inactive histone tail modifications.

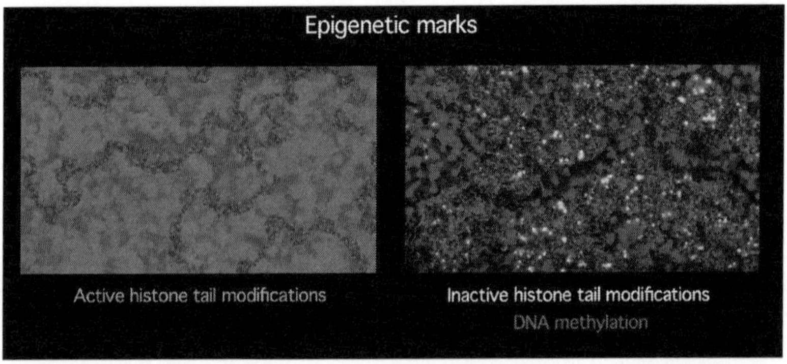

Figure 13. *X Inactivation and Epigenetics.* Active histone tail modification and DNA methylation are examples of epigenetic marks. They are modifications that distinguish between active and inactive chromosomes.

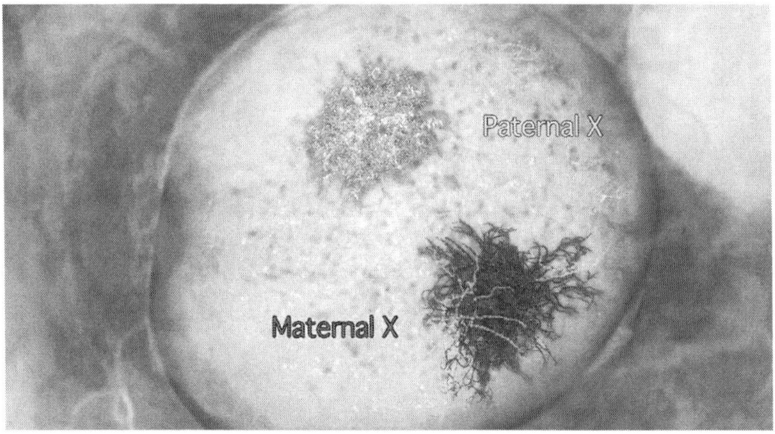

Figure 14. *X Inactivation and Epigenetics*. Single cell from a 100-cell embryo. At this early stage of development, the female embryos still have two active X chromosomes.

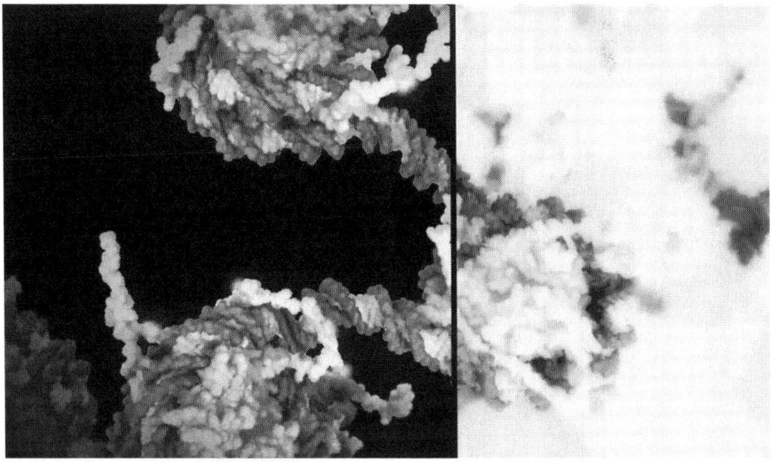

Figure 15. *X Inactivation and Epigenetics*. The inactivation of X chromosome.

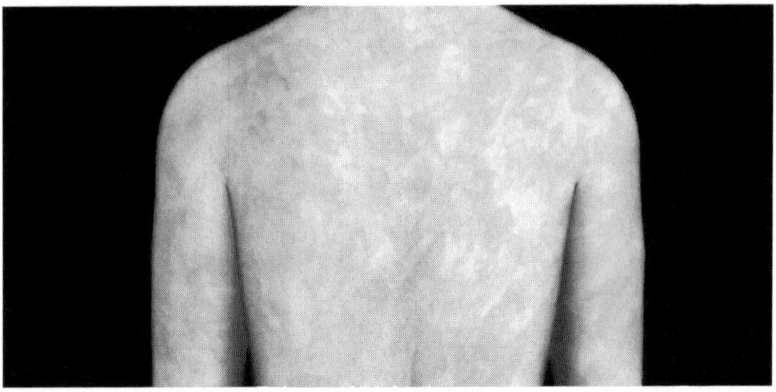

Figure 16. *X Inactivation and Epigenetics.* Hypothetical visualization of X inactivation on the skin of an adult female.

Institute of Medical Research in Australia, Etsuko Uno and Drew Berry examine the process by which one copy of the X chromosome in female cells (females carry two copies) is transcriptionally inactivated or silenced due to an epigenetic mechanism that corrects for a "potentially toxic double dose of X-linked genes" (Ahn and Lee 2008). What's important has less to do with the theory of epigenetics put forward in the animation than *how* it is advanced by means of the variability of simulations of transcription in the active and inactive X chromosomes.

What stands out about *X Inactivation and Epigenetics* is how the film experiments with the visual display of epigenetic regulation of DNA on X chromosomes. Figures 8–16 give us a sampling of the wide range of possibilities for imaging DNA (Figure 8) in different transcriptional settings (Figures 9–16). Not only do Uno and Berry experiment with scale—10,000,000x magnification (Figures 13 and 14), 1,000,000x magnification (Figure 13), 5,000x magnification (Figures 9 and 14), and 1:1 (Figure 16)—but they also experiment with the layering of different background and foreground relationships, as well as multiple color schemes, in order to showcase the potentially unlimited possibilities for visualizing the same object. And unfortunately, what these stills don't capture is how each of these experiments unfolds in time, with different speeds and rhythms of DNA transcription corresponding to each of the spatial arrangements. *X Inactivation and Epigenetics* gives the viewer a sense of the limitless possibilities for varying the spatial and temporal assembly of the same phenomena.

For Manovich, the variability of the media object, along with its automation, modularity, and mathematical existence, cannot be isolated from the final principle of new media: "transcoding." This principle describes how a media object is a part of "human culture": on the one hand, it is an artifact imbued with meaning and significance, and on the other hand, it is a "computer file that consists of a machine-readable header, followed by numbers representing color values of its pixels" (2001, 45). What this means is that molecular animations are at once a part of a system of human values (indeed, they are shaped at the intersection of various historical and epistemological transformations) and a part of the "computer's own cosmogony" (46). It is important to note that although Manovich (ostensibly) leaves room in his theory to evaluate the way in which human values shape contemporary media systems, this is ultimately not his interest and takes a back seat to his more immediate concern: the formal analysis of computerized media. This is true for both *The Language of New Media* and *Software Takes Command.* In the latter work, he admits that "we should not forget that the practices of computer programming are embedded within the economic and social structures of the software and consumer electronics industries. These structures impose their own set of constraints and prerogatives on the implementation of hardware and software controls, options and preference." However, what is "more interesting for media theory," Manovich continues, is that "software devices aimed at consumers significantly automate media capture and editing in comparison to their professional counterparts" (223–24). In other words, the logic of software is more interesting and important to media theory than the political economy in which it is embedded. I will have the occasion to return to many of these issues later in this chapter and throughout the book (see especially chapters 6 and 7).

But in Manovich's taxonomy, there is no ontological distinction between the digital images used in the laboratory setting for scientific research and those consumed online for entertainment, education, or other purposes. Digital images are an effect of computer software mediating user experience. Ontologically, digital images are of the same species; they are no more or less in touch with real/raw data, since they are all a product of software interfaces subjecting data to similar processes of mediation. It is the software and not the data we encounter (Manovich 2013, 154). What this means is that molecular animation, like any other digital media practice, is an experiment with software. Not weighed down by the real limitations of human perception, molecular and cellular animations freely experiment with images of transcription at 10,000,000x magnitude that are physically impossible to secure. Likewise, the steady movement into the concealed (virtual) world of the neuron in the opening scene of *Protein Packing* is produced in the "computer's own cosmogony" and is not constrained by experimental data about the spatial and temporal relationships of the neuron. This is a *digital* neuron and not a biochemical one.

If molecular animations are seen as experiments in software, then it seems fitting that *The Inner Life of the Cell* was first screened at SIGGRAPH and not in front of an audience of biologists. It also seems appropriate that scientist/animators, such as Drew Berry, David Bolinsky, and Janet Iwasa, would give TED Talks with millions of views and see their animations shared and consumed on social media sites. The entertainment value of molecular animation also explains why molecular animations are prominently featured in the 2012 BBC documentary *Our Secret Universe: The Secret Life of the Cell.* In this feature-length film, DNA transcription (see Figure 17a) is fully dramatized and bears a striking resemblance to a sequence from *Alien,* or some other Hollywood sci-fi film, complete with characters, protagonists, alien enemies, chase sequences, and a suspenseful soundtrack. If Figure 17a depicts the hero of our narrative—the molecular machinery in the cell, whose coordinated efforts ensure the faithful production of proteins—then the next shot (see Figure 17b) portrays the body's "ancient enemy": the virus. Marshaling stunning computer graphics and a host of sci-fi clichés, the virus is depicted as an "alien" invader that comes from elsewhere—outside of the body—and is "bent on destruction." The white blood cell (see Figure 17c) is called upon to save the cellular universe from alien invasion.

Outside of Hollywood-style graphics and narration, there are other possibilities for molecular entertainment. For instance, Ribosome Studio is producing some of the most visually stunning molecular animations in the field. Their stated mission is "Make science beautiful."[32]

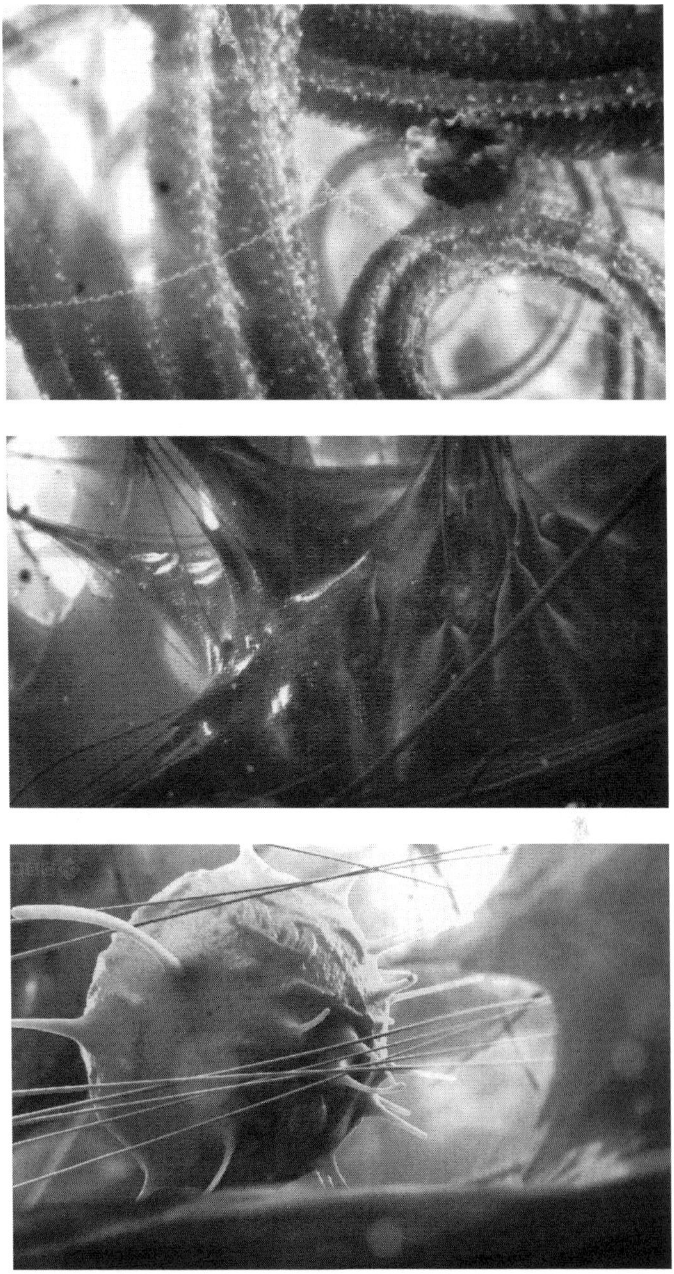

Figure 17. *Our Secret Universe: The Secret Life of the Cell*. Directed by Mike Davis. Wide-Eyed Entertainment. Distributed by the BBC, 2012. *(a)* DNA Transcription. *(b)* The Virus. *(c)* The White Blood Cell.

They "take your [client] scientific and medical discoveries and developments, and translate them into a beautiful medical animation, an impressive scientific illustration, or a stunning mechanism of action movie." What is striking about Ribosome's graphic style is how it simulates a Claymation of old physical models of biomolecules. But in simulating these earlier aesthetic forms, their films also announce that they do not aspire to cinematic realism. It's almost as if they celebrate the fabrication

Figure 18. *The Molecular Interactions of the MAPK Pathway*. Directed by Ribosome Studio, 2018.

Figure 19. *The Secret Life of Levodopa*. Directed by Ribosome Studio, 2018.

Figure 20. *Plasma Membrane*. Directed by Ribosome Studio, 2018.

of models and the use of earlier visual techniques (stop-motion) to capture it. This could not be further from the graphic styles used in most other molecular animations.

And yet, many of Ribosome's animations make use of interactive tools that few other molecular animations do. Short films, such as *Cell Membrane* or *Plasma Membrane* (see Figure 20), are 360-degree spherical videos that immerse the viewer in the world of the molecular model. These short films can easily be displayed on a phone using the device's accelerometer and magnetometer. And in these instances, Ribosome's animations come much closer to simulating a VR or AR experience of physical models than they do to replicating low-tech forms of visual capture. No distance separates the viewer from the object: she can explore and navigate her way through a simulated world of the physical model. Thus Ribosome's work is not about technical or aesthetic restraint or even a commitment to low-tech animation as we might see in Hayao Miyazaki's work. Quite the contrary, their animations announce their accommodation of both ends of the technical–aesthetic spectrum: from ultra-low-tech biochemical modeling and stop-motion filmmaking to the experimental spaces of AR and VR cinema. Computer animation can simulate physical models, stop-motion techniques of filming, and immersion into the world of physical models. Despite appearances, Ribosome's animations are just as sophisticated as films like *The Inner Life of the Cell*—perhaps even more. They too are experimenting with the modularity and variability of digitized molecules, but in a way that celebrates and immerses the viewer in its fabrication of molecular environments.

MOLECULAR NOISE AND ANIMATE ARTIFACTS

It should come as no surprise that many life scientists would find animation to be less than "scientific." As the biochemist and cell biologist Peter Walter noted in an interview for the *New York Times,* "Some animations are clearly more Hollywood than useful display," which means that "it can become hard to distinguish between what is data and what is fantasy."[33] It seems that for many working biologists, the more that molecular visualizations show themselves to be media objects not "held back by the drag of ordinary space and time" (Andrew 2010, 30) and experiment with the affordances of modularity, variability, and automation of the numerical image, the less confidence they have in their value for scientific research. While computer animations may be useful for popularizing biology, for disseminating information to mass audiences, and for educational purposes, a good deal (if not all) of them seem ill-suited for research. In fact, this view is not only popular among computer animation's detractors but it also reflects how many scientists and animators working in the cellular and molecular animation community feel about their own work—namely, the value of computer animation derives from its ability to educate students and the wider public about molecular and cellular processes that occur in time (Jenkinson and McGill 2012; Tasker and Dalton 2008). For an increasing number of researchers working in 3D time-based visualization, it is important for computer animations not to be confused with those technologies used in experimental settings to determine the structure and function of the biochemical world. Rather, animations help "to make sense of observations in the laboratory" and thus "stimulate the imagination, bringing a new dimension to learning chemistry"; however, at the end of the day their main use is to illustrate what is already known about the molecular and cellular phenomena in compelling ways (Tasker and Dalton 2008, 116, 117).

Although this view characterizes how a number of scientists feel about molecular animations, it is important to get clear on what exactly seems unsuitable about the technology for bioscientific research. That many computer animations are fantasy, or that they distort what is known about the biochemical world, does not tell us much. And what's more, the answer cannot be as simple as the following thought: molecular animations are digital simulations of molecular interactions, and so they do not index the "real world." In the world of twenty-first century biology, producing computer models, simulating molecular structures and processes, and managing large online databases is part and parcel of doing research in the biosciences. As Catelijne Coopmans, Janet Vertesi, Michael Lynch, and Steve Woolgar point out in their introduction to *Representation in Scientific Practice Revisited,*

> Computer screens have pride of place in laboratories and scientific offices, where researchers' attentions are as likely—or more likely—to be focused on colorful digital images, simulations, software suites, databases, or lines of code as on unruly specimens or instruments. Biomedical imaging enrolls fMRI and PET scans alongside X-rays, which themselves are frequently digitally manipulated to produce new modes of vision. (2014, 1)

For many scientists there seems to be a fine but important line that separates computer simulations bound for YouTube or the high school classroom and those intended for the discerning eyes of scientists trying to determine the structure and function of a macromolecule. But what is this line? How is it determined? And is it stable or a moving target?

To get some clarity, I want to follow Natasha Myers's discussion of an online debate that erupted after David Bolinsky's TED Talk about molecular animation and *The Inner Life of the Cell* (Myers 2015).[34] She notes that Bolinsky is harshly criticized by scientists for "fictionalizing" and even "fetishizing" the molecular world by anthropomorphizing very complex processes and making it seem as if molecules have intentionality. Frank Mason proclaims that "what David is really doing is 'Hollywood-izing,' i.e., glamorizing or applying advertising techniques to biology. Everything has perfectly adjusted colors . . . glows, etc. All of which is fake. Much like most advertising."[35] The chemist Isaac Yonemoto then remarks, "Yeah, but what you're doing is actually lying.

In the case of animations where you are showing microtubule/microfilament polymerization, you have made it look like monomers zip onto the growing tubule, magically 'knowing' where to go. . . . This is completely bogus. . . . You have subsumed truth to display something of 'beauty,' creating a fiction in the process." (Myers 2015, 196). Bolinsky, for his part, does not defend the referentiality of his "Mickey-Mouse science," but insists that fiction is not necessarily a bad thing. He embraces the subjective nature of "facts"—indeed, biological knowledge, he goes on to insist, is always a kind of fiction. Animations are incredibly useful *science fictions.*

What is suggestive about the online debate is that it reveals deep-seated commitments to referentiality in the visual culture of science. The colors, variations of space and time, clean edges, clear relationships, and distinctive shapes—or in short, all of those things that are the effect of computerized animation—are distracting and distort what is experimentally known about the physics and chemistry of molecules. This is perhaps why the standard notation in Molecular Imaging Maps (MIMs) in Figure 5 is often preferable to animation. And even the cartoonish graphic in Figure 4 makes no claim to "realism." One might argue that it is just our sense of realism and reference that a film like *X Inactivation and Epigenetics* plays with. By including text in the corner of the image about magnification scales (10,000,000x, 1,000,000x, etc.), the film makes it seem as if DNA were being imaged using a high-power microscope, when it is merely a graphical simulation of a magnified molecular process. What's more, such practices conform to much earlier photomagnification conventions, especially those used in the microcinematography of Jean Comandon and others in the early twentieth century (see chapter 7).

There is an important footnote to make here concerning whether the conception of scientific epistemology these and other critics have is too limited. Are they subscribing to a naive epistemology of nature? Myers suggests that such critics fail to appreciate that scientific knowledge is constructed in an experimental setting. They overlook that knowledge is produced and does not depend on a direct apprehension of nature "out there." She writes that "Mason's charge of fetishism, however, neglects to take into account that 'facts' also have at their root an association with 'things made,' and so facts and fetishes share the same 'ambiguous etymology.'" "It is Bolinsky," Myers continues, "who is able to acknowledge the fabricated nature of facts" (2015, 195). While I fully agree with this intervention, and I am sympathetic to drawing on Bruno Latour's and Isabelle Stengers's use of the "factish" to support it (Latour 2010; Stengers 2010), I also don't think those who are opposed to using computer animation for scientific research are necessarily committed to the easily refutable position of naive realism.[36] And what's more, I also don't think they have to be opposed to the constructedness of scientific knowledge in order to find problems with molecular animation. The real issue has less to do with scientists' anxiety over losing access to knowledge of a world "out there" and more to do with their concern over the loss of a common reference. (In this sense, the anxieties felt by cinema and media scholars are quite different from those felt by scientists.)[37]

To unpack what this might mean, we only need to refer to Ian Hacking's work on microscopy and entity realism. For Hacking, scientists are not watching nature when they peer into microscopes. In his seminal essay from 1981, "Do We See through a Microscope?" he dispels any pretensions we might have concerning our ability to actually see *through* a microscope. Generally, he aims to correct the assumption that microscopes are not interesting or relevant to philosophy, when in fact they bear directly on our understanding of the "reality" of scientific entities (Hacking 1981, 133). But beyond this, his goal is to build a theory of microscopic vision that challenges the widely held belief that scientists are passive watchers of nature. As Hacking explains, the untrained eye (of a philosopher) cannot just "see" through a microscope: "He will certainly not be able to tell a dust particle from a fruit fly's salivary gland" (136). This leads to the "first lesson: you learn to see through a microscope by doing" (136). Observation is not passive; it is an active skill that one learns how to do. We learn how to see "only after learning what it is like to move around in that world and intervene in it" (136).

Another crucial lesson in microscopic pedagogy is that what is seen through a microscope is not the same thing as the specimen put under the microscope. This is because the light rays are not reflected directly from the object but are diffracted (141).[38] As Karen Barad explains in relation to Hacking, although geometrical optics are adequate to account for the optics of the eye and magnification (refraction), different physical principles (diffraction) are necessary for explaining the optics of light and electron microscopy (Barad 2007, 50–53). For example, she explains how using the Scanning Tunneling Microscope (which I discuss in more detail in chapter 2) is nothing like the physical process involved in taking a photo. The quantum physical phenomenon of tunneling is akin to "touching" or "feeling" the surface of a specimen through a current (instead of "seeing" it) that "enables particles to traverse energy barriers that should be, at least according to the laws of classical Newtonian physics, impossible to cross" (52). Measurements of the current are obtained, which is what allows a computer image to be generated. What this means, according to Barad, following Hacking, is that in order to "see" anything under a microscope, a great deal of preparation is involved. "Obtaining a reliable image free of all artifacts," she continues, "entails experimental know-how, intuition, ingenuity (all of which are acquired through practice), a good deal of tinkering, [and] the honing of tactile techniques in tune with the specificities of the instrumentation" (51).

But then if you do not actually see the specimen, how do you gain any confidence that what you see is not simply an artifact of the instrument? How can scientists be sure that what is registered by the microscope is actually a part of the specimen? Some version of this question has troubled generations of scientists, and in the early years of microscopy it led many of them to ban the use of microscopes in the laboratory (Hacking 1981, 1983; Keller 2002). Hacking, for his part, neither suggests that all microscopic objects are artifacts of the instrument—not all microscopic images are aberrations—nor does he claim, and naively, that the microscope allows us to see the object itself.

Instead, he relies on his own experience in the laboratory to demonstrate how a part of learning-to-see-through-doing involves selecting out aberrations. The most obvious way to do this is comparatively. Hacking gives the following example: a low-powered electron microscope reveals small dots in red blood cells, and these dots may turn out to be important, electron-dense bodies in the blood. Alternatively, they might just be artifacts of the microscope. The only way to test this is to see if an unrelated physical process—the use of high-resolution light microscopes—detects the same structures on micrographs. It turns out that they do, and so it would be an incredibly "preposterous coincidence" that two unrelated physical processes detect the same structure (Hacking 1981, 144). We can assume with some confidence that these dense bodies are structures in the cell, all the while not committing ourselves to the view that we are observing the structures as they are in themselves.

What's important is that microscopes give scientists fairly good "maps" of structures that are in the specimen.[39] But this is only because scientists have removed aberrations and artifacts and then intervene in what is left—namely, microscopic entities that scientists can test and manipulate.[40] These visual maps are not the specimens themselves, so scientists must "learn to move around in the microscopic world" produced by an instrument instead of in a world of actual specimens (Hacking 1983, 209). Hacking argues that these microscopic entities exist as *real* experimental entities, and this serves to demonstrate a much larger point about realism that he elaborates in his seminal work, *Representing and Intervening* (1983). Positioned as a rejoinder to antirealist attacks, Hacking's book argues for an "entity realism," which is the idea that scientific entities—many of them unobservable—exist to the extent that they can be manipulated experimentally and used to generate knowledge about other things. Hacking's thesis is pragmatic—or to be more precise, a form of "New Pragmatism"[41]—inasmuch as it is committed to the reality of scientific entities based on whether or not scientists can do things with them; and as it turns out, a good deal of these entities are theoretical.[42] The point here is that many of the biological entities that life scientists see, discuss, and manipulate are real, but they are not natural phenomena sensu stricto. They are *real* micrographs that scientists experimentally manipulate.

Returning to those scientists who are either explicitly critical of molecular animation or do not think that their work should be used in the laboratory setting, I would suggest that their views are rooted in legitimate concerns about losing a common reference in the production of scientific knowledge. The issue is not that computer animation constructs its objects and that it maintains no indexical relation to the external world. Rather, the issue is that far too much of what is presented has not been experimentally verified (and could not be). Scientists have not come to an agreement about the shapes, timescales, relationships, to say nothing of the alluring colors, that computer animations present to audiences. Even if we agree that scientific knowledge is produced in highly artificial environments called laboratories, and the products of these environments are a mix of fact and fetish—a "factish"—then it is essential that the scientists who participate in these ritualized experiments agree upon the constraints and obligations within the community of practitioners in order to bring entities into existence (Stengers 2010). Scientists, not unlike other knowledge practitioners, demand a common reference, something to hold in common. And what we find is that molecular animation overwhelms the sensorium with objects, relationships, speeds, slownesses, and colors that have not been submitted to the demanding tests of scientists and therefore cannot be held in common as entities of scientific knowledge.

Thus, and in keeping with Hacking's vocabulary, the vast majority of the entities and relationships that molecular animations display turn out to be artifacts of the digital instrument. They are artifacts because they do not admit of experimental verification. And what has been experimentally verified in these animations (often) blends so seamlessly with what has not gone through these demanding tests that it is virtually impossible to tell the two apart. Just as Manovich speaks of live-action footage in computer animation in terms of one more element to manipulate (2001, 302), so too experimental data are transformed into components for endless manipulation and recombination in molecular animation.

What I am suggesting here is not the full story. There are certainly other possibilities for thinking about the relevance of computer animation to researchers, and this is what I discuss in chapter 3. This is simply one way to frame the concerns of scientists who do not think that computer animation is an instrument that should be used to determine what we know about the structure and function of the biochemical world. Instead of dismissing these concerns as reductive or detached from constructivist accounts of scientific knowledge production, I have tried to place them in the context of Hacking's entity realism. Such a perspective invites us to see how important it is to distinguish between a technological artifact and an experimentally determined entity. When molecular animation is put under this microscope it seems that its images of cellular and molecular systems produce far more artifacts of the media system than agreed-upon scientific content. This becomes even more explicit when molecular animations are read alongside Manovich's taxonomy of image media and the longer history of the cinema and animation of which they are a part. Seen in context, what is unique about molecular animation has more to do with the affordances of the technology, and the possibility of generating this or that image combination out of a dataset, than it has to do with generating new scientific content. There seems to be little ontological difference between animated molecules and other digital media content, and for many scientists this is enough to keep computer animations from helping to determine what is known about the biochemical world.

TENDENCIES OF ANIMATION AND CINEMA

Before I move on to offer a very different account of molecular animation in the next two chapters, I want to conclude by stating certain reservations that I have about the ontology of digital media that Manovich offers. Even if there is still something to be gleaned from his analysis of computerized media, it certainly does not exhaust the possibilities for characterizing digital media practices and infrastructures. (Kittler, for instance, argues that hardware is perhaps even more important than software; Thacker, along with many others, maintains that networks are at the base of our digital practices, while Beller contends that the post-Fordist economy underwrites the computerization of our age, and Parikka proposes that geological forces are at the base of planetary mediation today [Galloway 2011; Parikka 2015; Beller 2006; Kittler 1999].)

What's perhaps more worrisome for us is that Manovich's formalist conception of media presupposes a form of technological determinism that cannot accommodate the pluralism of media assemblage that working scientists seem to require. Of course, Manovich ostensibly critiques technical determinism (in both *The Language of New Media* and *Software Takes Command*) and argues that databases are not to be associated with specific technologies, and instead with a specific cultural form (2001, 233). Nevertheless, the modern computer seems to express the cultural form of the database most clearly. And as such, there is a tendency in *The Language of New Media* to valorize what digital computers can do—express the logic of the database—and hence to read digital cinema as the culmination of earlier forms of moving-image expression ("The Prologue: Vertov's Dataset," etc.). What this means is that the cinema realizes its potential in computer animation or digital cinema.

And in *Software Takes Command,* software becomes the transcendental metamedium for image production. The logic of softwartized media filters (in a very Kantian sense) all that is seen, which means that the plane of software conditions the phenomenal realm of digital image fabrication (Kohut 2013).[43]

In this schematization, the totalizing logic of software subsumes digital image culture and leaves questions of political economy, the body, discursive and epistemic practices, and so on subservient to its transcendental operations.[44] And while software metamediation may appear to capture what Coopmans and her colleagues write about the laboratory setting today (namely, that "computer screens have pride of place in laboratories and scientific offices . . . [and] researchers' attentions are as likely—or more likely—to be focused on colorful digital images, simulations, software suites, databases, or lines of code as on unruly specimens or instruments" [2014, 1]), there is another sense in which scientists are keenly aware of the multiple organizational logics that are possible for digital images. In other words, the technical determinism inherent to software metamediation is such that it leaves little room for the pluralism that would allow scientists to discriminate between and among digital images (see above). It seems unlikely, then, that all computer images are mere phenomenal manifestations of the same organizing logic, and much more likely that there are different modes of organizing the digital (corresponding to political, economic, and social regimes) and that molecular animation expresses a logic that is more in line with popular media consumption than it is with scientific knowledge production. (As we will see in chapter 3, this does not mean that all scientists object to this organization of the digital for research purposes, and in fact, there are good historical reasons for their interest in computer animation.) It is important to acknowledge in any case that this is not the only mode of digital image organization. What we require, then, is a nondeterministic theory of the digital image that can attest to the mode of organization that molecular animation expresses while also acknowledging other potentials for the digital image.

By way of conclusion, and in an effort to open up a different framework for understanding the organization of technical matter, I want to draw attention to Thomas LaMarre's book *The Anime Machine: A Media Theory of Animation* (2009). In his welcome and refreshing vision of animation, LaMarre backs away from all forms of media determinism. This would include Manovich's position (even if he does not critique it explicitly) as well as the more commonly known "apparatus theory." According to the latter thesis, which has become less and less popular over the years, and to which I return briefly in chapter 6, the materials of cinema and animation—for example, the movie camera, the animation stand, the celluloid, the paint brush—determine the visible, which is to say, they prescribe the possibilities for cinematic perception.[45] For LaMarre, apparatus theory, as well as associated theories of media determinism, lock technical materials into one set of possibilities for producing moving images instead of unearthing the ways in which those materials are organized (2009, 33–34, Introduction).

To develop his thesis, LaMarre draws on Félix Guattari's notion of the "abstract machine." In particular, he contends that Guattari's machine theory demonstrates how the materials of the apparatus are open, in principle, to different organizations, to different arrangements (Guattari 1995). The machine is not dependent on the technology, then. It is just the opposite: the abstract machine is what Guattari calls a "diagram," or the pattern of relations into which materials of the apparatus enter. The machine is what determines the apparatus in a particular instance without foreclosing other determinations. "A machine," writes LaMarre, "is not a structure. It does not totalize or totally determine every outcome" (2009, 33). This is why he claims that determinations of the apparatus are not *determinisms,* but rather *underdeterminations,* since other diagrams are possible (xxvi, 34). The key for LaMarre is to "find the machine on which the apparatus depends" (xxvi). What is theoretically at stake is uncovering the *animetic* and *cinematic* machines that organize diverse technical materials. These are the abstract diagrams that draw heterogeneous materials together, with no inherent relation to one another, for the purposes of producing one kind of image as opposed to another. Matter does not prescribe its organization, and hence the cinema and animation are not locked into specific devices, materials, and apparatuses. The cinema and animation, or rather "cinematism" and "animetism," are (virtual) tendencies of the image that can be actualized in a wide variety of materials.

Borrowing his notion of cinematism from Paul Virilio (1997, 1989), LaMarre shows how this tendency has its origins in late nineteenth-century mobile apparatuses of perception (unfolding at the intersection of perception from a moving train and in the film theater), but it is really meant to capture a much wider modern "technological condition" that aims to colonize every aspect of the visible world. This is a tendency of the image that is "deadly and inhuman." In cinematism, "the spectator becomes an apparatus-subject," writes LaMarre, "whose eyes and other senses are aligned with the apparatus, with the speeding train," so that "the eye becomes kine-eye, desirous of greater velocity and mobility, bent on its own destruction" (2009, 5). And for Virilio, this pervasive, optical logic has been militarized and has aligned itself with the "bomb's-eye view." Cinematism is therefore and at once capable of giving the apparatus–subject a sense of "control over the world" as well as the sensation that "the distance between viewer and target" has collapsed (5). Cinematism is thus a tendency of perception that seeks to master every aspect of its surroundings through greater mobility and flexibility, and in doing so eliminates all distance between subject and object. This tendency is actualized in a wide range of moving-image media, from classical Hollywood features to Disney classics, but it is also pervasive in today's digital media culture. LaMarre explains how digital animation and CGI have intensified the experience of cinematism to such an extent that we can begin to speak of a "hypercinematism" (74). It is at this point that LaMarre cites Manovich and alludes to the fact that hypercinematism in some ways corresponds to Manovich's notion of digital cinema.

Animetism is the other tendency of the moving image. Although historically animetism is most often harnessed in the technologies of cel animation (the animation stand), this wasn't always the case, and it certainly does not need to be the case (LaMarre 2009, 124).[46] According to LaMarre:

> Insofar as animetism arises in the same world as mobile apparatuses (the world that, in Virilio's view, gives rise exclusively to cinematism) animetism not only implies a different way of *perceiving* things in an accelerated world but also promises a different way of *thinking* about technology and of *inhabiting* a technology-saturated world. (6)

If infinite mobility and mastery are the guiding forces of mobile perception in (hyper)cinematism, then animetism puts the subject into a very different perceptual space: distance, interstice, hesitation, and disorientation are dominant. Instead of seamless movement into depth (which we see on full display in *Protein Packing*), animetism offers a different orientation: one where objects move by sliding across the surface of the image in uneven and unpredictable ways (as in a Hanna-Barbera cartoon or in a Hayao Miyazaki film). This is the effect of a multiplanar system where layers of the image (which exist in cel animation as much as they do in Photoshop) are openly (instead of tightly) composited and so do not give the viewer the experience of traveling through volumetric Cartesian space. The spectator is a part of a world that resists being consumed and mastered; this is a world that seems antithetical to the one expressed in molecular animation. The former is a perceptual system that offers a line of defense against the imperial/colonial perceptual practices of cinematism (Miyazaki). But because animetism and cinematism are tendencies of the image instead of affordances of particular technologies, they "[appear] across divergent series of cinema and animation" (34).

Without pursuing this classificatory scheme any further, I want to conclude with the following observations. LaMarre's meditations on cinematism and animetism clarify how the organization of technical matter does not exhaust its possibilities for expression. If the modularity and infinite variation of vision are expressed in molecular animation in ways that correspond to expressions across the popular mediasphere, then this is because they share a specific organization of matter. It is their shared diagram or machine that accounts for their shared expression, and not their technical matter. It is a specific organization or logic of the image that some scientists find objectionable for scientific research. But if this organization of the visual can at times seem inimical to scientific knowledge practices, then it is not because it is an *animate* organization. Molecular animations seem to express cinematism, or rather, hypercinematism, much more than they do animetism. What this means is that those scientists who object to using computer animation for scientific research are actually protesting the *cinematic* tendencies of the image rather than the animetic ones. This leaves us wondering whether and to what extent molecular animation could be reimagined to express the animetic tendencies of the image and how this (re)organization of the image would fare within various domains of bioscientific research.

In other words, might the epistemologies of the life sciences be more animetic than they are cinematic?

But for the time being, here is what we know: molecular animation expresses a mode of visual organization whose genealogy seems closer aligned to popular entertainment culture than to bioscientific research. And while this leaves open the question of whether molecular animation could be reimagined and redesigned according to other visual logics (which I discuss in the Postscript), I think the more pressing question at this point is whether we have a full picture of the visual culture of the life sciences and its knowledge practices. Although the objections leveled against using molecular animation for research seem to originate from the experimental culture of biology, and there are good reasons for them (which is why I turned to Hacking), they are by no means representative of today's life science research in toto. As I argue in chapters 2 and 3, when the biosciences are opened up to both their experimental and theoretical cultures, molecular animation's organization of the visual field seems to be equally at home in a genealogy of epistemic practices in the biosciences. And in fact, it is precisely those characteristics that make molecular animation an entertainment medium that also resolve essential questions in experimental and theoretical epistemology in the life sciences. This will leave us wondering how molecular animation could at once be a medium for entertainment and for scientific knowledge production.

2

VISUALITY AND EXPERIMENTAL KNOWLEDGE PRACTICES

> The "eyes" made available in modern technological sciences shatter any idea of passive vision; these prosthetic devices show us that all eyes, including our own organic ones, are active perceptual systems, building in translations and ways of seeing, that is, ways of life. There is no unmediated photograph, or passive camera obscura in scientific accounts of bodies and machines; there are only highly specific visual possibilities, each with a wonderfully detailed, active, partial way of organizing worlds.
>
> —Donna Haraway, "Situated Knowledges"

This passage appears about midway through Donna Haraway's seminal essay "Situated Knowledges: The Science Question in Feminism and the Privilege of Partial Perspective" (1988). In the context of the essay, her words offer a welcome rejoinder to what she calls the "god trick" of "infinite vision" that modern science too often champions. As an example of this "trick," she cites the volume dedicated to the hundredth anniversary of the National Geographic Society, which presents two complementary chapters: the first documents outer space with stunning images that celebrate the space race and other planetary wonders; while the second chapter, "Inner Space," puts the incredibly small on display—photographs of t-cells and viruses from scanning electron microscopes, and so on. Scientific perception is presented here not only as potentially infinite—able to capture the infinitely large and the infinitely small—but also as neutral and objective: "These fabulous objects come to us simultaneously as indubitable recordings of what is simply there and as heroic feats of techno-scientific production" (Haraway 1988, 582). For Haraway, such confident visual displays conceal the fact that they are the products of modern science's colonial/patriarchal presumption that nature can be mastered.

The narrative I construct in the next two chapters is interested in a similar form of technoscientific mastery. It is this narrative that lends support to the idea that molecular animation not only culls the enthusiastic backing of an increasing number of scientists but is also primed for playing the "god trick" Haraway cites. If chapter 1 demonstrated that molecular animation is an outgrowth of popular media culture, and for this reason seems more appropriate for mass communication and entertainment (and even incites criticism among scientists), then chapters 2 and 3 illustrate just the opposite: this technology is also taken seriously by a growing number of life scientists and sets a high bar for future innovations to the field of molecular and cellular visualization. More than this, in these chapters I argue, and in seeming contrast to the narrative just constructed, that molecular animation is a part of another genealogy that stretches back to the ancient Greeks, passes through the history of microscopy and visualization technologies in the nineteenth and twentieth centuries, and culminates in computer-animated molecules. This is a history that underscores the central importance of visual perception to Western thought, and to the life sciences in particular, and emphasizes how the objects of knowledge produced in the experimental culture of biology are inseparable from its visual practices.

This history is then complemented in the next chapter by an account of theoretical and mathematical biology throughout the twentieth and twenty-first centuries, and how the abstract models of biochemical processes it produces requires visual and experimental verification. Drawing on a series of case studies, I show how molecular animation is a bridge between these two epistemic cultures in biology (experimental and theoretical). What's more, I argue that it is the "hypercinematic" organization of the visual we discussed in chapter 1, which so clearly fits into the history of entertainment media, that is poised to yoke the theoretical and experimental domains of the life sciences together. It is at this point that the fundamental problem that anchors this book will come into full view—namely, how do we understand computer-animated molecules as simultaneously objects of popular media consumption and of scientific knowledge? This question will prime us in the following chapters to think seriously about how to construct a theory of animate molecular media that does not treat these histories and relevancies as opposed or irreconcilable.

To orient our discussion in this chapter, the first section draws on scholarship from the history of science that illustrates the primacy of visual perception in the experimental life sciences and how this primacy emerges out of a much longer conversation in the history of Western epistemology about the intersection of "vision" and "knowledge." And while I anchor my discussion in Western thought and then use this abbreviated history to chart the emergence of particular tendencies in the visual epistemic practices of the experimental life sciences, there are clearly many other histories that can be told. The intention in this section is to expose the importance of visual perception in the life sciences by grounding it in a set of persistent, if always mutating, questions about knowledge and vision in the history of Western thought. This history then helps to frame a series of essential moments in the nineteenth and twentieth centuries where the technicities of visual perception and knowledge production collide in the biological sciences.[1]

In this spirit, I show how visual and epistemic practices came to be entangled in the life sciences and how microscopy emerged at the center of knowledge production in this domain. Through a close examination of microscopic knowledge practices throughout the nineteenth and twentieth centuries, one of the central claims of this chapter is that the very object of the life sciences—namely, "life"—eludes the visual epistemic practices that are entrusted to solidify knowledge claims about it. Beginning in the mid-nineteenth century, I demonstrate how there is always an aspect or dimension of life that eludes the visible and knowable, which is what makes a coherent and complete visual epistemology of life's essential processes—heredity and development—impossible for experimentalists. I draw on exemplary moments from visual epistemic practices in the biological sciences—that traverse cell and chromosome theory, as well as classical and molecular genetics—as evidence for this claim. Taken together, these examples illustrate how the *space* and *time* of the living have each functioned as forms of excess (conceived as virtual conditions in Deleuze's sense) to visual knowledge practices in the experimental life sciences. Where the spatial inscription of life's essential processes are what eluded visual epistemic capture up until the mid-twentieth century, the temporality of life's molecular processes are what later came to resist the capture of visual epistemology. This chapter functions as the first half of a genealogy of visuality in which experimental methods generate excesses that are then supplemented by the theoretical operations of the biosciences (in chapter 3). What makes this supplementation possible is molecular animation. Molecular animation is a genealogical and conceptual bridge between experimental and theoretical methods in the biosciences.

ON SEEING AND KNOWING

In order to frame our discussion of visual epistemology, I turn to work in the history of science on the importance of visual perception in the life sciences. In particular, I refer to Evelyn Fox Keller and Christine R. Grontkowski's claim that visual perception has come to (1) delimit what is knowable—what is knowable must also be visible—and to (2) distinguish the life sciences from the physical sciences in the history of Western science (Keller and Grontkowski 1983). Even though their argument may now seem dated in light of more recent work on visual epistemology in the social and historical studies of science (Daston and Galison 2007; Wellmann 2017b; Myers 2015), in what follows I contend that there are still many conceptually relevant aspects of their analysis of the primacy of vision for Western scientific knowledge. This discussion will set up Keller's more recent work on visuality in the experimental life sciences, which frames many of the questions animating our own investigation in this chapter.

In their article "The Mind's Eye," Keller and Grontkowski argue that the logic of Western thought is inextricably bound to the priority given to visual perception. In ancient Greek, for example, the verb *oida* (οἶδα) means both "I know" and "I have seen." They explain that in the *Timaeus* [61d–68e], Plato "describes the creation of the sense of sight in the same context as the creation of soul and intelligence in human beings; all of the other senses are described in the context of the creation of man's material nature" (1983, 210). The preeminence of sight for

Plato is also tied to his theory of light as a "medium of perception" and the critical importance of the sun among the "heavenly bodies" (212). Keller and Grontkowski show how the special status given to visual perception can be found throughout Plato's dialogues, including the *Republic* and the *Phaedrus,* and argue that this privilege has significantly influenced Western epistemology, to the detriment of the other senses (Haraway 1988; Irigaray 1985).

They go on to suggest that modern science's presumption that nature can be known objectively can be traced, at least in part, back to Plato. However, the supposition that nature offers herself up for objective knowledge is only really accomplished with Descartes's radical separation of the subject from the object (Keller and Grontkowski 1983, 212–13).[2] This separation entailed the "dematerialization of knowledge" in scientific modernity, which meant that knowledge of nature's eternal truths was divorced from all sense organs, including Platonic vision, in order that abstract ideas could be known through "pure intellection." Despite these dematerializing tendencies in scientific modernity, which ushered in a dualism that we still haven't wrenched ourselves from, Descartes, Newton, and Leibniz remained fundamentally interested in optics and sense perception. Descartes, for example, famously wrote *La dioptrique* (or *Optics,* one of the three essays attached to the *Discourse on Method*), a short treatise on vision and the laws of refraction that was essential to his natural philosophy. Meanwhile, Newton claimed that the "eye was the first scientific instrument," a fact that can be seen throughout Newton's work on theology, psychology, and metaphysics (Frank Maunel, cited in Keller and Grontkowski 1983). And Leibniz, for his part, contends that monads are governed by "perspectivism" (Krämer 2012).

Historians and theorists of science and culture tend to attribute this abiding fascination with visual perception among modern rationalists to a metaphorical sense of vision instead of vision through a physical eye. Martin Jay, for instance, contends that "Cartesian perspectivalism" is synonymous with the triumph of the modern, disembodied subject, whose perspective is not that of a situated and embodied subject but that of pure and abstracted intellect (1988, 10–11, 17). Jay's reading is emblematic of a wider tendency to interpret references to the "eye," "vision," and "perspective" in modern scientific rationalism as metaphors for knowledge—the "mind's eye"—instead of references to the tactility of visual perception (Krämer 2012). Keller and Grontkowski (1983) are not completely satisfied with this explanation. They explain that there is still some residual Platonism in Newton's understanding of vision: the mental eye remains tied to the physical eye in the intellect's "communion" with God's vision—to see as God sees. This in turn has sparked much debate about "God's sensorium," which Leibniz criticized Newton for (Connolly 2014). Likewise, Sybille Krämer argues that in the mathematics of Descartes (analytical geometry) and Leibniz (infinitesimal calculus) there remains a commitment to a mode of "tactile seeing," which carves out a central place for the physical eye in rational epistemology (Krämer 2012).

What this means more generally is that the division between inner and outer sight in modern rationalism is not as clear and distinct as it is often portrayed in the critical literature. Even so, modern rationalism tended to identify with disembodied abstraction. However, rationalism was not the only domain of knowledge interested in erecting a border between inner and outer sight: indeed, the life sciences were also invested in drawing lines in the sand. In her 2002 work *Making Sense of Life,* Keller argues that the two greatest achievements of the seventeenth century were the invention of the calculus and the popularization of the microscope (2002, 210–11). If inner sight was indispensable for grasping immaterial truths (via mathematical calculation), then understanding stubborn and messy living systems—for example, the problems of development and generation—seemed to require something else: namely, *outer sight.* By the seventeenth century, "students of living forms countered the primacy of the mind's eye with the primacy of the corporeal eye, claiming epistemological certainty for direct observation rather than for deduction, relying on instruments of sight rather than on those of calculation" (210). Anton van Leeuwenhoek's popularization of the microscope in the late seventeenth century was just as significant a development for modern science as Newton's and Leibniz's invention of the calculus. Keller's claim is that technologies of inner and outer sight corresponded to the twin tendencies of modern science: toward mathematical abstraction on the one hand and direct observation on the other. Vision is at the very heart of modern scientific epistemology, even if it is so fundamentally embedded in modern presumptions about knowledge that it often goes unrecognized (Keller and Grontkowski 1983, 213).

Today, most scholars of science and technology would insist that inner and outer sight are deeply interwoven capacities that have no clear and distinct separation in scientific practice, even if there are some differences. Typically, these differences are articulated in terms of evidence and explanation (evidence is what is observed and either helps support or debunk an explanation or theory),[3] and scholars have gone to great lengths to demonstrate that there is no such thing as "naive observation"—namely, observation that does not already presuppose some theoretical framework. All observation is "theory-laden" according to Thomas Kuhn, Paul Feyerabend, and numerous other philosophers of science. And although I won't examine the plausibility of theory-ladenness here,[4] a topic that has preoccupied analytical philosophers for decades now, suffice it to say that the intersection of theory and observation is a hallmark of social and cultural studies of science as well, including those variations advocated by feminist STS scholars such as Haraway, Sandra Harding, and Karen Barad. Indeed, the example of "infinite vision" plucked out of Haraway's "Situated Knowledges" at the outset of this chapter presumes the legitimacy of a very particular theoretical framework.

In view of this, Keller addresses the tight coupling of observation and explanation in the history of the experimental life sciences. She argues that observing has frequently been considered to be a kind of knowing in biology: "Are there not circumstances," she asks, "when the mere observation of a phenomena is itself so satisfying and compelling that no further explanation seems to be required? When seeing itself is a kind of understanding?" (2002, 206). "Is clarity in thinking," she goes on to inquire, "always and necessarily of a higher epistemological value than clarity in seeing?" (207). In other words, observation is often enough, and this is nowhere more apparent for Keller than in the experimental life sciences. Channeling Lorraine Daston's work on the purity of evidence—"unequivocal in its interpretation and irresistible in its persuasive power" (Daston 1991, cited in Keller 2002, 207)—she reads the early history of embryology, roughly between the eighteenth and nineteenth centuries, as evidence for the identification of visual perception and knowledge in the history of the biosciences.

There is an important point of clarification that needs to be made here. Keller does not advocate a naive theory of visual representation.

She does not claim that visual technologies (microscopy) make it possible for scientists to see nature as it is in itself, which is explanation enough. Although she is clumsy with her language at times, and even uncritically references the "direct observation" of nature (229), which cannot help but remind us of Haraway's "god trick," Keller has been critical of ocularcentrism throughout her career (e.g., 1983), and this work is no different. In an important endnote, she explains that her "focus in this chapter on visual access is not intended either to discount or to undermine the insights from recent critiques of ocularcentrism in modern science but rather as an occasion to meditate on the implications of that very tradition for what has counted, and for what continues to count, as explanation" (Keller 2002, 333n2). And just below this note she reminds us that there is "no such thing as naïve seeing" (207). Keller's main point seems to be that technologies of visual perception have been the main source of explanation and knowledge production in the history of experimental biology, and in embryology in particular. Thus, certain modalities of seeing have come to have epistemic worth. Visual practices have become *epistemic practices.*

This interpretation is neither ahistorical nor impervious to the critical tradition in the humanities and social sciences. According to Keller, she offers a "meditation" on ocularcentrism at work in the sciences. Ideally, this meditation would function as a platform from which a critical theory of visual perception could be launched, not unlike the one conceived in Lisa Cartwright's *Screening the Body.* Different regimes of visuality usher in different modalities of knowing in biology, both of which are instruments of the dominant regimes of power (Cartwright 1995; Haraway 1988; Foucault 1994). The takeaway here is that within the experimental life sciences, technologies of seeing are not separate, or in any way exterior to practices of knowing: they are one and the same thing (Keller 2002, 207–8). It is through historically situated visual epistemic practices that biologists have come to know what life is.

The history that Keller goes on to tell is rich in suggestiveness: the microscope is the most important prop, and the characters are some of the most well known in the history of microscopy; all of them conspire to get us to where we are today in the twenty-first century: scientists seem to be "watching" life unfold before their very eyes, an event that does not seem to require any further explanation (212, 231–33). But unfortunately, all the critical insight that one would hope to follow from Keller's claim that visual practices are knowledge practices in experimental biology is either implied or overlooked entirely. In the remainder of this chapter, I weigh in on Keller's thesis about the identification of vision and knowledge and examine not only how it becomes apparent in exemplary moments in the history of the life sciences but also how these moments raise important questions about the limitations of knowledge. In particular, I investigate episodes in the history of the life sciences when the spatial and temporal organization of living systems are in excess of what is visible and therefore knowable.[5] These moments stand out because biological research is converging on an object (e.g., the location of the gene on the chromosome or the temporal unfolding of genetic material), but the object in question is nonetheless invisible, even if it remains important to the determination of what can be known. I recount these episodes, which cover territory as diverse as nineteenth-century cell theory, classical genetics, and time-lapse florescence microscopy, in order to trace the logic of the visible and the invisible in the knowledge practices of the experimental life sciences.

THE SPATIAL INSCRIPTION OF LIFE

I begin by focusing on the era in modern science when the cell was first identified as the primary unit of life. This is the moment when life itself was first spatially located in a microscopic entity. As many historians of biology note, the development of "cell theory" throughout the mid- and late nineteenth century redirected scientific attention onto a single microscopic entity: the cell. And this tiny entity promised to offer answers to the mysteries of life, chief among them, development and heredity (Coleman 1977, 171). In other words, the fundamental characteristics of biological life—namely, development and reproduction—were spatially inscribed through the perceptual economies of nineteenth-century microscopy.

Microscopic observations in the years 1838–39 are what made cell theory possible. Of course, the concept of a cell had been around for centuries before the formal articulation of cell theory in the nineteenth century. Robert Hooke was the first to identify plant cells under microscopic observation, though he was unable to say anything concrete about structure or function due to poor magnification (Inwood 2002). The historian William Coleman explains that in the early 1800s, botanists, such as the French botanist Charles Brisseau-Mirbel, became once again fascinated by the microscopic structure of plant cells and by 1831 were able to identify the cell nucleus.[6] Much of the renewed fascination in cell structure by the early nineteenth century can be attributed to advancements in the microscope. From the time of their popularization in the seventeenth century, simple microscopes—which use one lens to magnify a specimen via angular magnification—proved to be the most reliable visual technology, though they had low magnifying abilities. Compound microscopes, on the other hand—which made use of a compound lens system involving a magnifying lens (objective lens) compounded or multiplied by the ocular lens (eyepiece), and which permitted far greater magnification and resolution than the simple microscope—were full of aberrations (Keller 2002, 212–13). But by 1830 great improvements had been made to the compound system—essentially, by combining lenses with different refractive potentials—and most of the anomalies had been corrected. What this meant is that far greater faith could be placed in the observations made through the compound magnifying system, and this ultimately led to the identification of the cell as the primary unit of life.

Matthias Jacob Schleiden (1804–81) and Theodor Schwann (1810–82) were the co-creators of cell theory.[7] Cell theory was far from the first time that life had been inscribed in a unifying substance, however. Such unifying substances of organic phenomena were cornerstones of German *Naturphilosophie,* despite the fact that cell theory was supposed to break away from the metaphysical (vitalist) assumptions of the former.[8] Indeed, it was precisely in order to distance themselves from the vitalism advocated by *Naturphilosophen,* which Schwann's own mentor Johannes Peter Müller supported, that Schwann and Schleiden devised their principle of cell generation: *cytoblastema.* According to this doctrine of plant and animal formation, the cytoblastema was a "structureless substance" that was capable, from within its own organization, of giving rise to cells (Schwann, cited in Coleman 1977). The idea is that cells do not beget cells (our theory of cell division); rather, all cells emerge from a "structureless substance," which is "sometimes quite fluid, at others more or less gelatinous" (Coleman 1977, 28). For Schleiden and Schwann, cell formation was supposed to be strictly analogous to crystallization, an identification that guaranteed that their theory of organic phenomena had eliminated all the last vestiges of vitalism. In this way the formation of plant and animal life could be accommodated within a physicalist explanation. The irony here is that their theory of cytoblastema is shockingly close to Lorenz Oken's vital substance—the very thing that cytoblastema was supposed to overcome (28).

Despite the wide dissemination of the cytoblastema hypothesis of cell formation in the mid-nineteenth century, microscopists were among the most vocal skeptics (30). By 1860 it had been observed by independently working microscopists that cells were formed from preexisting cells (cell division), a discovery which not only disproved the cytoblastema hypothesis but also gave cell theory a new kind of significance: the cell did not need to refer to anything outside of itself in order to reproduce itself—there was reproductive continuity (30).[9] Thus, there was no need to seek principles beyond the organic functioning of the cell—the non-organic cytoblastema, for instance—to find the unit and unity of life. The cell is sufficient unto itself.

As many historians note, the mid-decades of the nineteenth century were a fruitful time for microscopal observation: the cell was believed to unify plant and animal life and it was also what bound successive generations of cells. And in 1876 microscopists determined the basic structure of the cell—including the nucleus, which contained more specialized structures, such as the chromosome—and even suggested that the chromosome (although the term was not actually coined until 1888 by Heinrich Waldeyer) was responsible for the generation of new cellular life (35). But what the molecular structure of the chromosome was and moreover how intergenerational continuity was established through chromosomal activity was beyond the purview of direct empirical observation and so remained a space of conjecture (40).

However, a number of crucial microscopic observations during the late nineteenth century would help scientists hone in on the importance of the chromosome. For example, in 1876–77 Oscar Hertwig and Hermann Fol observed that the fertilized cell contained two nuclei (copies from the parents) that became one in the daughter cells.[10] This observation helped disprove the prevailing hypothesis that the sperm merely provided "excitation" to the egg, and in this way paved the way for new questions regarding the transmission of molecular material from one generation to the next. Additionally, the German anatomist Walther Flemming explored the fibrous network within the nucleus, which he called the "chromatin" (or the chromosome, a term which had yet to be coined), and observed that during nuclear division, or mitosis, the chromosome became threaded-like; from there, he described the behavior (the precise sequence of movements) of the chromosome during the duplication of nuclear material. What these and other observations at the end of the nineteenth century confirmed for many biologists is that the chromosome was the essential element controlling inheritance. Even work in descriptive and comparative embryology began to zero in on the importance of the chromosome. Thanks to the comparative studies of Casper Friedrich Wolff and Ernst Von Baer (42–43), nineteenth-century embryologists decided in favor of epigenesis over preformationism (Lenoir 1980). And while this "victory" paved the way for later findings that would fit well with the latest developments in cell theory (namely, the priority of nuclear activity), the precise nature of the nucleus's control over individual development remained unknown. In other words, despite these and other observational successes, what the molecular structure of the chromosome was and how hereditary units were distributed according to its operations fell beyond the scope of empirical observation.

My interest in recounting this history, which is neither exhaustive nor particularly original, is to underscore how the fundamental characteristics of biological life—namely, reproduction and development—were spatially circumscribed by nineteenth-century perceptual technologies. Growth and reproduction, phenomena that plagued philosophers and scientists for millennia, not only had a material basis (something that had been postulated for centuries) but also had a specific location within the biophysical architecture of the organism. The technologies of microscopic perception created the conditions under which many biologists in the late nineteenth century could agree that the contents of the nucleus, and more specifically, the chromosome, were the source of life's most essential activities. For these experimentalists, *to see* chromosomal activity was also *to know* where reproduction and growth took place (Keller 2002, 207).

Despite the convergence of many experimental efforts on the chromosome, and the emerging consensus that this "stainable material" (as Flemming called it) was the site of life's developmental and hereditary operations (even filling in the heredity gap in evolutionary theory [Müller-Wille and Rheinberger 2012]), it was also the source of deep epistemological frustration: it functioned like a gatekeeper to a molecular world that would remain impervious to empirical observation for the next half century. The chromosome was a promise that a material answer to the structure and function of life would be available one day soon. So while these nuclear contents were thought to hold incredible explanatory power, the material responsible for chromosomal activity was beyond the scope of observation. It's as if the chromosome was Janus-faced: one side turned toward human perception, signifying that an answer to many scientific questions about life was possible, and the other side turned toward its molecular workings that refused to answer the demands of human perception.

What's remarkable about the era of experimental biology that spanned the late nineteenth and mid-twentieth centuries is that it gave a wealth of determination to the unobservable conditions of cellular and nuclear activity. What was beyond the seen and known did not go without structure and determination. And this structured excess had a name: the "gene." Following Staffan Müller-Wille and Hans-Jörg Rheinberger's suggestion in *A Cultural History of Heredity,* the era known as classical genetics, which followed the rediscovery of Gregor Mendel's laws around 1900, was not one thing or set of ideas. Instead, it was "marked by conceptual ambivalences and by research practices that changed continuously" and is therefore better understood as "a loose ensemble of research strategies and technologies that . . . helped to forge a disciplinary identity" (Müller-Wille and Rheinberger 2012, 130). Following Müller-Wille and Rheinberger, as well as a number of other historians of science, including Elof Axel Carlson (2004) and Phillip Thurtle (2007), I argue that the ambivalences and uncertainties are crucial: what the gene was for classical geneticists was constantly under negotiation. The gene was not one thing investigated using one set of techniques. Rather, the gene was multiple, capable of taking on many different meanings, resonances, and instantiations depending upon the experimental practice in which it was situated. But for all the ambivalences and divergent strategies, some of which I recount below, the gene remained an invisible abstraction, a theoretical proposition, that was irreducible to the visible features in the cell. Despite this, I contend that the gene is an essential element in the logic of the visible and the invisible in the history of the experimental life sciences: it gives sense and meaning to observable characteristics in the cell.

In what follows, I draw on some of the key figures from the era of classical genetics, especially Wilhelm Johannsen, Carl Correns, and Thomas Hunt Morgan. I use them as touchstones for understanding what the gene meant to geneticists in the early twentieth century (Müller-Wille and Rheinberger 2012; Gaudillière and Rheinberger 2015) and suggest that the epistemological and ontological space opened up by classical genetic research lends itself to being read through new philosophical lenses.

GENETIC OUTSIDES

In their study of classical genetics, Müller-Wille and Rheinberger explain how the various conceptions of the gene that surfaced after the rediscovery of Mendel's laws were adaptations and not direct translations of Mendel's original experiments published in 1866. Mendel (1822–84) was an Augustinian monk who conducted hybridization experiments with peas and proposed that heredity was discontinuous, instead of continuous. Availing himself of sophisticated statistical analysis, Mendel discovered that certain traits were "dominant" while others were "recessive" due to the distribution of heredity units, or "factors" (that would later be called genes), which could be predicted. These experiments led to Mendel's laws of inheritance—segregation, independent assortment, and dominance (see especially Carlson 2004).

It is crucial to keep in mind that although Mendel's laws are thought to be foundational for classical genetics, they were reinterpreted as "epistemic objects" good for the study of heredity generally, instead of restricted to the narrow domain in which they were applied. "When Mendel published the results of his experiments with peas," Müller-Wille and Rheinberger explain, "he saw them as a special law that applied to hybrids only" (2012, 137, 128). But when Hugo de Vries, Carl Correns, and Erich Tschermak rediscovered them in 1900, the time was ripe for them to be taken as "general laws" (138–39). Whatever the situation that may have led up to their re-/mis-interpretation, Mendelian "factors" constituted a new "knowledge regime of heredity" that spread quickly throughout Germany, Scandinavia, Great Britain, and North America between 1900 and 1910 (138). What the content of this knowledge regime was remains somewhat unclear. Or in any case what the gene was, where it was located, and what it did lacked empirical verification. This is why the gene is such fertile ground for us: it was a hypothetical entity that was an organizing principle for research instead of a verifiable object known through observation and experimentation.

Following Müller-Wille and Rheinberger's history, this abstract conception of the gene is evident in the post-Mendelian experiments of the Danish botanist Wilhelm Johannsen.[11] In the annals of science, Johannsen is remembered for coining the genotype–phenotype distinction in 1909.[12] Although genotype and phenotype are idealized notions that separate genetic and developmental spaces in a way that is no longer tenable, Johannsen's thesis was generated out of the constraints of his experiments on pure plant and animal lines (138).[13] Based on these experiments, he surmised that the genotype was the "sum total of all genes" and that it did not evolve or change but was "ahistorical" (132–33). Moreover, genes are not empirically verifiable, for Johannsen, and so they do not have a chemical makeup—he even resisted Mendelian segregation into individual genes (145).[14] Instead, the genotype is an abstract organizing principle that gives sense and meaning to the narrow confines of experimentation on pure lines. What was left outside the purview of Johannsen's experiments, however, was hybridization.

Carl Correns, on the other hand, experimented extensively with hybrids, which yielded a conception of genetic factors and traits that was quite different from Johannsen's. As a co-rehabilitator of Mendel's laws, and one of the founding fathers of experimental genetics in Germany, Correns made an astounding number of contributions to experimental biology.[15] But among his most important contributions was his distinction between "dispositions" and "characters." After years of experimentation with hybrids, Correns became "increasingly aware of the statistical regularities that manifested themselves with respect to particular characteristics in the offspring of his hybrids" (141). Between 1894 and 1899, he crossed countless corn and pea varieties and came to several important conclusions regarding the relation between dispositions (or the unobservable units of heredity, similar to the genotype) and characters (the observable and physiological spaces of development, similar to the phenotype): although hereditary and developmental spaces are related, they are fundamentally separate and behave independently. While Correns agreed with the Mendelian speculation that heredity was located in the chromosome (leading to a chromosomal theory of inheritance that was fleshed out by Sutton and Boveri), this also meant that developmental spaces are located outside of the nucleus. In the end, Correns identified two separate spheres, with their own logics, functions, and locations in the cell. Where Johannsen's experiments on pure lines led him to the conclusion that organization was an expression of the genotype (145), Correns's experiments, by contrast, led him to believe that there was an independent developmental space altogether that was located in the cytoplasm. This space somehow interacted with dispositions (via a "mechanism responsible" for the development of the dispositions), but it remained irreducible to them.[16] Characters are the microscopically observable features through which we come to learn something about the unobservable dispositions in the nucleus, although not in any direct or causal way, since they are only knowable through characters.[17]

If Johannsen and Correns were skeptical about identifying the constitution of hereditary dispositions, then the eminent embryologist and geneticist Thomas Hunt Morgan pushed what we can say and know about genes (without actually observing them) to the epistemological limit. Unlike Correns, who worked with numerous plant species at once, Morgan famously conducted large-scale breeding experiments at Columbia University with a model organism: fruit fly *Drosophila melanogaster*. Working in the famous "Fly Room" from the early 1910s through the late 1920s, Morgan and his group experimented on the mutants of the fruit fly to construct maps of its genotype (genetic maps) (Kohler 1994). The major breakthrough came when Morgan and his group discovered that genes could be linearly grouped in particular locations on the chromosome, like "beads on a string" (Morgan 1926), which "proved" the so-called chromosomal theory of inheritance. Crucially, Morgan and his group were able to study the relative position of genes on chromosomes based upon the frequency of genetic recombination. He asserted that the frequency of genetic recombination—essentially, that paired chromosomes could "cross over" to exchange genetic information—was somehow linked to the distance of genes on a chromosome. What he found is that the closer two genes were to one another on the chromosome, the more likely it was that they would be inherited together. Genes that are farther apart are more likely to be split up during recombination. Genetic linkage therefore depends upon the relative distance genes share to one another on the chromosome. Thus, genes are spatial and recombinable units that have a determinable location (Morgan et al. 1915; Müller-Wille and Rheinberger 2012).

What's remarkable about Morgan's experimental program is that even as it enabled scientists to isolate the gene's relative location and thus spatialize heredity in a way that had never been achieved before, it was unable to make the molecular structure of the gene visible in a microscope. The physicochemical gene was still invisible and thus, strictly speaking, unknown, and so it could only be studied indirectly through visible traits. However, Morgan and his group knew, for instance, that more than one gene was often responsible for a given trait (and vice versa), which is why they resisted the idea of a one-to-one relation between gene and trait. Instead, they proposed a differential notion of the gene, which meant that they studied the change in gene–trait relationships instead of either one in isolation from the other (Rheinberger, Müller-Wille, and Meunier 2015; Schwartz 2008). Despite these experimental successes, in Morgan's 1933 Nobel acceptance speech, he notes that "at the level at which the genetic experiments lie it does not make the slightest difference whether the gene is a hypothetical unit, or whether the gene is a material particle" (Morgan 1935, 3; cited in Rheinberger, Müller-Wille, and Meunier 2015). In other words, the "gene abstraction" had real experimental effects (Morgan 1935, 3).[18]

My interest in Johannsen, Correns, and Morgan has more to do with the fact that they formulated gene theories from out of the specific demands of their experimental practices than it has to do with the accuracy of the gene theories they proposed. The gene became a remarkably effective research tool despite having no visible physicochemical reality. But it is precisely because the gene lacked this reality, since no microscope could visually capture it, that it could take on so many different meanings, resonances, and functions; it was a multiplicity, irreducible to an empirically verifiable entity with the power to disqualify claims about it. What's equally significant is that the many different meanings attributed to the gene were determined by the experimental circumstances that they were entrusted to give sense and meaning to. So, for example, Johannsen's genotype and phenotype distinction gave intelligibility to the fact that groupings of characteristics were stronger determinants of inheritance than the observed traits of parents. The genotype gave meaning to the phenotypic tendencies Johannsen observed in his pure-line experiments. Correns, on the other hand, devised a notion of genetic disposition based upon the fact that cytoplasmic and nuclear spaces seemed to be independent of one another. He then used the notion of genetic disposition to understand what the functions of the cytoplasm and observable characters had to be. Finally, Morgan's notion of relative genetic location based upon genetic recombination was derived from his microscopic observation that linked traits often separated. Morgan's gene theory made frequencies of traits intelligible. With their genetic abstractions in place, each of the scientists could give meaning to their empirical observations.

What this also meant is that observations were rendered intelligible by means of what had no demonstrable physical reality: it was a hypothetical entity and not an object of knowledge that scientists could be confident in. The sense and meaning of microscopically visible traits were guaranteed by what could not be seen or known as such. The invisible gene was the condition of intelligibility for what was visible. It is in this sense that the gene is the invisible excess at the very heart of the visible (not unlike what the opening sequence of *Protein Packing* presents). The gene is more virtual than actual in Deleuze's sense of the term.[19] This is because the gene indexed an abstract realm that is the condition of possibility for the production of sense and meaning. But this realm is not pregiven or determined in advance (it is not a Kantian transcendental) but generated out of what is required by the biologist's experimental setting. In other words, virtual conditions (genes) are not determined in advance of what they are called upon to condition. When I referred to the classical gene as a multiplicity (above), it was just this idea of a genetic condition firmly anchored in the requirements of an actual practice that I had in mind. The gene is not one thing in this era of experimental genetics, it is a virtual multiplicity, an invisible excess, capable of generating sense and meaning in a variety of experimental contexts. In Deleuze's vocabulary, the virtual and actual reciprocally determine each other (Deleuze 1994, 171). *Qua* outside, the gene was not exterior to the visible in any absolute sense; on the contrary, it was virtually exterior to experimental research in genetics: it is what made experimental observations intelligible. The gene was the invisible excess that ordered what could be seen.

Although historians emphasize the importance of classical genetics, in hindsight molecular biologists are quick to dismiss the classical gene as a stopping point on the way to discovering its molecular structure: the DNA molecule. From our perspective, however, the classical gene is made possible by ongoing negotiations between the visible and invisible that do not go away once the molecular structure of the gene is articulated. As I demonstrate in the next section, James Watson and Francis Crick's identification of the double-helical structure of the DNA molecule in 1953 was in no small part due to advances in visualization technologies. But this does not mean that the gene, and by extension the essential structures and functions of life itself, are visually captured. The relation between the visible and the invisible simply undergoes a transformation and shifts to new and more complicated terrain. If the previous sections of this chapter demonstrated how microscopic perception was unable to image the genetic agent responsible for life's heredity and development, then the next sections show how life's time-based processes are what elude visual perception.

PHOTO 51 AND THE STRUCTURE AND FUNCTION OF THE DNA MOLECULE

In the annals of science, Watson and Crick's discovery of the structure and function of the DNA molecule in 1953 goes down as one of the crowning achievements in all of science. In two seminal papers they demonstrate how the DNA molecule exists as a three-dimensional double-helical structure and are then able to deduce its function (Watson and Crick 1953a, 1953b). By doing so, they set to rest generations of debate over the physical and chemical basis of genes. "It has not escaped our notice," Watson and Crick explain at the end of their first paper, "that the specific pairing we have postulated immediately suggests a possible copying mechanism for the genetic material" (Watson and Crick 1953a, 737). In their elegant model, nucleotide pairs are structurally similar—A:T and C:G—and fit equally into the sugar–phosphate backbones. The pairs are held together by hydrogen bonds, which are not difficult to break apart, and so are easy to imagine "unzipping" in replication. Thus the structure of DNA suggested its function as genetic material. The second paper, "Genetical Implications of the Structure of Deoxyribonucleic Acid," describes how DNA replicates.

Historians of science are quick to note that Watson and Crick's achievement did not occur in a vacuum. The identification of the DNA molecule with the genetic material had been in the making for some time. DNA was first identified by the Swiss chemist Friedrich Miesher in 1869 (then called "nuclein," and later, nucleic acid), and then years later the Russian biochemist Phoebus Levene suggested that nucleic acids were made up of nucleotides. And in a seminal paper from 1944, Oswald Avery and his colleagues hypothesized that genes were made up of DNA. Erwin Chargaff then expanded upon Avery's insights and proposed that A and T bound together just as C and G did—though he had not formalized how these bonds work in the molecular structure of DNA (Kresge, Simoni, and Hill 2005). Maurice Wilkins's and Rosalind Franklin's X-ray crystallographic work was also crucially important to Watson and Crick's model (they shared the Nobel Prize with Wilkins); and they were also said to be competing with Linus Pauling, who had proposed an incorrect model of DNA just before Watson and Crick announced theirs (Pauling and Corey 1953).

Watson and Crick were thus anything but lone researchers who hit on something big. In the many stories told about how Watson and Crick arrived at their model, Wilkins's and Franklin's X-ray crystallographic work at Kings College is often mentioned as a key factor. Watson and Crick briefly mention in their paper that "X-ray evidence obtained by the workers at King's College London, and presented at the same time, gives qualitative support to our structure and is incompatible with all previously proposed structures" (Watson and Crick 1953b, 965).[20]

What is rarely discussed, however, is how this work also situates Watson and Crick's discovery within the visual epistemic practices of the life sciences. X-ray crystallography, or X-ray diffraction, is a technology used for determining atomic configuration within a crystalline molecule.

In the early twentieth century, physicists found that X-rays have much smaller wavelengths than visible light and that they could easily pass through certain objects, but other objects, like crystals, were obstructions and caused the waves to diffract. "Diffraction," as Natasha Myers eloquently puts it, "is a phenomenon produced when any wave, such as a sound wave or a wave of radiation[,] passes through a regular array of objects. When waves encounter obstructions, they get deflected and change course, often forming patterns as one wave overlaps and interferes with another" (2015, 52). It was William H. Bragg who realized in 1913 that he could use crystals to diffract X-rays, and from this insight he built a "technology for investigating the very atomic configuration of molecules that were packed inside crystals" (52). Dorothy Hodgkin then applied X-ray crystallography techniques to the visualization of biomolecules. To perform a crystallographic experiment in which a visualization of a complex biomolecule is obtained, a crystal must first be formed. This is an incredibly difficult task, given that the crystal must be appropriately uniform and just the right size.[21] Once a crystal is obtained, an X-ray beam is shot through it that hits the electron clouds of atoms, which diffracts the beam. The scattered X-rays then produce a diffraction pattern on a screen behind the crystal. From these patterns, scientists can determine electron density and the position of atoms within a molecule.[22]

In the case of the DNA molecule, it was Rosalind Franklin's PhD student Raymond Gosling who produced the photograph referred to as "Photo 51," which was so significant for Watson and Crick's double helix model. Although Wilkins is the one who originally determined that using a solution of sodium thymonucleate obtained high-quality X-ray patterns from DNA, Franklin, who was hired by Wilkins, perfected the method; and then Gosling is the one who produced Photo 51. What's questionable is how Watson and Crick came upon the photo. As one commentator put it: their acquisition of the photo has all the elements of a good scandal (Cobb 2015b).[23] Wilkins, who was well known for not getting along with Franklin, showed Watson the photograph without Franklin's knowledge. Upon seeing it, Watson noted that "my mouth fell open and my pulse began to race" (n.p.). Availing themselves of Franklin's private notes and data about the experiment that accompanied the photograph (again without her permission), Watson and Crick

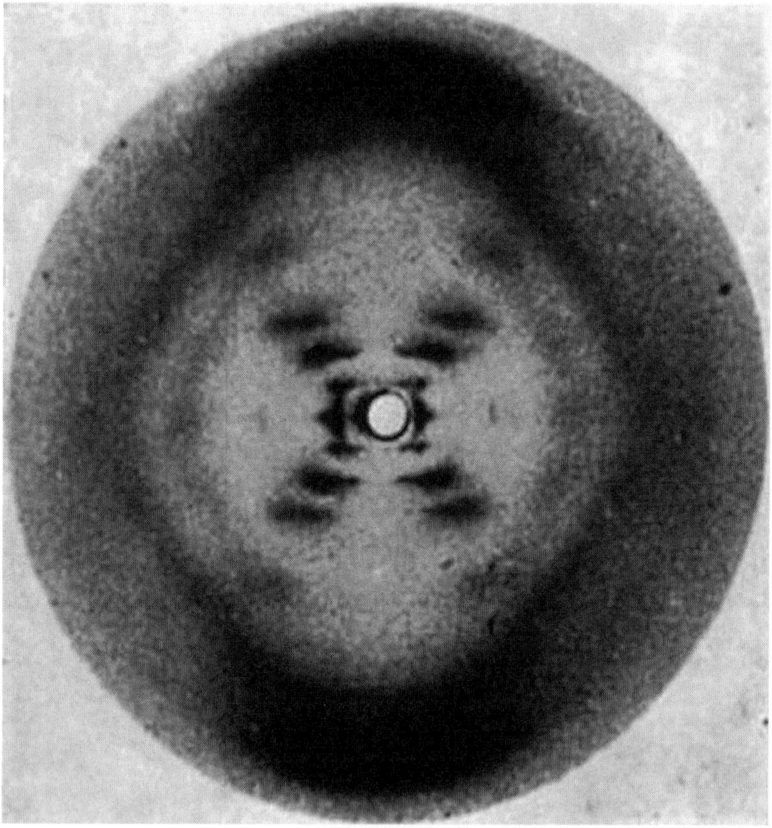

Figure 21. Photo 51.

were able to finally work out the double-helical structure of DNA. In fact, it was only one year prior to their seminal paper that Watson and Crick proposed a three-strand model of the molecule that was severely criticized by Franklin. It was ultimately the photograph that provided the visual confirmation that Watson needed to determine the chemical organization of the molecule and propose its genetic function.

While there is no shortage of historical and critical work on the molecularization of the gene, there is also an underrecognized visual dimension to this history. Prior to the famous Photo 51, Pauling, Watson, and Crick all proposed mathematical models of the DNA molecule; but they all had theories that lacked empirical verification. Watson and

Crick's first attempt at modeling the molecule in 1951 was deemed a serious failure (responding to certain theoretical problems but not others), and they were even told to stop working on it (Cobb 2015a; 2015b). It was only after discovering that the American scientist Linus Pauling was also trying to crack the structure of DNA that Watson and Crick were able to resume their work. Pauling, for his part, proposed a triple-stranded DNA model,[24] but it too was theoretically untenable, and what's more, it was based on blurry diffractive patters of DNA taken by William T. Ashbury.[25] Gosling's image, by contrast, was the lynchpin inciting the double-helical model of DNA. When Keller insists that there are moments in the history of the experimental life sciences when seeing is a kind of knowing, I think this is precisely the kind of incident she had in mind: knowing the structure and function of DNA proved impossible without seeing some trace of its helical form. Although it was not the molecule itself that scientists were seeing, the diffraction patterns nevertheless gave scientists the empirical hold they needed. A coincidence of seeing and knowing had to occur in order for Watson and Crick to publish their seminal papers. But if seeing diffraction patterns is what gave scientists the confidence to locate the gene in a molecule, then the gene's resistance to visual capture does not disappear for experimental biologists: it simply shifts to a different register. It is time instead of space that resists visualization.

DRIED-OUT LIFE UP CLOSE

In an opinion piece written for the *Guardian* in 2012 titled "You Can't See DNA Unless You Look Properly," Stephen Curry, a professor of structural biology at Imperial College London, took it upon himself to correct a flurry of bad science reporting surrounding an article published by a group of Italian scientists led by Enzo di Fabrizio (Gentile et al. 2012). For Curry, what's worrisome is not what the scientists claim in the article but what the journalists suggested about it: that DNA could be seen for the first time. In a piece written for *New Scientist,* Roland Pease (2012) argues that the Italian scientists had used "an electron microscope [to capture] the famous Watson-Crick double helix in all its glory"; and he even suggests that the image "would allow researchers to see how other biomolecules interact with DNA." Other reporters made even stronger claims about the article, suggesting that this was "the first ever microscope image of an isolated DNA molecule" (Wild 2012).[26]

Curry is eager to set these reporters straight in a way that is instructive for us. First, he notes that the researchers did not obtain the first image of DNA using an electron microscope. Electron microscopy had been imaging DNA molecules long before the 2012 article. Second, X-ray crystallography, Curry's preferred method of obtaining information about biomolecules, had been imaging DNA since the early 1950s (see above), and this method yields more precise information about molecular structure (according to Curry). The issue for some is the way the information is obtained: scientists must calculate how the molecule visibly diffracts an X-ray beam, and this is not as satisfying to the lay public. What is new about the research, according to Curry, is that a high-contrast image of DNA fiber was obtained. But the researchers obtained a high-contrast micrograph of tightly packed *bundles* of double helices, not a "single DNA" molecule. In other words, the individual building blocks of DNA—the smaller units called nucleotides—were not captured in the image.

The issue is that the micrographs do not give us information about how DNA interacts with protein molecules, like the reporters insist. The reason for this is twofold: (1) tightly packed bundles of double helices do not occur naturally in the living cell (intricate preparation is required—see discussion below), and (2) electron microscopy must dry out its samples in order to obtain images, which makes it impossible to understand how proteins *actually* interact with DNA. In other words, while the micrographs showcase the impressive magnification and resolution power of the electron microscope, they do not get us any closer to understanding *how* DNA actually works in a living system. What this anecdote from science writing underscores is that, since the 1950s, life scientists have become remarkably good at obtaining images of DNA molecules, even high-contrast micrographs that garner tremendous excitement from the public. Despite these successes, scientists have struggled to image our other key intuition about what genes do—namely, unfold in time.[27] Molecular biologists have been able to refine their understanding of the structures of DNA and RNA molecules, and advances in optical and electron microscopy, as well as X-ray diffraction, have made this work possible. However, what experimental biologists still struggle to observe is how genes function, that is to say, how they engage in complex interactions within the cell over time. To illustrate the challenge that experimental biologists face, I draw attention to what is gained and lost with the tremendous resolution and magnification power available for imaging biomolecules. While these visual technologies offer an exciting range of affordances—for example, atomic-scale imaging (~0.05nm)—our question will be: At what cost?

Imaging the structure of a biomolecule like DNA requires a level of magnification and resolution (the former is worthless without the latter) that is generally considered impossible to achieve with the light or optical microscope. While optical imaging systems were all that classical geneticists had available to them, molecular geneticists have had electron-imaging systems for over half a century (Rasmussen 1997). An electron microscope (EM) emits beams of electrons, instead of rays of light, to create an image of a specimen. Because the wavelength of an electron beam is much shorter than light photons—up to 100,000 times shorter—the EM has a much higher resolving and magnification power than an optical microscope. In fact, an EM has a resolving power that is approximately 250 times greater than an optical microscope (0.001μm compared to 0.25μm to 0.3μm) and a magnification power that is up to two million times, compared to 500–1,500 times in an optical system. The first prototype of an EM was built in 1931 by two Germans, the physicist Ernst Ruska and the electrical engineer Max Knoll, and today it is used in a wide variety of research and industrial contexts (although still based on the early prototype), including the imaging of biological materials, metals, and crystalline structures, and the features of different molecular surfaces; it can also be used in the construction of silicon chips, forensics research, and a host of other industrial and research settings (Freundlich 1963; Ruska 1987; Ackermann 2011). The introduction of the EM into life science research is far too complicated to recount here (Rasmussen 1997),[28] but suffice it to say that after the EM's adoption by the biological sciences, which would have to wait until the end of the Second World War, biological imaging was never again the same (Rasmussen 1997). I say this for at least two reasons.

The first reason has to do with the sheer power of the imaging system. Structures that had never before been delineated were now brought into view. With the early EM—the Transmission Electron Microscope (TEM)—it was possible to determine intracellular structural details all the way down to 10nm; this was a level of magnification and resolution that light microscopy simply could not deliver. By the mid-1940s, resolution had been brought down to 2nm (thanks to a number of advancements, such as voltage acceleration and lens technology), and today, with advances in TEM technology, it is now possible to image at the atomic scale (~0.05nm) and have it reconstructed using 3D tomography. And with a slightly later iteration of the EM, the Scanning Electron Microscope (SEM), it is possible to get 3D images of surfaces. The SEM scans an electron beam across the specimen (instead of through it) and then electrons bounce off it to produce an image of the surface structure in 3D (McMullan 2006).[29]

Let's return to Di Fabrizio's and his colleagues' work on imaging bundles of double helices. The images published in their article are illustrative examples of both TEM and SEM imaging (Figures 22a and 22b). Images *a, b, c* in Figure 22a are obtained using the SEM. As the authors explain, these are SEM images of DNA bundles suspended over superhydrophobic pillared devices: *a* is the DNA bundles hanging on top of pillars, *b* is a close-up, and *c* is a top view. Image *d* provides a computer sketch of the concept behind the direct TEM imaging used in *e:* "The passing through etched hole allows the suspended DNA to be orthogonally crossed by the microscope electron beam" (Gentile et al. 2012, 6454). In Figure 22b, we see TEM images of DNA fiber in *a* and *b*; the arrows in the inset of photo *a* indicate a 2.7nm pitch of double helix, and the white rectangle in image *b* corresponds to where the intensity profile of the image was measured (or the set of intensity values taken from regularly spaced points in an image).

These images are especially good at capturing how TEM and SEM technologies image structural details in biomolecules. But they also illustrate how these details are manufactured by a physical system. To obtain micrographs, the DNA specimen has been stripped of what makes it a macromolecule worthy of our attention—namely, its function within a cell. Stephen Curry makes a similar observation in his article for the *Guardian,* but I think it's a point worth driving home.[30] The molecule is deprived of in situ complexity, which the authors refer to as "background noise": the DNA fibers are "suspended and background free," and under these conditions "the background noise is removed and the

a

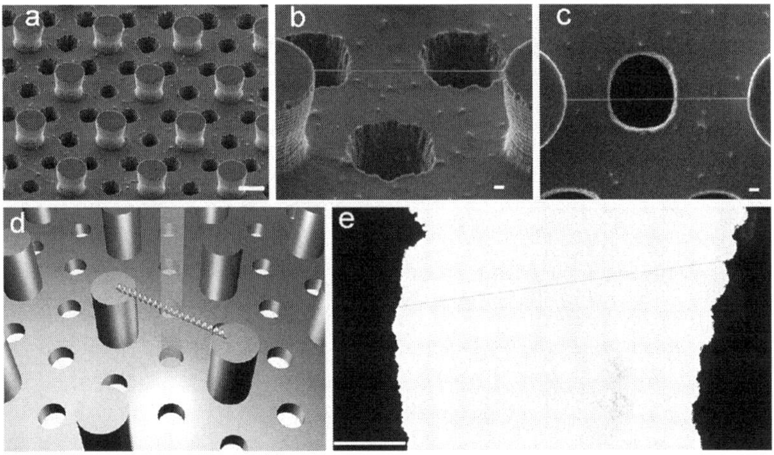

b

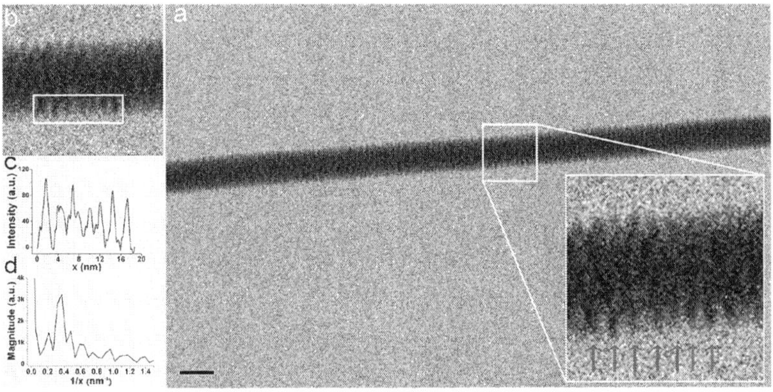

Figures 22a and b. "Direct Imaging of DNA Fibers: The Visage of Double Helix." Published in Gentile et al. 2012.

spatial resolution is primarily dictated by the intrinsic scattering and absorption cross section properties of the molecule itself" (Gentile et al. 2012, 6453). By purifying bundles of their environment, they become visible to human perception. The bundles have to be elaborately prepared before they can be imaged: dried out and suspended over a layer of pillared silicon. In short, the visibility of the fibers depends upon making their environment invisible.

This brings me to the second reason why biological imaging was never the same after the introduction of the EM: specimens became dried out, dead, and thus lifeless objects for the scientific gaze. How nucleotides interact with their intracellular environments is left to abstract calculation instead of empirical observation. As a matter of fact, any in situ environment seems to pose a serious challenge to the high-resolution EM. In 1985 the celebrated microscopist Robert Day Allen wrote that after the introduction of the EM in the 1950s, many medical students "gave up the study of living processes, except for indirect evidence that came to them fortuitously from electron micrographs" (cited in Keller 2002, 218). Edward L. Chambers remarked as early as 1961 that

> these techniques [of the EM] seemed so fruitful of results that the aim of their originators—to visualize live cell structures—was lost sight of. More and more attention was focused on killed and fixed cytological structures, despite the fact that many of the details, now so clearly visible, might have well been due to the coagulating action of the particular fixative used." (Chambers, quoted in Keller 2002, 336)

At least part of the reason behind the rapid development of TEM and SEM technologies was to correct for the fact of the elaborate preparation of specimens. The cryo-EM (cryo-electron microscopy), ETEM (Environmental TEM), and ESEM (Environmental SEM) all damage the specimen far less than traditional TEM and SEM imaging. Nevertheless, imaging DNA molecules in situ is beyond the scope of current EM technology.

Advancements in light microscopy, on the other hand, have come a long way in the last couple decades to eradicate the so-called vital barrier.[31] By the early 2000s, advances in biotechnology, microscopy, and cinematography converged in florescence microscopy to make significant improvements to the imaging of live cells (Stephens and Allen 2003). The green florescent protein (GFP) is largely responsible for these developments. First isolated from the jellyfish *Aequorea victoria* in the 1960s and then used in modified forms as a reporter of expression beginning in the 1990s, the florescent protein has become critical to live-cell imaging (Keller 2002, 219–21). Once the protein is inserted into the organism using genetic modification techniques, scientists are able to observe the expression of those proteins tagged with GFP in situ. What makes GFP so important to microscopists is that it is far less harmful to the organism than other florescent molecules. GFP has "revolutionized live cell imaging," according to David J. Stephens and Victoria J. Allen, because it makes it "feasible to perform multicolor imaging of living cells" (2003, 82). Stephens and Allen also note other crucial factors to consider when performing live-cell imaging, including microscope choice—for example, widefield, scanning confocal, spinning disc confocal, phase contrast, and differential interference contrast (83, 84)—as well as light exposure, speed of acquisition, and room conditions. They also mention that the resolving power of the light microscope has broken the diffraction limit (down to 100nm of resolution) (86). And inasmuch as diffraction limits the resolving power of light microscopy, "sub-diffraction" imaging of live cells holds out the promise of closing, at least partially, the tremendous gap separating the resolution power of optical and electron microscopy (Patterson 2009). These and other elements contribute to the authors' concern for maintaining the health of the specimen as they obtain high-resolution images of it.

Another crucial feature of live-cell imaging using florescence microscopy is the use of time-lapse video. Stephens and Allen note that as processes in living systems "occur in three dimensions over time . . . to get a complete picture we need to image cells in four dimensions" (84). Time-lapse video is an important technique for capturing the changes in living systems that are too slow, fast, or subtle for human perception to register. The authors explain that most confocal systems and epi-illumination microscopes either already are or can easily be adapted to generate four-dimensional images. And while combining moving-image media and microscopy dates back to the early twentieth century, its importance for bioscientific research today is readily apparent because it can visually capture *molecular* processes. Where EM technologies strip away the relational context of living systems, time-lapse techniques give dynamism back to the imaging of molecular bodies. With time-lapse microscopy, scientists see what relations emerge over what periods of time.

For example, Muzzey and van Oudenaarden (2009) use time-lapse experiments in conjunction with two fluorophores in a single yeast cell to resolve a long-standing confusion over how the proteins are composed in the cisternae (the flattened membrane disks that make up the cis Golgi network) of the Golgi apparatus (Losev et al. 2006; Muzzey and van Oudenaarden 2009). The Golgi is the organelle responsible for sorting and processing macromolecules within the eukaryotic cell. Essentially, it's the cell's "post office." What was unclear to scientists is how the networks within the Golgi are formed: Do they maintain their protein composition for a long period, or do they transition from cis to trans, the two main networks in the apparatus? Using quantitative time-lapse microscopy, the order of molecular events was ultimately determined, which allowed scientists to conclude that "rather than retain exclusively cis- or trans-specific proteins for a long time, cisternae rapidly mature, with secretory-processing factors progressively shuttling through them" (Muzzey and van Oudenaarden 2009). Muzzey and van Oudenaarden use this and other experiments with time-lapse microscopy and computer image analysis (Matsuura-Tokita et al. 2006) to "demonstrate the power of quantitative time-lapse microscopy to distinguish among possible mechanisms, extract important information about their kinetic properties, and generate new hypotheses" (Muzzey and van Oudenaarden 2009). Developments in light microscopy, biotechnology, and time-based imaging have worked to ensure that the image expresses the dynamism of larger intercellular structures, such as the Golgi body. Historians, such Janina Wellmann, have picked up on this and have shown how "live imaging," or "in vivo imaging," combines the methods of time-lapse microscopy with 3D digital imaging to capture living specimens, and how this in turn has "animated" biological knowledge in new ways.

Interlude: A Note on Excluded Histories of Time-Based Media

Given the prominent place that time-lapse florescence microscopy has in our genealogy of molecular animation, it may seem strange to omit early twentieth-century time-lapse microcinematography from our story. As Hannah Landecker, Lisa Cartwright, Thierry Lefebvre, Oliver Gaycken, and a host of others argue, time-lapse experimentation with live cell cultures plays an essential role in the visual culture of the life sciences throughout the twentieth century. As I discuss in chapter 7, microcinematography in France, England, Germany, and the United States in the early 1900s provided a new way to think about the development of living systems: in time (Landecker 2007; Cartwright 1995; Lefebvre 1993; Gaycken 2015). For this reason, microcinematography seems like a hallmark of any history of visual epistemology that includes contemporary time-lapse microscopy and ends in molecular animation. There seems to be a shared concern among these visual practices: capturing the dynamic (rather than static) unfolding of living systems.

The issue that needs attending to, however, is the following one: by the time that microcinematographic practices matured in the 1910s–1920s cytology was no longer of interest to most working biologists. At that time, genetics was gaining serious momentum, and questions about cellular development were taking a back seat because sights were literally set elsewhere: on the molecular world of the gene. Experiments with time-lapse microcinematography did not pave the way for the "new cytology" that the French surgeon and microcinematographer Alexis Carrel hypothesized (Landecker 2007, 87–90). This knowledge was largely sidelined in favor of genetics and then later molecular genetics.[32] As Landecker writes, "If we look at the historical literature, it would seem that the cell was of little importance after the rise of genetics and molecular biology in the 1920s and 1930s" (6). Of course, today, visualizing how genes function in complex cellular environments (using in vivo imaging, for example) is essential to building maps of organismic development and heredity in time. But this was not always the case and only became a priority for researchers long after the heyday of microcinematography.

The challenge, then, is to understand the relevance of this cinematic knowledge for a genealogy of the biosciences that leads to molecular animation. In other words, in what way and by what means did these knowledge practices exert their influence? In the main, histories of visual culture now weave early scientific filmmaking into the cultural history of life science imaging. For instance, Landecker charts the epistemic conditions under which life became technological. And as it turns out, cinematographic experiments are an essential part of this history (see chapter 7). This is the spirit of the project I am invested in as well. But if I depart from projects like this, then it is because I also aim to account for how microcinematographic knowledge was largely on the margins of bioscientific knowledge practices in the early twentieth century.

My hypothesis is that we need a fuller picture of the social and political conditions of image-making during this era in order to understand how microcinematography exerted influence over the design and fabrication of bioscientific knowledge practices. This is why I focus on what is visible and knowable in the era of classical genetics, arguably the preoccupation of experimental biologists during the early twentieth century. I am not neglecting microcinematographic knowledge in the biosciences, rather I am waiting to layer it into our genealogy once we have a conceptual apparatus capable of articulating its full meaning and relevance for a history of scientific knowledge that makes molecular animation possible.

TEMPORAL EXCESS

When we return to twenty-first century bioscientific imaging, it's clear that time-lapse optical microscopy is a widely used and valued technique for imaging live cells. However, even if this practice allows scientists to "see" molecular events, and thus bring molecular imaging together with organismic development in time (in a way that was impossible with classical microcinematography), then the magnification and resolution power of these optical systems still stand in the way of bringing the complex structures of the molecular world into view. (For instance, where a single strand of DNA has a diameter of 2.5nm, the cisternal membranes of the Golgi apparatus, by contrast, range from 0.5 to 2.0 μm in diameter [Dupree and Sherrier 1998].) This may not always be the case. With the resolving power of the light microscope reaching below the diffraction limit and EM technologies getting better at accommodating molecules in situ, we might wonder whether watching transcription using time-lapse microscopy is not too far off. In other words, might *X Inactivation and Epigenetics* be a precursor to the future of microscopy?

But even if we assume that resolution and in situ imaging do not pose the challenge we may have once assumed, then we confront another crucial problem: imaging the environment that enables and supports gene expression is next to impossible. Where the "Central Dogma" endorsed by Francis Crick in the early years of molecular biology may have led some to believe that gene expression was a linear and ultimately comprehensible process (Keller 2000), this view has well and truly been cast aside in favor of far more rich and complex gene–environment relationships. Much of this will be discussed in chapter 3, especially in relation to the history of theoretical and systems biology.

But suffice it to say that it is wider gene–environment relationships, especially in the postgenomics era of life science research, which have slowly come to dominate the research agendas of most working biologists. This is not only true for systems and theoretical biologists who design mathematical models but it is also true for the experimental work in evolutionary and developmental biology (evo-devo) (Mallarino and Abzhanov 2012; C. Goodman and Coghlin 2000). Evo-devo biologists have shown that morphological complexity in organisms does not necessarily correspond to genomic complexity, and that it is gene regulation, instead of individual genes, that accounts for many of the novel forms within a population (Carroll 2005). With the discovery of the Homeotic Hox gene family, or the "homebox" DNA sequence, in the 1980s, biologists have been able to research and experiment on the importance of the temporal unfolding of gene regulation networks rather than the activity of single genes within networks. What this means is that gene expression cannot be understood according to the action of a single gene—the Hox gene alone is said to be made up of around 180 base pairs—but according to how a gene is a part of a much larger distributed network of genes that have complex temporal rhythms. From this perspective, understanding the timescales operative in the activity of DNA transcription is a problem that needs to be sorted out mathematically (Balaguera et al. 2017).

What this and other work in experimental biology underscores is how imaging genetic material in situ seems more like a fable or a useful fiction rather than something that could ever happen in practice. Even when micrographs capture something that resembles the "natural state" of a biomolecule, they have had to eliminate either temporal or spatial complexity. For example, although EM technologies have made it possible to obtain micrographs with tremendous spatial detail, time has to be abstracted from them; they are static and lifeless snapshots. Time-lapse light microscopy, on the other hand, gives temporality back to the image of living processes; but its images lack spatial detail. And what's more, even if resolution and magnification no longer pose a serious challenge to light microscopy—or could rival EM in any case—experimental and theoretical biologists have demonstrated that there are numerous concurrent timescales involved in gene expression, and many of them have yet to be identified or measured mathematically (Orlando et al. 2010).[33] For this reason, time poses a serious challenge to the visualization of molecular processes in the cell: these processes unfold at timescales that technically augmented visual perception simply cannot register (see also McGill 2008). Molecular imagists may have to remain content with snapshots, abstractions, and partial views of living systems.

By way of conclusion, I want to insist that cellular and molecular imaging in the twenty-first century does not simply eliminate the visual epistemic excesses that haunted earlier eras of bioscientific visualization. Where nineteenth- and twentieth-century biology may have been limited by its ability to locate genetic material on the chromosome, today's biology, by contrast, is limited by its ability to bring the multiplicity of timescales involved in the operation of genetic material into view. As a consequence, I would like to suggest that it is time, instead of space, that functions as the excess to visual knowledge practices in the era of molecular genetics. Although time is not excluded from the visualization of molecular processes, visualization practices in the biosciences struggle to articulate complex molecular architectures and concurrent timescales.

One of the central claims of this chapter is that visual epistemology is central to knowledge production in the life sciences. From the outset I suggested (drawing on Evelyn Fox Keller, among others) that the prominence of visual perception is what distinguishes knowledge in the experimental life sciences from the physical sciences (and theoretical biology—as we will see). This prominence was evidenced in a variety of ways in the history of cell theory and early microscopy. It took seeing some trace of the helical structure of DNA for scientists to shore up knowledge about the structure and function of genetic material. And throughout the twentieth and twenty-first centuries, spatial knowledge of biomolecules became more precise through enhanced visualization technologies, but the same cannot be said of their temporal dimensions. There is a significant temporal blind spot in the visual field of biomolecular processes.

One might wonder whether this matters. In other words, does the invisibility of time (or the inability to visually capture it) matter to life science researchers today in the same way that the invisibility of the gene's location mattered to twentieth-century researchers? Many scientists argue that most of these timescales, if not soon all, can be understood mathematically, and that's what matters. In short, we don't need visualization, because mathematics has well and truly made the temporal complexity of molecular systems comprehensible, or at least in principle. Time is not "virtual" in the way that space was; time can be known through advanced mathematics. This attitude toward the mathematization of the cellular and molecular processes has to do with a struggle at the very heart of biological knowledge, which I will discuss in the next chapter. But at stake here is nothing less than the status of visual epistemology in the life sciences as such. For if mathematics is not only necessary but also sufficient for knowledge production in the life sciences, then visualization is little more than an add-on, something that may be vital to the communication of ideas but not knowledge production as such.

This tension frames the beginning of chapter 3 and anchors my claim that visualization is still crucial to knowledge production in the life sciences. I contend that molecular animation practices are what promise to fill in the temporal blind spot in the visual field: they bring complex molecular architectures together with multiple scales of temporality. In doing so, I also claim that these visualization practices matter to knowledge production insofar as they allow scientists to build molecular and cellular systems back up—that is, give sense and meaning to what lacks spatial and temporal referents (Thurtle 2018). Thus computer animation is a technology that fills a gap in the visualization practices in the biosciences, and in doing so, helps to reassert the relevancy of visualization for knowledge production. What will nevertheless haunt the next chapter's reflection on molecular animation's relevancy to biological knowledge is Donna Haraway's critical assessment of the scientific gaze and the "god trick." The truth of Haraway's insights will become apparent once the parallel between molecular animation's role in visual epistemology and in popular culture (chapter 1) is made plain. This will provide the groundwork for seeking a deeper, critical genealogy of the animated image in the life sciences.

3

A FEELING FOR THEORETICAL BIOLOGY

> There seems to be no limit to the amount of information that we can accumulate, and today . . . we face the question of what is to be done with all this information.
>
> —Sydney Brenner, "Theoretical Biology in the Third Millennium"

I ENDED THE PREVIOUS CHAPTER with the provocation that molecular animation finds its relevance amid the increased mathematization of molecular biology. This claim was made at the end of an extended discussion about the interdependence of visualization and knowledge production in the experimental life sciences, and with specific reference to the fact that the complex temporalities of the molecular world can be mathematically modeled, but they too often remain invisible. It is in this context that I proposed that molecular animation becomes an integral visualization tool: at once grafting itself onto the computational and mathematical trends in molecular biology and filling in the blind spots of the visual economy of the experimental life sciences.

But to understand how mathematical and computational approaches to the life sciences have come to sit alongside experimental methods, and in animation practices no less, we need to unpack, both historically and conceptually, the resistance to mathematical and computational biology among experimentalists and the reason for their uptake in the last few decades. What we discover is that the relevance of mathematical modeling in the life sciences has less to do with its reducing experimental approaches to mathematical models (as they once were) and more to do with each epistemic culture benefiting from the other: computational biology has become more experimental just as much as experimental biology has become more computational. To appreciate this epistemic shift, and how animation is situated at its intersection, the first half of the chapter reviews some of the major attempts to bring mathematics and biology together in the twentieth century—including those made by D'Arcy Wentworth Thompson and Alan Turing and by cybernetics and information theory—and how systems biology in the postgenomic era has managed to find common ground with experimental approaches to molecular biology. In the second half of the chapter, I turn to two case studies from molecular animation and demonstrate how they at once bring experimentalism into systems biology and also make it possible for temporally and spatially determined systems to be a part of the visual economy of experimental biology.

This chapter makes it possible to think about molecular animations as a part of two very different genealogies. This will set the stage for chapter 4, which demonstrates how the characteristics that tend to make animation a valuable tool for contemporary biologists are also the ones that make it an expression of popular media culture. I use this collision of popular and scientific worlds in molecular animation as an opportunity not only to showcase the complexity of this media object but also to argue that there is no available discourse that addresses this complexity. It is in an effort to fill in this gap that the rest of the book is dedicated.

MATHEMATICAL IDEALIZATIONS: THOMPSON AND TURING

Perhaps one of the most well-known and frequently cited attempts to fuse mathematics and biology in the twentieth century comes to us from D'Arcy Wentworth Thompson's seminal book *On Growth and Form,* which was written in 1917 (second edition in 1942) and is over one thousand pages in length.[1] Thompson, a zoologist, anatomist, and recreational classicist, sought "to study the inter-relations of growth and form, and the part which the physical forces play in this complex interaction; and, as part of the same enquiry, to use mathematical methods and mathematical terminology to describe and define the forms of organism" (D. W. Thompson 1942, 1026). And what is more, he endeavored not only to redescribe the biologist's field to her in mathematical language but also "to shew the mathematician a field for his labor" (1026). For this reason, Thompson was critical of his contemporaries' resistance to physical and mathematical descriptions of biological form—he even suggests that morphology is the last stronghold of irrationality—and endeavored to demonstrate that morphology could indeed be mathematized.

The book's relation to mathematics is a strange one, though. This is not a typical mathematical treatise. As Stephen Jay Gould remarks in his 1971 review, *On Growth and Form* does not contain any of the "differential equations and mathematical statistics that adorn modern work in ecology or population genetics"; instead, we find elegant descriptions of the "partitioning of space, the tetrakaidekahedron, the Miraldi angle, the logarithmic spiral and the golden ratio" (1971, 236). According to Gould, this is because Thompson was a zoologist who was also a classicist: it is Aristotle, Pythagoras, and the later Plato who inform his mathematization of morphology. This is not to say that Thompson holds to a strictly Pythagorean or Platonic doctrine of number—for instance, "things are numbers"—instead, he channels their "general attitudes: that solutions to the world's mysteries should be sought in 'the geometrical aspect of number'; that simplicity, regularity, symmetry, harmony and truth would be found conjoined; that the world of 'science' could be approached and comprehended by poetic vision" (Thompson, quoted in Gould 1971, 237). It is through quasi-mystical intuitions of number and form that biology is capable of moving beyond the contingencies of particulars to the universal principles of organization.

According to Gould, there is no "anachronistic Greek biology" in Thompson's work. Ancient mathematics inspires his efforts to bring wildly disparate and seemingly incongruous systems together under abstract physical laws. This synthetic work accounts for Thompson's magnificent and varied analyses of the correspondences between the shape and size of natural and physical systems.[2] Such correspondences between organic and inorganic forms are more than mere analogy: they are the result of the same efficient cause at work in the shaping of matter (241). As Thompson writes:

> The waves of the sea, the little ripples on the shore, the sweeping curve of the sandy bay between the headlands, the outline of the hills, the shape of the clouds, all these are so many riddles of form,

> so many problems of morphology, and all of them the physicist can more or less easily read and adequately solve: solving them by reference to their antecedent phenomena, in the material system of mechanical forces to which they belong, and to which we interpret them as being due. . . . Nor is it otherwise with the material forms of living things. Cell and tissue, shell and bone, leaf and flower, are so many portions of matter, and it is in obedience to the laws of physics that their particles have been moved, moulded and conformed. (1942, 10)

That Thompson worked across so many fields and with such great acuity accounts for *On Growth and Form*'s widespread praise and its secure place among other classics in the annals of science. In fact, after the publication of the second edition in 1942, Nobel Laureate Peter Medawar even proclaimed that Thompson's book was "beyond comparison the finest work of literature in all the annals of science that have been recorded in the English tongue" (in Keller 2002, 61; Medawar 1962, 232). Such high, if slightly exaggerated, praise for Thompson's work should have meant, under normal circumstances, that *On Growth and Form* had a tremendous impact on twentieth-century science. But it is precisely *On Growth and Form*'s lack of clear influence that makes it such a curious work in the history of science.

Despite the high praise it received from noteworthy members of the scientific community, the work remained difficult to characterize. Was it a work of art, science, mathematics, poetry, or all of them (Keller 2002, 62)? Gould, for his part, called it an "unusable masterpiece" that was plagued by "excessive length and difficulty of application" (1971, 253). In reference to this latter difficulty, he quotes Medawar, who, notwithstanding his admiration for the work, insists that Thompson's method was "analytically unwieldy" (253). What Medawar is referring to is that Thompson's mathematical morphology depended upon multivariant calculations of form—which is to say, more than two variables—when the mathematical study of morphology at the time used bivariant analysis (two variables). That bivariant mathematics was insufficient to study the growth of a whole organism was not in doubt. Rather, the problem was that the multivariant analysis of growth and form was difficult, if not impossible, before the electronic computer. Hence, the brilliance of Thompson's work, according to many, lies not so much in what it was able to achieve—that is, with actual equations—but more in what it predicted: namely, the superiority of multivariant calculation for the analysis of morphology. This is why Gould writes that "D'Arcy Thompson's vindication has come only now, and it has just begun" (254).

Looking back, it is hard to tell whether there is any truth to what Gould prophesized about Thompson's work. Gould gives us a taste of what may be in store by citing a couple of examples of scientists who have drawn on Thompson's coordinates to understand the morphology of chimpanzee and human skulls (Sneath 1967). And indeed, Thompson's work has found an audience among some adherents of computational and mathematical biology in the twentieth century, as well as in fields outside of the life sciences. My own introduction to Thompson's work did not come from the history of biology or even science but from contemporary biocomputational practices in architecture and design (Lynn 1999). Thompson's work is frequently cited in practices of computational architecture that find inspiration from morphologies that traverse organic and inorganic matter. But even if *On Growth and Form* continues to find readers and inspire new mathematical apprehensions of form, it has had a paltry influence on the readership it sought to affect most: the "naturalist" who might be influenced by a "few mathematical concepts and dynamical principles" (D. W. Thompson 1942, 1026). In her discussion of Thompson's reception, Keller writes that "a scan of the technical biological literature shows little evidence of an impact on the main trends of biological research as these continued to unfold over the course of the century" (2002, 70). While Gould, Medawar, and other biologists offer high praise of Thompson's work, and even prophesize its impact on the life sciences, it was unable to steer the course of biological research in the twentieth century.

Keller offers a concise explanation for the book's failure to have a larger impact on biological research: it failed to match the empirical realities of biological phenomena. If Thompson wished to pierce through the messiness of particulars to quantify them in terms of general physical principles, then twentieth-century biologists by and large remained tied to the empirical realities of *actual,* instead of the ideal, constructions of form. According to Keller, "The strengths that Thompson saw in the methods of mathematics and physics lay in their economy, in their focus on idealized examples," but biologists, in contrast, took these "characteristics" to be "not a strength but a weakness" (74). Where Thompson asserted that "this is how it *could* happen," he never considered "how it need happen, and certainly not how it *does* happen in any particular instance" (74).

That there was such a focus on the experimental realities of living phenomena in twentieth-century biology is more than evident in the work discussed in the previous chapter—the cell theorists, Mendel, Johannsen, and Correns were all experimentalists. It's not difficult to see how Thompson's disregard for experimental biology led him to write off genetics altogether, perhaps the most significant preoccupation of the twentieth-century life sciences. In a clear and deliberate affront to classical geneticists, Thompson writes that "the question whether chromosomes, chondriosomes or chromidia be the true vehicles or transmitters of 'heredity' is not without its analogy to the older problem of whether the pineal or the pituitary body were the actual seat and domicile of the soul" (1942, 268). Chromosome and gene theories were anachronistic, they were the specters of a Cartesianism that Thompson believed we were, or at least should be, well past (Keller 2002, 71–72). Thompson's mathematical and physical characterizations of biological phenomena were thus at odds with the dominant experimental culture of biological research for most of the twentieth century. And for many biologists, there were no physical and mathematical idealizations capable of dreaming up the structure and function of the DNA molecule; this could only have been discovered through messy empirical research (106).

This is not to say that there is no place for mathematics in twentieth-century biology, that mathematical biology has not found a home in the last few decades of life science research. Rather, the preoccupations of biologists for much of the twentieth century were such that even deeply relevant forms of mathematical research had to wait until later decades to find a hold in biological research practices. According to Keller, a shining example of this is Turing's 1952 paper, "The Chemical Basis of Morphogenesis." While Turing's work is most often associated with computer science and AI research, his detour through developmental biology "had a deep and lasting influence," according to Keller, "on virtually all subsequent attempts to model these processes mathematically" (2002, 89). Turing was not skeptical of genetic research in the way Thompson was. Instead, he, like many biologists at the time, was troubled by the fact that the classical gene had yet to offer a satisfactory account of development. Although embryology tended to be subsumed under genetic research at this time, there was still little known about how hereditary material directed the development of the organism.[3]

Turing sought to solve this problem. Although he references very few sources from biology (one being Thompson), he provides a mathematical model of embryogenesis. In it, he proposes "to discuss a possible mechanism by which the genes of a zygote may determine the anatomical structure of the resulting organism" (Hodges 1983, 641). He postulates the existence of chemical substances called "morphogens," which are called upon to account for the nonlinear development of the organism from out of disturbance of the homogenous system (Turing 1952, 37). Morphogens had not been chemically isolated; instead, they were theoretical "form producers" that could be anything that "diffuses into a tissue" (Keller 2002, 93) and "somehow persuade[s] it to develop along different lines from those which would have been followed in its absence" (Turing 1952, 38). Through a pair of differential equations, Turing modeled how two hypothetical morphogens would be capable of generating pattern and structure "due to the instability of the homogenous equilibrium, which is triggered off by random disturbances" (37).

I will not delve into the numerous issues surrounding Turing's chemico-mathematical model for working biologists at the time—for example, why not look at actual chemical reactions instead of only hypothetical ones?—I will simply note that it was slow to gain recognition, despite its relevance. As Giuseppe Longo notes in his "Letter to Turing," Turing's 1952 paper on morphogenesis is of an altogether different nature than his previous work on the Discrete-State Machine (DSM). With the DSM, Turing sought to emphasize "the physical structure of its hardware, with discrete states, and the discrete-writable, alphabetical nature of its software." But with morphogenesis, he was able to "grasp the role of the continuum, of the interplay between non-linear dynamics and physical measurement, which is always an interval, always approximate."[4] And it is precisely this conception of continuity that opens the door to thinking about nonlinearity in a non-Laplacian frame (something that Henri Poincaré worked on but remained undeveloped), the importance of which is only beginning to be understood in the domain of biological systems (Longo and Montévil 2014). Among the few to recognize the merit of Turing's paper at the time of its publication was C. H. Waddington, whose mathematical acumen far exceeded most biologists'. Waddington recognized that there was no explanatory framework for embryogenesis, and so he found Turing's model appealing (Keller 2002, 103–4).[5] In particular, Waddington acknowledged that what was missing in biological research was a mathematical–theoretical framework for understanding the generation of biological form. Although he thought that Turing had overestimated the significance of morphogens, they may be useful for understanding pattern formation due to "chance variations around an equilibrium," but not embryogenesis more generally (Waddington 1953, 122–23; Keller 2002, 105).

Returning to the slow uptake of Turing's model, Keller adds a not insignificant historical detail: there could not have been a less hospitable time for his research. Just a year later, Watson and Crick would publish their seminal papers that determined the structure and function of the DNA molecule. What this meant was that "DNA, translation mechanisms and codes took center stage, and even the attention of mathematicians and physicists looking for greener pastures . . . was captured by the new excitement" (Keller 2002, 106). It took decades for Turing's work to see the light of day, and anything that looked and smelled like mathematical biology received little to no recognition. Keller contends that the triumph of Watson and Crick's work, as well as the subsequent successes in molecular biology, just reconfirmed for biologists the value of its experimental culture (106–7), and in particular the priority of determining molecular constituents of living organisms through experimental methods (see also Boogerd et al. 2007).

As Crick would claim in 1966, "The ultimate aim of the modern movement in biology is to explain all biology in terms of physics and chemistry" (2004, 10). This statement is exemplary of what came to be known as the "reductionist" approach in molecular biology that would dominate the life sciences in the latter half of the twentieth century. While we already encountered this research paradigm in chapter 2, it's worth emphasizing that an experimental methodology that presupposes that the determination of isolated molecules and their structure have "sufficient explanatory power to provide an understanding of the whole system" is fundamentally at odds with the work proposed by Thompson, Turing, and other mathematical biologists, which seeks to provide general explanatory frameworks for morphology, embryogenesis, and other biological processes (Van Regenmortel 2004). In short, what molecular biologists sought was experimental specificity, not theoretical generality.

FURTHER IDEALIZATIONS: CYBERNETIC BIOLOGY

But doesn't this story conflict with some of our basic intuitions about the rise and dominance of molecular biology in the twentieth century? For example, is there not a significant body of scholarship in the theoretical humanities that shows how cybernetics and information theory had a tremendous influence on molecular biology? Are we not often told that the turn toward molecular biology, and many of its associated concepts (e.g., genetic code, program, information), is due to the success of applied mathematics, modeling, and idealizations—or those things for which Thompson and Turing were also found guilty (cf. Thacker 2005, 2004; Doyle 1997)? Scholars often point to key scientists, theories, as well as the Cold War political and economic climate in order to account for how "information," "code," "feedback," and "control" were introduced into biology and even helped spark the molecular revolution in the life sciences. Norbert Wiener's cybernetics, John von Neumann's cellular automata, Claude Shannon and Warren Weaver's mathematical theory of communication, as well as François Jacob and Jacques Monod's lac operon model all conspired in the context of the postwar military–industrial–academic complex to determine, both directly and indirectly, the material and symbolic spaces of molecular biology.

Eugene Thacker, for instance, attributes a great deal of importance to Jacob and Monod's work on "genetic regulatory mechanisms" (Jacob and Monod 1961). Thacker contends that their notion of the "genetic program" is explicitly indebted to early work in cybernetics (chock-full of control, feedback, etc.) and it is what makes it possible to conceive of the cell as a kind of computer as well as the attendant intuition that the computer can be hacked. This work, albeit nascently, is thought to anticipate the rise of computational biology and biological computing in the following decades (Thacker 2004, 98–99). Alexander Galloway (2004) makes similar points about the intersections of "wet" biology and "dry" code in the latter half of the twentieth century. And social scientists, such as Kaushik Sunder Rajan, suggest that the cybernetic influence on molecular biology paved the way for life itself to be commodifiable information or code in advanced capitalism (Rajan 2006, 22). Still others play fast and loose with the idea that a conflation exists between three- and two-dimensional systems (molecule and code) and then attribute a whole host of semantic and conceptual slippages to the field. In these views, cybernetics and information theory are constitutive for molecular biology and later developments in biotechnology and bioinformatics.

What these glosses often fail to recognize are the complex historical and epistemological circumstances that shaped the material and symbolic spaces in which Cold War information theory, engineering, and mathematics engaged with the experimental culture of biology. They generally fail to account for the fact that "feedback" and "regulation"—hallmarks of cybernetics and information theory—had been in circulation within experimental biology well before cybernetics introduced them into the field; they also neglect to appreciate how their meanings are derived from the specificities of experimental findings concerning developmental pathways. While these distinctions tend to be appreciated by historians of science, they are less often acknowledged in critical-humanities scholarship, which is often more interested in theoretical provocation than historical robustness.

Phillip Thurtle has recently written on this in *Biology in the Grid* (2018). There, he draws on William Bateson's notion of variation and the regulation of gene expression as well as Ed Lewis's work on the bithorax mutants in order to demonstrate how modularity and regulation are already fully embedded within the experimental culture of biology. (In fact, Thurtle even reverses the causality, suggesting that cybernetics got their ideas about regulation from biology.) Keller also discusses how Waddington, Frederick Warburton, and Edward Umbarger were writing about feedback in the mid-1950s and how none of these early citations can be directly tied to the invasion of cybernetics and information theory into biology (2002, 153).[6] Together their work suggests that feedback, regulation, and control mechanisms were not simply imported from cybernetics and information theory into biology. While there is little doubt that cybernetics and information research in the context of the Cold War military-industrial-academic complex had a significant impact on the interest in and circulation of control mechanisms in biology, their uptake is less of a historical and an epistemological rupture in the culture of experimental biology than it may be a shift in emphasis or focus (Thurtle 2018).

To take stock of the impact cybernetics and information theory had on biology, Lily E. Kay addresses the citational, discursive, and epistemological practices of scientists during this era of biological research. And in what follows, I trace some of her steps through this history. What's significant, if not entirely surprising, is that the greatest contributions to the cybernetic vision of the organism come from mathematicians, physicists, and computer scientists instead of biologists (Kay 2000, 73–127). Crucial to this vision was Wiener's seminal *Cybernetics: Or Control and Communication in the Animal and the Machine,* which suggests that the principles of control and communication engineering are applicable at the cellular and molecular levels and even proposes a model of "cybernetic heredity." Von Neumann, for his part, worked closely on biological automata, focusing in particular on neuronal and gene systems, and concluded that system function is independent and irreducible to the parts of a system (Kay 2000, 109). The potential impact of these theories on bioscientific research became well known as they circulated among mathematicians, scientists, and industry throughout the mid-twentieth century. Mathematicians and engineers were not alone in advocating the use of cybernetic and information systems to understand biological phenomena. There were countless biologists and biomedical partners who were compelled to apply control and feedback mechanisms to their research. But more often than not these forays into cellular automata and other information-theoretic models yielded very little in the way of biological knowledge, and for this reason did not make the kind of splash in the life sciences that Wiener and others would have hoped. According to Kay, the main reason for this is that these theories, by and large, failed to offer biochemical specificity, when specificity was precisely what would make cybernetic theories useful to biologists working with actual specimens.

A case in point is Joshua Lederberg, a geneticist from the University of Wisconsin, who was intrigued by von Neumann's work on cellular automata, and in particular how von Neumann's work on self-reproduction could be applied to his research on genetic reproduction (113). Lederberg wondered what it would mean for a nucleus or a gene to "reproduce" in the way von Neumann imagined, especially when "most intracellular particles will not reproduce at all outside of the context of a given kind of cell" (Lederberg to von Neumann, quoted in Kay 2000, 113). Despite Lederberg's intrigue, and his correspondence with von Neumann, he was ultimately troubled, among other things, by the lack of attention paid to the individual components of a reproducing system. Indeed, "genes, or even nuclei, are incapable of producing anything, much less copies of themselves, when isolated from the whole machine" (113). While self-reproduction applies to systems, they are quite independent of their parts. Hence, von Neumann's model could not attend to the specificity of genetic material, which is precisely what an experimental geneticist demands. For this reason, Lederberg writes in one of his letters that "I have the occasion to feel rather uncomfortable at the adequacy of the genetical concept of 'self-reproduction'" (113). Additionally, Lederberg was uneasy about the conflations made between digital and non-digital coding, as well as between "information tape" and genes. All of this led him to the conclusion that biologists and information theorists were "working at very different planes" (Lederberg, quoted in Kay 2000, 114). Kay summarizes Lederberg's concerns:

> Molecular biologists were concerned with *realistic* models of reproduction. Perhaps they could design a chemical analog to a punch-card reproducer, and thus go "a long way toward the experimental initiation of life," if only they possessed an equal knowledge of the model's parts. They could simulate an autocatalytic system by punch-card reproduction with more than one or a few bits on it, Lederberg pointed out, but none of the chemical machines they could devise would produce results anywhere near the complexity of an organism. (114)

There were still other, less disenchanted biologists who worked closely with the models of cybernetic organisms that von Neumann and Wiener sketched. For example, Wiener's *Cybernetics* was well received by biologists such as J. B. S. Haldane and H. Kalmus. Using Wiener's notion of a "cybernetic heredity," in conjunction with Shannon's theory of redundancy, they worked out a theory of genetic transmission and mutation (e.g., a mutation is "noise which gets incorporated into a message").

These endeavors were nonetheless based on the outdated (though popular at the time) "protein paradigm of the gene" (78, 86). In fact, Kay points out that this was the operative genetic paradigm for most "cybernetic efforts to represent heredity and life in terms of information and communication," as they had yet to be exposed to the DNA-based paradigm of the gene (78).

The Austrian-born physician and radiologist Henry Quastler, on the other hand, attempted one of the most literal translations of cybernetic and information-theoretic notions into the life sciences: he peddled an "information-based biology" that was even equipped (eventually) to accommodate the DNA-based genetics (115, 125; Quastler 1953). What's instructive about Quastler's work is that it was not for lack of resources and attention that his foray into information theory failed to yield a new paradigm for biological research. According to Kay, his work came into prominence in the 1950s as he spearheaded many public efforts to make Shannon, Wiener, and von Neumann's ideas about information and self-reproducing machines relevant to biologists, and more specifically address how information content could deliver biological specificity. Working in close company with cybernetic luminaries, and backed by the Control Systems Laboratory at the University of Illinois by the end of 1940s, Quastler embarked on an all-out "quantification of information flow in biological control systems" (Kay 2000, 118). During this time he worked closely with the theoretical physicist Sydney Dancoff (Dancoff and Quastler 1953) to generate a Shannon–Wiener theory of the gene, which proposed a "chromosomal error control" system that served as a "checking device" for the "high-error" process of replication (Kay 2000, 118).

After Dancoff's death, Quastler continued to rally support for his cybernetic biology and attend to the specificity of biological control systems. In a paper titled "The Measure of Specificity," he argued that "when properly qualified, an information content would supply the exact measure of specificity. Establishing an estimate of 'enzyme specificity' based on binary codes of 'right' and 'wrong' substrates, he showed that the expression resembled Shannon's entropy equation" (120). By the early 1950's, Quastler's expertise in the area of biological control systems was even acknowledged by the eminent cybernetician Heinz von Foerster, who presented Quastler's work at the Ninth Macy Conference in 1952.

Despite these many efforts to produce an information-based biology, which taken together was supposed to yield a new epistemological framework for the life sciences, there were a number of underlying weaknesses with the agenda. According to Kay, it was "devoid of predictive capability, means of theory-testing, and experimental agenda" (123; Apter and Wolpert 1965).[7] The vulnerabilities came to a head by the end of the 1950s, when biologists and cyberneticians alike grew frustrated with the inability of an information-based biology to yield any new facts, or even to create an experimental research agenda more generally. Apparently, Shannon and von Foerster even came to accept the limited scope of information, command, and control, and grew skeptical of their usefulness to working biologists (Kay 2000, 126–27).

In attending to the failed attempts to import cybernetics and information theory into the life sciences, Kay is careful to point out that the discourse of information nonetheless managed to seep into biology (see also Kay 1997). But when it did, notions of information, regulation, and control were taken to be "metaphors" that were "emptied of [their] technical content" instead of indices of a true information-based biology that had taken hold (Kay 2000, 127).[8] Even Jacob and Monod's lac operon model, which is often taken to be a shining example of "cybernetic biology," is far from just "applied cybernetics." While their work is often credited with popularizing notions of feedback and control in molecular biology, Keller argues that they adapted these notions to meet the demands of experimental biology in very specific ways. Their lac operon model provided a revolutionary way to think about genetic regulation of the production of the enzyme beta-galactosidase in *E. Coli* through negative control. But as Keller makes plain in her careful consideration of their work,

> even while popularizing these concepts among their colleagues, they [Jacob and Monod] also tailored them in ways that minimized conflict with the dominant framework of the time. Where others had taken the essence of the cybernetic metaphor to lie in the absence of any form of centralized control, and had seen feedback as providing the mechanism by which global regulation could be achieved, Jacob and Monod redefined both feedback and regulation (and even epigenesis) to refer to genetically controlled processes. (Keller 2002, 170)

For Keller, although cybernetics and information theory managed to influence the discourse of molecular biology, they were filtered through the experimental culture of biology. "Information," "control," and "feedback" came to have different meanings for biologists than for mathematicians and information theorists. Many of these notions had already been in circulation within experimental biology (Thurtle 2018; Keller 2002) and so were not information-specific concepts (Judson 2004); and when they were applied to biology from information theory, they were applied with very limited success. The larger point is that cybernetics and information theory, like the mathematical and physical models of biology that preceded them, were met with resistance by the experimental culture of biology in the twentieth century.[9] Despite the appeal of applying simple and elegant models with explanatory power to biological phenomena, there was no upending biologists' deep commitment to the experimental analysis of messy living systems.

SYSTEMS AND EXPERIMENTATION

But something very different seems to be happening in late twentieth- and early twenty-first-century molecular biology. Computer simulations and mathematical modeling are part and parcel of life science research today. As Catelijne Coopmans and her colleagues insist, "researchers' attentions are as likely—or more likely—to be focused on colorful digital images, simulations, software suites, databases, or lines of code as on unruly specimens or instruments" (2014, 1). They go on to suggest that the laboratory setting has fundamentally expanded to "other spaces and places via collaborative ventures, shared data centers, and information and communication technologies" and that "this expansion challenges the very distinction between laboratory and field" (1–2). Similarly, Keller contends that the distinction between the virtual and the real organism has been blurred so that "our perception of 'real time' comes to be more and more closely assimilated with our perception of 'reel time'" (2002, 233).

Given this, it becomes necessary to ask: How and why have information and communication technologies, mathematical modeling, and computer simulation found their way into molecular biology? If experimentalism was the reigning epistemological framework for twentieth-century biology, then how did mathematical and computer models become viable tools for researching molecular phenomena today? Did experimental epistemologies finally concede to the reductive simplicity of physico-mathematical modeling? Or is there something else at work? In what follows, I contend that the biological sciences' embrace of mathematical modeling in recent decades has less to do with conceding to reductive simplicity and more to do with finding adequate means for dealing with extraordinary complexity. As we will come to see, it is in the epistemological space now shared by experimental and systems biology that molecular animation finds a home.

In the last few decades, there are a number of biologists who have taken to historicizing their field. They acknowledge that even though systems thinking was in the air in the latter half of the twentieth century and had a tremendous impact on countless other fields—physics, engineering, and so on—biology was slow to catch on (Trewavas 2006).[10] But today, systems theory has found a secure foothold in mainstream biology. Looking back, many biologists attribute the lack of systems thinking in the history of molecular biology to the fact that there just wasn't enough molecular data to make systems-level analysis useful to biologists (Bruggeman and Westerhoff 2007; Alberghina and Westerhoff 2005; Westerhoff and Palsson 2004; Kitano 2002; Stevens 2004). In the days of Wiener, molecular biology was still very much in its infancy, and large datasets dealing with system-level molecular function were not yet available. As the systems biologists Frank Bruggeman and Hans Westerhoff explain, "Systems biology approaches did not connect to the more rapidly expanding experimental molecular biosciences, largely because of the lack of sufficient quantitative molecular data" (2007, 45).[11] And while this supposedly reductive approach offered an impressive catalogue of genome parts, it did not, and perhaps could not yet, deal with how genes and gene products function together as a part of gene regulatory networks (Davidson 2006; Yuh, Bolouri, and Davidson 1998). In short, molecular biology was in the business of collecting experimental data, but it was not in a position to comprehend how genome-wide data worked together (Bruggeman and Westerhoff 2007). So while systems theory made serious headway in areas such as nonlinear dynamical systems in mathematics, as well as in metabolic engineering, and helped to characterize self-organization via nonequilibrium thermodynamics, molecular biology has been relatively slow to avail itself of systems theoretical analysis.[12]

Scientists often attribute the rapid spread of genomics, and omics-based research more generally, to the popularity of systems perspectives in the biosciences. According to the editors of *Systems Biology: Philosophical Foundations,* the "development of the various high-throughput technologies used in genome sequencing, transcriptomics, proteomics and metabolomics have enabled the comprehensive analysis of complete living systems in terms of the identity and concentration of all their components" (Boogerd et al. 2007, 6). In other words, high-throughput, genome-wide experimentation made new tools for analyzing, synthesizing, and experimenting on the wealth of data necessary (Tanay et al. 2005). As Boogerd and his colleagues put it, systems biology takes the molecular and cellular data gathered from omics research and models how it works together in a living organism, how the whole comes together to form a system. It is for this reason that genomics and systems biology have deeply intertwined histories: "Genomics provided a critical boost to the popularity of systems biology," since genomics requires systems-level analysis if it is going to say anything about the dynamic organization of cellular life (Conesa and Mortazavi 2014; Auffray et al. 2003; Boogerd et al. 2007). Sequencing an entire genome, Bruggeman and Westerhoff continue, is nothing but a long list of parts unless it is pieced together into a systematic whole.[13]

Systems biology is not one thing or practice but a diverse set of practices, technologies, and expertise that share certain goals. According to Hiroaki Kitano (2002), head of the Systems Biology Institute in Tokyo, and president and CEO of Sony Computer Science Laboratories, systems biology aims to describe (1) "system structures," including the "network of gene interactions and biochemical pathways"; (2) "system dynamics," or how a system behaves over time under varying conditions; (3) the "control methods," or what controls the "state of the cell"; and (4) the "design methods," which are those techniques that could be used to intervene in or modify systems in order to yield desirable system behavior. A similar description of systems biology is offered by Bruggeman and Westerhoff; but they emphasize that systems analysis provides an exhaustive account of the constituents of emergent behavior. They write that systems biology seeks

> (i) a (complete) characterization of an organism in terms of what its molecular constituents are, with which molecules they interact, and how these interactions lead to cell function; (ii) a spatio–temporal molecular characterization of a cell (e.g. component dynamics, compartmentalization, vesicle transport); and (iii) a thorough systems analysis of the "molecular response" of a cell to external and internal perturbations. In addition, information from (i) and (ii) must be integrated into mathematical models to enable knowledge-testing by formulating predictions (hypotheses), the discovery of new biological mechanisms, calculation of the system behavior obtained under (iii), and finally, development of rational strategies for control and manipulation of cells. (2007, 46)

This shift in emphasis—where molecular constituents play a primary role in systems biology—is important to note and will be revisited below. But suffice it to say that in both descriptions, systems-level analysis seeks not only to describe system behavior but also to control, manipulate, and design it. I do not want to dwell on this point here, but such an emphasis on design brings into focus how the discourse and practices of engineering have infiltrated mainstream biology.[14]

The transition to systems-level analysis has not always been met with open arms. Much of what passes for systems biology has done little to mitigate the concerns of mainstream molecular biologists. As Charles Stevens of the Salk Institute explains in 2004, "For many molecular biologists, a systems-level explanation leaves them feeling that, until the underlying molecular mechanisms are known, the approach is hopelessly superficial" (2004, 51). Unless you have a list of genes and their interactions, some molecular biologists insist, you cannot adequately understand function. Historically, criticisms such as these are leveled against what is referred to as "top-down" systems biology. As the name suggests, a top-down approach to systems biology entails gathering data on molecular function from a "bird's-eye view" and then formulating hypotheses about the molecular interactions that could have led to the behavior in question (Boogerd et al. 2007; Bruggeman and Westerhoff 2007, 46). In other words, you move from the whole to the parts. One advantage to this method is that it is particularly good at determining the behavior of large datasets; it then uses it to induce molecular organization based upon correlations.[15] However, one serious limitation to this approach is that this "rarely (if at all) leads to the formulation of relations between properties in terms of molecular mechanisms. Although the emphasis formally lies on inductive discovery science, such discoveries rarely lead to molecular knowledge" (Bruggeman and Westerhoff 2007, 49).

Bottom-up systems biology is the other approach to the system-level analysis. Unlike the top-down method, the bottom-up method leverages experimental studies that determine both the "kinetic and physicochemical properties" of molecular parts in order to predict functional behavior (47). The strength of the bottom-up approach is that it bases its models (ideally) upon actually determined properties of molecular constituents (experimental data). There are many risks involved in this approach as well, especially when modeling kinetic parameters: "As soon as the model is enlarged and the same data are again used for parameter estimation, the parameters will assume different values" (47; see also Moles, Mendes, and Banga 2003).

To correct for this problem, Bruggeman and Westerhoff cite the silicon cell program as an example of bottom-up systems biology that serves to illustrate the importance of working with "actual" experimental data instead of data "fitted" to a model.[16] The program is made possible by a global consortium of researchers who aim to work with experimental data on pathways in order to build models of emergent reaction networks. Researchers predict system-level properties that emerge from experimentally determined molecular constituents by comparing the behavior of the in silico model to what is actually observed at the system level; the discrepancies are accounted for and then fed into the next generation of the "silicon cell." Crucially, the discrepancies between observed and mathematically predicted behavior are not "fitted away," but they are the basis for building better and more robust models. In the end, the silicon cell program serves as a platform for online in silico experimentation with biological systems (see also Alberghina and Westerhoff 2005). Ultimately, Bruggeman and Westerhoff argue that we need to bring both approaches to systems biology, top-down and bottom-up, together: "It should then be realized that bottom-up systems biology cannot tolerate possible unknown factors and, hence, will need to integrate with top-down, genome-wide systems biology to ensure completeness" (Bruggeman and Westerhoff 2007, 49). This integrated perspective aims to bring the experimental data into genome-wide analysis to ensure better and more robust modeling of system behavior. This work functions less as a critique of molecular biology than it does an updating of it, so that it may interface with the wealth of data gathered from high-throughput measurements. Such efforts to integrate the experimental work of molecular biology into systems-level analysis actually reflect a broader tendency of systems biology to become more experimental. Keller explains that in the twenty-first century, systems modeling and mathematical biology are far more "grounded in [the] experimental realities" of organisms than they were in the days of Thompson, Turing, and Wiener. She writes that

> of the many differences made by this turn to computers in biology, one of the most prominent has been to render this conceptual work performed by these models easier to recognize as theoretical. Yet the new kinds of mathematical models biologists develop and explore are far more closely grounded in experimental realities and hence far more acceptable to working biologists than any of the other mathematical models that had earlier been possible. (2002, 237)

To illustrate this point she draws on two examples in contemporary biology, one from Eric Davidson's work on gene regulatory networks and the other from Garrett Odell's work in developmental biology (239–64). With respect to Davidson's work at the California Institute of Technology, Keller points to his extensive research on the promoter of the developmental gene *Endo16,* a gene that encodes a protein important for sea urchin development (Yuh, Bolouri, and Davidson 1998). Crucially, Davidson and his colleagues developed a model for the regulation of *Endo16*—which made sense of the deluge of experimental data—that became a tool for thinking with experimental data that dealt with the regulation of this particular gene instead of an abstract model that could be applied to gene regulation in general. In short, their model was not generalizable, and for this reason did not meet the demanding expectations of physicists, for whom models must be widely applicable. But this did not pose a problem for experimental biologists, since it told them something important about variability: "It [made] one realize how different [gene regulatory systems] can be" (Davidson, quoted in Keller 2002, 247). Odell's work at the University of Washington, on the other hand, models the temporality of gene networks and gene products associated with the segmentation of insect bodies, such as *Drosophilia.* Odell and his colleagues start with experimentally known gene interactions and then use models to understand where they need to update their experimental data. Mathematical models are interwoven with experimental realities of organisms and provide a program for experimental work (von Dassow et al. 2000).

The takeaway is that the mathematical modeling of system behavior is no longer disconnected from experimental biology—and hasn't been for sometime now. Or at least mathematics no longer seems detached from experimentalism in the way that it was for most of the twentieth century. For this to have happened, two very different epistemological cultures had to be brought together. In chapter 2 we examined the experimental culture of biology, and in particular paid special attention to how visualization is a cornerstone of knowledge production in this domain. But by the end of the analysis we came to appreciate how systems of gene regulation are far too vast and complex to be isolated and visualized in situ. By contrast, in this chapter we encountered a very different mode of producing knowledge about organisms—namely, through the mathematical modeling of system behavior. In this mode, scientists generate models of organismic behavior believed to have emerged from molecular interactions, but they often lack experimental verification for their calculations. Even when we take into account the many successes of computational systems biology (silicon cell, *Endo16,* etc.), the epistemic cultures of systems and experimental biology still seem far from seamlessly woven together. Nevertheless, what we're witnessing in the early decades of the twenty-first century is that mathematical and experimental biology no longer prescribe opposed or competing epistemological frameworks (or at least in the way that they once did). Indeed, the data-rich landscape of postgenomic biology seems to necessitate some interweaving of the two.

In the remainder of this chapter I explore the intersection of mathematical and experimental biology using molecular animation. Where chapter 1 situated molecular animation in terms of popular media, and more specifically suggested that it shares many of the same features as contemporary screen media, in what follows I leverage the history of biology recounted in chapters 2 and 3 to examine how molecular animation is equally an outgrowth of the intersection of experimental and theoretical cultures in biology. Molecular animation is indebted to these different epistemological cultures finding common ground. Such a perspective not only invites a fresh view of computer animation but also encourages us to reevaluate the modalities through which systems and experimental epistemologies come together. My contention is that computer animation is one such point of intersection.

ANIMATING SYSTEMS

I want to preface this discussion with a quick note about a popular use of molecular animation that I am not particularly interested in discussing in this chapter. In recent years many scientists/animators have begun addressing the educational uses of animation in biology classrooms. For example, Jodie Jenkinson and Gaël McGill (2012) discuss how molecular animation overcomes various learning challenges in the undergraduate biology class. Likewise, Hsin-Yi Chang and her colleagues (2010) discuss the importance of evaluating animations based on how well they allow middle school students to "understand the particulate nature of matter." And even iGem, the world's premiere synthetic biology competition for undergrads, has touted the virtues of molecular animation for depicting synthetic constructions.[17] What these discussions illustrate is how animation is a tool for communicating already agreed-upon facts to mass audiences.

And while this is an increasingly common way to discuss the function of computer animation within the life sciences—especially when many scientists do not think that its "fictionalizations" are suited for the laboratory—I am not concerned here with the dissemination of biological facts to wider audiences (see chapter 1 for discussion). Rather, in this chapter I'm interested in those scientists who use computer animations as a crucial part of their research practice. In opposition to those objecting and/or cautious scientists we encountered in chapter 1, what I'm suggesting is that computer animation, like the microscope, has emerged as an integral part of knowledge production at the intersection of experimental and systems biology. To unpack this claim, I draw on two case studies taken from the world of molecular and cellular animation. The first one comes to us from Jason Sharpe, Nicholas Woolridge, and Charles Lumsden in their extensive work, *In Silico: 3D Animation and Simulation of Cell Biology with Maya and MEL*. The other one comes from Janet Iwasa's work with the Reck-Peterson Group at Harvard University on the animation of motor proteins. What's instructive about these two examples is that they offer different ways of thinking about the significance of molecular animation for scientific research. However, and crucially, they also provide unique perspectives on bridging experimental and systems approaches through animation.

Case Study 1: Biocomputational Vision

Perhaps no research team has taken the computational stakes of molecular and cellular animation as seriously as Jason Sharpe, Nicholas Woolridge, and Charles Lumsden in their extensive work, *In Silico: 3D Animation and Simulation of Cell Biology with Maya and MEL*. What is intriguing about the study is that it situates molecular animation within the history of mathematical and computational biology. Drawing on the history of cybernetics and information technology—from the work of John von Neumann, Claude Shannon, and Norbert Wiener to recent developments in biochemical computing—they argue that biological systems are just complex computing machines. "There is no evidence," they conclude,

> for some mysterious, supernatural "life force" acting alongside the chemistry of living matter . . . Cells obey the laws of chemistry and chemistry the laws of physics. The laws of physics and chemistry are, in turn, intensely and fundamentally mathematical. One must conclude that mathematical calculating machines—capable of large scale computing based on enormous volumes of data and information—can be used to analyze, predict, and ultimately re-design the biochemical activity of cells and organisms. The machines can do this by invoking mathematical operations to precisely represent the biochemical mechanism and calculate its properties. (2008, 35)

This otherwise swift and reductive view of living systems is complemented by an equally strong emphasis on the importance of visual perception in scientific research. Sharpe and his colleagues give a brief history of the priority of visual perception in the West and offer an account of its significance for knowledge production in the biosciences more specifically. They then discuss the daunting task of visualization in the data-rich landscape of the contemporary biosciences:

> Phenomena at the cellular and molecular levels are the principle focus of modern medical and bioscience research. At these scales structures and events are often difficult or impossible to see in the lab, in real time, as they happen: the distances are too small, the times too short, the events too unusual. Instead, they are measured—mapped as numerical properties. But how can we envision what the numbers mean? New tools are needed that leverage the power of computer graphics (CG) to see into the complex web of structure and function in cells and tissues. We, and other researchers in the emerging field of computational biology, are meeting this challenge with CGI: using the computer as a visual information machine to harness the brain's enormous prowess for insight into the knowledge encoded in cellular data and computer models. We call this approach *in silico,* since it is focused on computer methods for research discovery that will complement traditional *in vivo* and *in vitro* methods. (4)

What's noteworthy is how the authors bring a data-intensive systems approach to the biosciences together with the experimental priorities of observation. And what's more, they even suggest that computational approaches to biology seem to be missing something if they cannot visualize cellular and molecular data. In short, biological data need to be given visual form in order to make sense, to mean something to scientists. And it is this dual emphasis on computation and observation, systems and visualization, that not only confirms Keller's intuition that there is a much more complex relation unfolding between systems and experimental approaches to biology than in previous eras but also paves the way for the development of new visual technologies that can bridge the gap between mathematics and visual perception. But for these researchers, visual perception is nothing more than a high-speed "pattern recognizer" that resembles a supercomputer behind our eyes (4). And for this reason, the best visual technologies are the ones that align the computing power of human vision with biochemical computers. What's significant about molecular animation in particular is that it is uniquely situated to bridge these two computational registers. Their wager is that molecular visualization is no longer about visual representation—namely, re-presenting one thing (nature) in terms of another (image) (Daston and Galison 2007). Rather, at stake is expressing the immanence of systems to one another through their common language: mathematics.

Overall, their work aims to demonstrate how the high-end proprietary software program Maya Autodesk is primed to close this computational gap. And while their research focuses on the affordances of Maya, they note at the end of their study that most of what they have described is also available using other "scriptable high-end packages for 3D animation and rendering" and may be available one day soon using lower-cost or open-source software such as Blender or the Visualization Tool Kit (Sharpe, Woolridge, and Lumsden 2008, 578). At the time that Sharpe and his colleagues were writing in 2008, neither Maya nor any of the other high-end 3D animation packages were designed specifically to animate molecular and cellular structures and processes. Until recently, animating molecular geometries was not a straightforward or easy process. Usually, a good deal of effort went into uploading molecular structures into Maya from the Protein Data Bank (PDB), an open-source platform where scientists upload experimentally determined 3D structures of proteins and nucleic acids sourced from X-ray crystallography, cryo-electron microscopy, and elsewhere.[18] As Maya is not outfitted for the specific requirements of modeling the geometries of proteins and nucleic acids found in the PDB, there was little that was intuitive or efficient about bending Maya's rendering capabilities to the specific geometries of molecular structures. Scientists would have to build representations in programs and then export the geometry to a file format recognizable to Maya (29). However, the geometries of PDB files were often not preserved in the translation to Maya.[19]

In recent years, the workflow has been significantly streamlined. Working with the biologist and programmer Campbell Strong, Gaël McGill customized a plug-in for Maya called Molecular Maya (mMaya), using MEL (Maya Embedded Language) and Python (high-level language for general-purpose programming). The plug-in allows scientists to upload PDB files directly into Maya, thus maintaining a direct connection between the geometries of the underlying PDB files and the geometries Maya is capable of building (see Figures 24–26). What's more, there are a series of "kits" available for mMaya that automate the

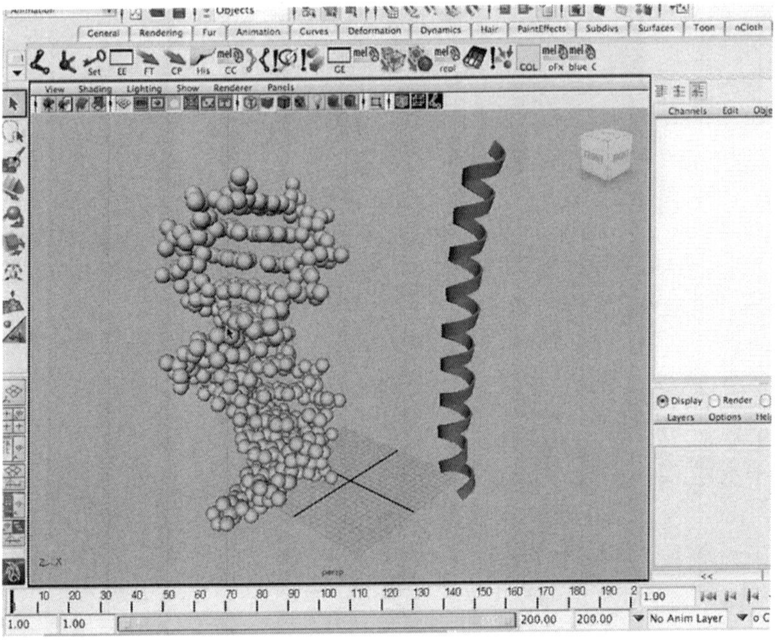

Figure 23. Screenshot taken from Molecular Maya Tutorial. On the left is an example of a DNA molecule brought in from PDBreader.mel. This image puts a sphere in the position of each atom; however, only one type of representation is available to the user with this script. On the right is a model built in a molecular graphics program; the tesselation of the geometry is very high (once you look closely) and so is less than optimal. The issue with both of these methods is that they lose connection to the initial PDB file.

animation process. For example, the "Double-Stranded DNA (dsDNA) Kit" automates the simulation of DNA sequences as well as other protein interactions with the dsDNA. Likewise, the "Cryoskeleton Kit" enables rapid modeling of actin or microtubule fibers.

In designing mMaya, McGill and Strong used MEL and Python scripting languages. While MEL is Maya's built-in language, Maya can be a host application for Python as well. There is some debate in the Maya community about whether Python will overtake MEL, as Python is more powerful, gives the programmer more control, and is not Maya-specific.[20] In this regard, Python is a general-purpose, object-oriented language (OOP), where MEL is not. MEL is a good example of a domain-specific language, which essentially means that it is built for a

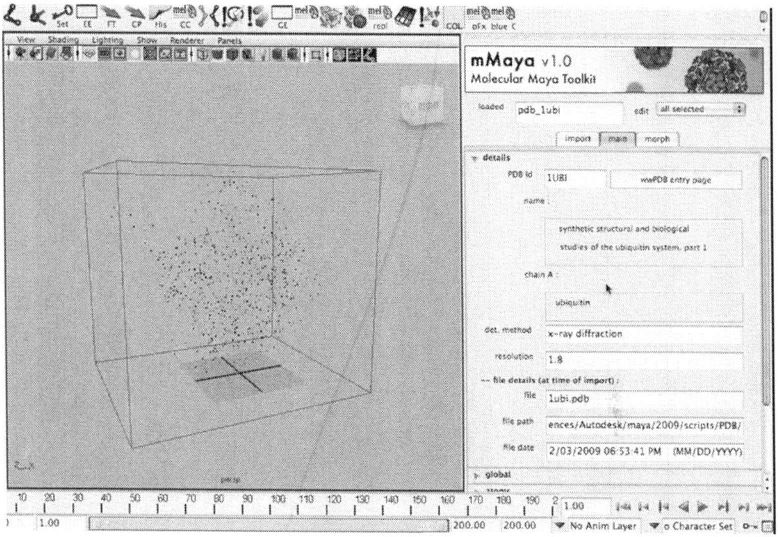

Figure 24. Screenshot of Molecular Maya Tutorial. Imported PDB file. On the right is information about the PDB file (the protein, ubiquitin) obtained from X-ray diffraction. On the left is the protein coordinate set, represented as a cloud.

specific application domain. Since OOP is arguably the dominant programming paradigm, and Python is an OOP language, Python has a natural appeal. Essentially, OOP is a programming methodology based on "objects" and their interactions, rather than a procedural list of actions (procedure-oriented programming). Objects are defined as data fields, which are instances of a "class" of objects (essentially, a blueprint that identifies the variables and methods that are common to a certain kind of object) and thus have similar contents and behavior as the other objects in the class. Each object is a self-contained unit with properties and procedures, known as "methods" (or what it can do), which can be altered without affecting the other objects. Objects exchange information with each other through "messages," and each object has information about which objects it can interact with.

Sharpe and his colleagues describe OOP as a "transactional" programming language, which seems to lend itself to the exchanges that happen in the biochemical world: "The organs and tissues of our bodies are built from cells that coordinate their activity by exchanging chemical

a

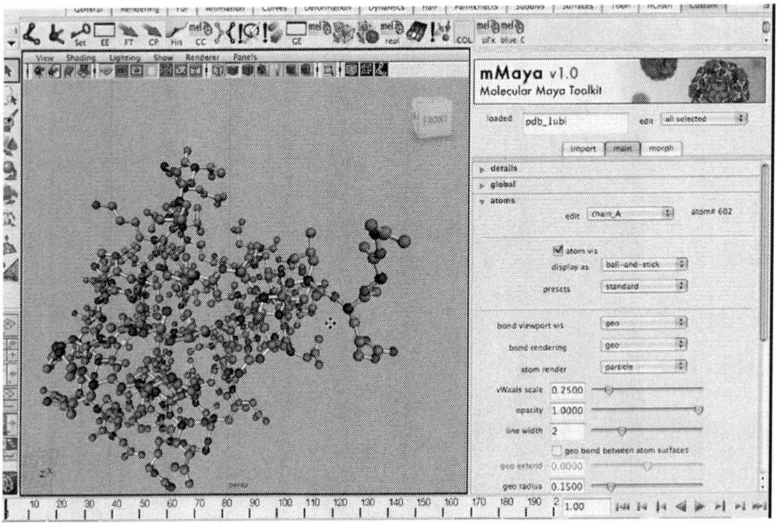

b

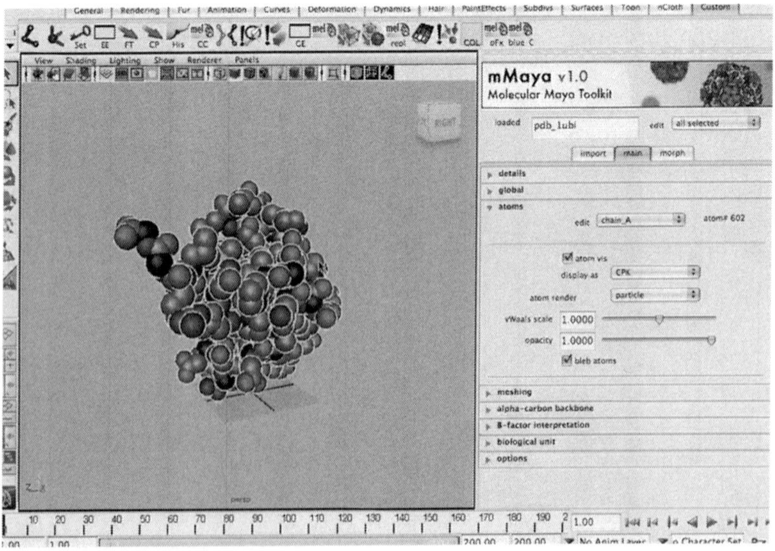

Figure 25. Screenshot of Molecular Maya Tutorial. Once the structure is imported, there are different options for representation that are built into Molecular Maya and automatically available to the user. *(a)* Ball-and-stick representation. *(b)* CPK representation.

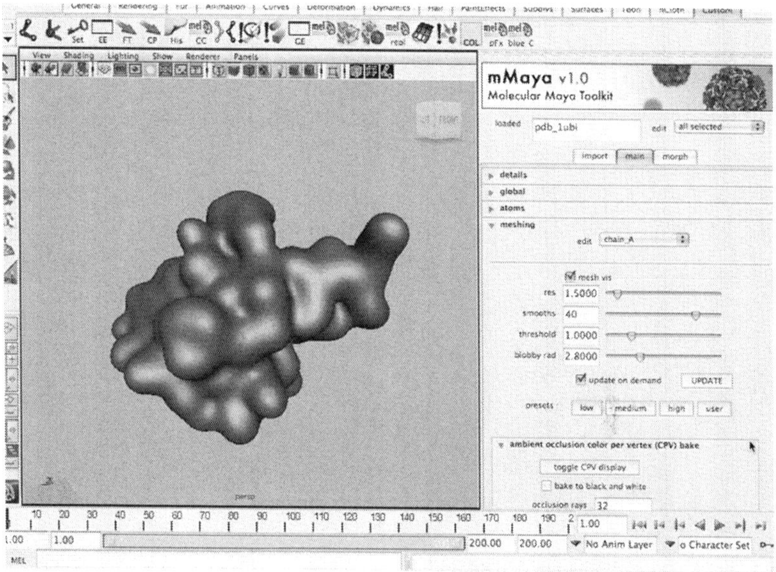

Figure 26. Screenshot of Molecular Maya Tutorial. Maya's built-in meshing capabilities can be used to fill different kinds of meshes of that point cloud; the user has a great amount of control over the parameters to "tighten the mesh."

and electrical signals" (Sharpe, Woolridge, and Lumsden 2008, 40). For this reason, OOP has frequently been used for the purposes of computational and systems biology, and in the last couple of decades Python has become an important programming language in these domains. Projects such as Biopython, an open-source archive of bioinformatics tools written in Python, illustrate the language's increasing relevance for the field (Chapman and Chang 2000).[21] In their research, Sharpe and his colleagues acknowledge the power and importance of these programming capabilities for MEL and Maya,[22] which Molecular Maya evidences, but they also don't minimize the significance of Maya's embedded language.

They suggest that MEL is a domain-specific language well suited to problems specific to the animation of objects. For instance, OOP scripting languages (C++, Python, Java, etc.) are not designed to deal with the mobility of the biochemical world. Under these conditions, objects function less like "nodes in a communication network" and more like self-directing, correcting, and modifying agents—molecules and cells are dynamic systems. Maya and MEL, with their built-in capacities to model 3D motion, are much better equipped to handle the programming challenges surrounding the dynamism of the biochemical world. For this reason, Sharpe and his colleagues suggest that an "agent-oriented-programming" (AOP) workflow is often preferable to an OOP one, and MEL remains an invaluable tool for bringing this workflow about (Sharpe Woolridge, and Lumsden 2008, 40; Shoham 1990).

But at the end of the day, these researchers are committed to leveraging the mathematical capabilities of programming languages to model increasingly complicated geometries, movements, and exchanges. And with this comes the enriched capacity to translate biochemical geometries and movement vectors into visual form. Thus they conclude that

> just about any structure or process that can be described mathematically—from simple molecular diffusion to a complex AI scenario—can be expressed using MEL. This means that, in addition to employing Maya's built-in physics capabilities, you can program the rules of cell activity relevant to a specific project, in essence writing your own *in silico* biology engine. Parameters and equations then can be adjusted to test hypotheses or to make predictions *in silico* in advance of expensive, time consuming real-world experimental work. (Sharpe Woolridge, and Lumsden 2008, 17)

Their contention is that anything, at least in principle, that can be described mathematically, including motion over time, can be expressed in Maya. While this may have seemed aspirational in 2008, with the development of mMaya and increasingly flexible plug-ins, animation software is able to blend seamlessly with bioinformatics software. As Graham Johnson and his colleagues explain, their open-source plug-in, Embedded Python Molecular Viewer (ePMV), goes inside any number of host 3D animation software packages and "creates both a common model and a common GUI (Graphical User Interface) environment to unite molecular modeling software and 3D animation applications, as well as algorithmic contributions from a variety of scientific software packages" (2011, 300). What this means, the authors argue, is that "ePMV can import data and adjust (simplify or enhance) the level of detail (e.g., from atomistic to geometric for PDB files) to suit specific audiences or to convey the confidence levels for each object shown in the scene" (298). The larger point is that the computational landscapes of biocomputing and bioinformatics are so interwoven with 3D graphics that Sharpe and his colleagues' provocation about the computational immanence of vision and biochemistry seems more realistic today than ever before. And what's more, the experimental priorities of many biologists can also be accommodated: predictions can be made and hypotheses can be tested, all the while they are carefully observed in silico. In short, the many advantages of computational systems analysis are now paired with the hallmarks of experimentalism: hypothesis formation, rigorous testing, and observation. But unfortunately, concrete examples of how animation contributes to experimentation are not provided. This aspect of molecular animation will be fleshed out in the next case study.

But for the purposes of this example, what stands out is that technologies of visual perception can be anchored to the mathematical complexities of biochemical systems. In chapter 1 we focused on the aesthetic capabilities of molecular animation in films such as *The Inner Life of the Cell, Protein Packing,* and *X Inactivation and Epigenetics,* and in particular addressed how human perception traverses numerous cellular and molecular scales in these films. Now we're in a position to appreciate how it is possible (in principle) for these perceptions to be mathematically fused to the geometries and movement of biochemical systems. We can also appreciate how this capability fills in a gaping hole in the visual economy of the biological sciences: much of what is invisible to researchers using light, electron, or confocal microscopy is brought into view. And at the end of chapter 2 I suggested that what is largely invisible to these researchers are the complex movements and temporalities of the biochemical world. The work of Sharpe and his colleagues provides a conceptual framework for how computer animation is equipped to address many of these challenges. The next case study examines how molecular animation can also engage with puzzles concerning movement over time through its unique capacity to facilitate experimentation.

Case Study 2: In Silico Experiments

Unfortunately, experimentation (in particular, hypotheses-forming and testing) too often gets overlooked in the scientific literature about computer animation. Animation's "hypothetical character"—where it is often difficult to determine where an experimentally known fact ends and "science fiction" begins—leads many scientists to distrust it as a medium for scientific research. Some of this skepticism stems from the fact that many of these dynamic 3D models have not been experimentally verified (as we saw in chapter 1). Janet Iwasa is a biologist and animator who embraces this aspect of molecular animation and contends that it is integral to experimenting with movement over time (2010, 699). This is why Iwasa situates animation in terms of physical 3D models of molecules that were once popular among structural biologists. "Historically," writes Iwasa, "physical 3D models of molecules have been used as thinking tools and have aided in scientific discovery. . . . In some cases, these models were created as an educational device, but were later brought into the laboratory and used to help researchers visualize and solve a problem" (699). In this same way, molecular animations can also function as "thinking tools" that help scientists formulate problems and test solutions.

As we noted in chapter 1, Natasha Myers (2015) examines 3D modeling in structural biology using ethnographic methods. She pays special attention to how the 3D computer modeling of X-ray crystallography is a part of the lineage of physical modeling and emphasizes how physical and computer modeling are embodied experimental practices. She writes that "modelers working with physical or digital media combine their keen visual and spatial sensibilities with their sense of tactility, . . . drawing these together with forms of proprioception . . . and kinesthesia. . . . In this sense, protein modeling is akin to artisanal production: it is a 'reflexive, anticipatory practice guided by synesthetic evaluation' at every stage of the model-building process" (2015, 75–76). Myers even suggests at one point that molecular animation may be a part of this genealogy, even if the technology is more susceptible to criticism (194–99). While I don't think Iwasa goes as far as Myers, and this is largely because she has different theoretical and practical commitments, the former embraces the idea that dynamic 3D modeling, like physical modeling before it, generates a space for thinking and experimenting with time-based phenomena that are too small or complex to observe in the laboratory setting.

However, distinguishing a proposal or hypothesis from what is already experimentally known proves to be a difficult challenge for animators. Iwasa underscores this and suggests that one strategy is to render structures according to the level of confidence that scientists have in their data. She explains that

> when communicating a molecular or cellular process through animation, different rendering styles can help explain the level of confidence and the amount of experimental data that support a specific model. It is important to be selective about how and when animation is used and to be aware of how a visualization is perceived by the audience. A process that is poorly understood will be less misleading to the audience when shown as a simple 2D sketch rather than a 3D animation, while conversely, a well elucidated molecular mechanism will be best served by a visualization where detailed information in time and space are clearly conveyed. Animation software enables animators to transition easily between different rendering styles; however, animators are sometimes tempted to render animations in eye-catching styles that can unfortunately result in distraction from the key message or even miscommunication. (2010, 702)

To avoid such miscommunications, she notes that high levels of surface detail in protein renderings might suggest "a higher emphasis on molecular structure," and thus a greater amount of confidence, while "smoother" or "cartoonish" renderings might point to proteins that have not yet been crystallized (702).[23] What's clear in any case is that the semiotics of animation is a crucial battleground for researchers working with computer animation.

A particularly illustrative example of the experimental uses of computer animation can be found in Iwasa's work with the Reck-Peterson Group at the Harvard University Medical School on motor proteins. Motor proteins transport proteins and vesicles in the cytoplasm and are a difficult class of molecules to get a handle on. This is mainly because scientists still do not understand how these proteins "walk" along microtubules. So, for instance, although there is a wealth of experimental data about the structure and movement of dynein, one of the three families of cytoskeletal motor proteins, "the mechanism by which dynein is capable of walking processively along microtubules remains an area of active investigation" (702). Anthony J. Roberts and his colleagues claim in their paper for *Nature Reviews* that "dynein is currently at the frontier of cell motility research at the molecular level, as its mechanism of movement is much less well understood than that of kinesin and myosin" (2013, 714). They go on to show that while the architecture of the dynein motor domain has been isolated using two-dimensional electron microscopy, X-ray crystallography, and three-dimensional cryo-electron microscopy (cryo-EM), crucial features of its movement still remain untested. Understanding this movement over time is a pressing concern for biomedical researchers, especially since defects in the dynein motor have been linked to Parkinson's and Lou Gehrig's diseases. And what's more, dynein is also responsible for transporting numerous viruses to the nucleus to infect the cell.

In an effort to study this movement, Iwasa built manipulatable 3D animations of dynein with the Reck-Peterson Group. In an article published in 2012 for *Nature Structural and Molecular Biology,* Samara Reck-Peterson and her colleagues summarize how they acquired their data for dynein movement and what it suggests:

> By combining two-color, single-molecule microscopy with high-precision, two-dimensional tracking, we have shown that dynein's stepping mechanism is distinct from all cytoskeletal molecular motors characterized to date. Although many of dynein's steps alternate in time, most stepping events do not switch the leading or lagging spatial identities of dynein's two heads, a distinction from the two-headed processive myosin and kinesin motors that use alternating and passing (hand-over-hand) mechanisms. Strikingly, our data suggest that dynein's steps are uncoordinated when the distance between motor domains is small, presumably because the intra-molecular strain is low. However, as the distance between motor domains increases, our data show that dynein becomes increasingly coordinated, likely through a tension-based mechanism. (Qiu et al. 2012, 8)

The researchers' data suggest that dynein's stepping patterns are variable, unlike kinesin and myosin—"hand-over-hand stepping" (5). Nevertheless, the uncoordinated movements seem to become coordinated as the distance between the motor domains increases. But this is a counterintuitive and even awkward way to think about movement over time in three dimensions. This is what Reck-Peterson struggled to express using 2D and physical models, which is how Iwasa became involved with the research project.

Figure 27. *Molecular Motor Struts Like a Drunken Sailor.* Directed by Janet Iwasa. Produced by Reck-Peterson Group, 2012.

Together, Iwasa and the Reck-Peterson team designed animations that prototype dynein stepping down a microtubule. What's more, these animations include various handles allowing researchers to manipulate the images to inspect different possibilities for movement. In the end, Reck-Peterson confirms that "the animations have made it easier to talk concretely about our ideas, both within the lab and with others in the field" (Lok 2011). Similarly, Iwasa notes how the manipulatable models help scientists "explore the structure of dynein, and it has been instructive in revealing the conformations of dynein that are possible or impossible given structural and steric constraints" (2010, 702). In particular, it became obvious to researchers that dynein's "drunken-sailor strut" is what makes the protein more flexible than other motor proteins, enabling it to step around obstacles and keep the cell alive (Qiu et al. 2012).

It's crucial that this animation is a tool for thinking about what's possible for movement over time, about what hypotheses can be confirmed, denied, and reformulated. For this reason, it was especially important for Iwasa to flag the exploratory nature of the animations to researchers. This is why she had to build multiple versions of it. The first animation was deemed too technical, and thus misleading, since viewers may have believed that model was based upon actual structural data. A second model was then built that displayed "softer, smooth surfaces" in order to flag its "hypothetical" character (Iwasa 2010, 702). As she put it, the model was then able to effectively help researchers in "formulating and visualizing different hypotheses" (701). The animations helped Reck-Peterson and Iwasa give sense and meaning to their data. 3D computer visualizations allowed them to turn their data into meaningful information. The "drunken strut" of dynein is no simple anthropomorphism, then: it actually gave scientists a way to think about what is possible and impossible for protein movement in the biochemical world.[24]

ANIMATION, SYSTEM, EXPERIMENT

In the data-intensive landscape of molecular biology, scientists are often at a loss for how multiple spatial and temporal scales work together, how they interact and combine to permit protein movement, for example. How, for instance, are dynein motor domains coordinated in such a way that they permit such a large protein to move along a microtubule? Observation-rich experimental biology had the advantage of being able to understand phenomena through observation. For this reason, it was suggested in chapter 2 that observation does not just test a preexistent theory (in fact, it has been argued that this does not happen at all in experimental biology), but it is what allows biologists to discover new phenomena in the first place (Keller 2002). But in the post-omics era of molecular biology, observation-rich experimental practices seem to be more of a rarity. It is not that biologists do not want or need these confirmations, it is that it has become increasingly difficult to imagine how they would come about. Epistemological priorities have to be adjusted.

Using these two case studies as our guide, it's clear that bioscientists use molecular animations to communicate more than agreed-upon scientific content. My hypothesis is that molecular animation is beginning to build experimentalism back into post-omics systems biology. These two case studies in particular serve to illustrate how, on the one hand, even the farthest reaches of the molecular universe can, in principle, be given visual form through the latest animation technologies (case study 1). Scientists are increasingly able to appeal to observation (which is central to experimental approaches) in order to understand systems of organized interaction. Where observation was once limited to individual parts and their interactions, now it seems to be expanding to entire systems. What this also means is that animation holds out the promise of filling in many of the "blind spots" in the visual field of molecular biology. In chapter 2 we saw how biologists working with micrographs have to assume, instead of witness, much of the temporal and spatial complexity in the biochemical world. In particular, movement over time was shown to pose a serious challenge to the visualization of molecular processes. Systems of interaction were, by and large, invisible to researchers. While systems of experimental parts could be mathematically modeled in differential equations, they could not be observed and understood in the same way. Molecular animation pulls experimental data from a variety of sources—including X-ray crystallography, cry-electron microscopy, florescence microscopy, and so on—to generate a visualization of interacting systems for researchers to observe and feel their way through.

But on the other hand, case study 2 shows how computer animation helps researchers get a handle on what is and is not possible for movement in the biochemical world. Using molecular animation, scientists perform in silico experiments to generate knowledge about biochemical systems. Animation not only expresses what is known but also incorporates what is not yet known. Computer animation builds proposals for the molecular world, performs experiments on them, and then determines what does and does not work. This is precisely what Iwasa does with the Reck-Peterson group. Computer animation can incorporate exteriority, what lies at the edges of experimental knowledge. It folds what is experimentally unknown (the exterior) into what is visible and knowable (interior).

What these two case studies serve to demonstrate is not simply that visualization is built into systems-level analysis, but that experimental methods are themselves folded into the visual epistemic work that molecular animation performs. Even though these studies seem to be doing different conceptual work, they also need be thought together, as complementary practices that illustrate the range of possibilities for the computer-animated image in the life sciences. Together, they express the affordances of molecular animation and should be treated as different aspects of an emergent domain of visual knowledge practices. And this domain is what serves to bring systems and theoretical research in the biosciences into closer proximity with the experimental methods that have dominated the life sciences for centuries.

PART II

The true method of discovery . . . makes a flight in the thin air of imaginative generalization.

—Alfred North Whitehead,
Process and Reality

4

ECO-SOCIAL MEDIA

PART I OF THIS BOOK engaged with numerous histories and practices associated with molecular animation. But in bringing this heterogeneity into view—from popular cinema and media to experimental and theoretical practices in the life sciences, cybernetics, and systems biology—the book also attempted to situate contemporary molecular animation in an increasingly complex genealogy. Over the course of Part I it became clear that there is no simple or straightforward history of molecular animation. The ensemble of materials, practices, and meanings that are most commonly associated with it appear to be an outgrowth of two very different visual cultures.

On the one hand, molecular animation finds itself situated within popular visual culture. We saw how molecular animations share a visual logic with other forms of popular screen media; they also circulate through social media channels like other popular media content. What's more, these connections are very often institutionally reinforced, since it is not uncommon for scientists to receive their animation training in Hollywood studios or through popular online tutorials. It is these connections between popular visual culture and molecular animation that inform many of the reservations scientists have about using computer animation for research purposes. And then on the other hand, computer animation has also proven to be crucial for reconciling two different epistemic cultures in the history of the life sciences. By bridging theoretical and experimental epistemologies in biology, computer animation is primed to become an essential visual technology for life science researchers. Still, there is little scholarship that takes stock of the competing genealogies of the technology, and certainly not in the way that I have outlined here. While there is often some acknowledgment that computer animation can be used for both entertainment/mass communication and for scientific research, no theoretical assessment currently exists for the strange fact that molecular animation somehow grows out of both domains.

There may be a straightforward answer to this problem: molecular animation can be historically and conceptually situated in either domain, or perhaps some combination of the two, because it is not one thing. What falls under the heading of "molecular animation" is a diverse set of practices that may be more or less suited to the requirements of entertainment/communication or scientific research, respectively. I imagine that this explanation would satisfy many STS-minded scholars inasmuch as it takes stock of the diversity of materials, agents, and discourses that constitute the various practices that get grouped under one heading. And what's more, this explanation would be equipped to show how different molecular animation *practices,* and not molecular animation as such, have different histories and uses.

I begin this chapter by showing why this explanation is unsatisfying. This is not because I assume that molecular animation is one thing. Quite the contrary: we have already seen how the practices of molecular animation comprise a wide range of methods and materials, operating in multiple institutional settings, with various histories and meanings. Rather, I make the more subtle claim that this heterogeneity does not preclude specific forms of coherence from emerging across these practices. In what follows, I show how there is a coincidence of *values* that occurs in molecular animation, despite the diversity of its composition. This is also why molecular animation is such an important visual technology for media philosophy: it exposes the shared values of the popular mediasphere and visual epistemologies in the life sciences. My proposal is to frame the heterogeneity of molecular animation in terms of a deeper convergence of values operating across the visual cultures of science and entertainment. This insight will be the basis for articulating the problem that anchors this book's larger speculative project—namely, what theory of media explains this convergence, and what mechanisms brought it into being, and for what reason?

It is in order to address these questions that Part II of the book makes the move from the ground of a particular observation (Part I) into the thin air of abstraction. Whitehead and Foucault will be the pilots for this portion of the speculative journey. In the second half of this chapter, I turn to Whitehead's work on value, social order, and ecology in *Process and Reality.* I review not only how his thought has gained attention in media theory in recent years, but also how his conceptions of value and social order tend to be neglected or glossed over in favor of notions such as "prehension" and "experience" that underwrite various strains of posthuman materialism. I show how Whitehead's work is especially well suited for addressing how disparate and even conflicting human and nonhuman materialities coordinate into forms of social order for the sole purpose of promoting *value.* As Whitehead insists, "Existence, in its own nature, is the upholding of value intensity" (MT 111). The production and promotion of value intensity in Whitehead's sense will anchor our own speculative engagement with how heterogeneous elements with various histories and meanings converge in molecular animation. In the next chapter, I demonstrate the necessity of supplementing Whitehead's work on the production of social order and value in human societies with Foucault's understanding of power and governance.

SHARED VALUES

I want to start this chapter by entertaining the following proposition: the reason that molecular animation can be framed in terms of diverse visual histories is that it is really just a heterogeneous set of practices misleadingly grouped together. In other words, molecular animation is not one thing, and any attempt to frame it as one misses its inherent multiplicity. So, for example, there are many instances where molecular animation aligns itself with a mode of entertainment or pop science that has clear ties to other popular screen-based visual practices. This is clearly the case for the animations in the documentary *Our Secret Universe,* as well as for Ribosome Studio's videos. But this is also true of *The Inner Life of the Cell.* As David Bolinsky insists, the film was not intended for biological research. In his TED talk, he claims that he wants to "spark" the "passions" and "imaginations" of students so that they will go on to "find out really how life works."[1] These animations are intended to make their audiences fascinated with the crowded and bustling operations of molecular machinery. In trying to convey this point, Bolinsky draws freely on metaphors of busy city life and even likens the operations of the motor protein kinesin to a FedEx delivery truck.

This seems to be in stark contrast to the genre of animation that Janet Iwasa and Samara Reck-Peterson use. For them, computer animation is a tool for research. Like other research tools and instruments in a laboratory setting (microscope, beaker, flask, Bunsen burner, physical model, mathematical model, etc.), computer animations (including the algorithms, software, and hardware that make them possible) collaborate with a wide range of other materials, actors, and expanding computational infrastructures to render phenomena intelligible to competent scientists who are able to come to an agreement about them. In their case, computer animation brings aspects of dynein mobility into sharp focus. Working alongside mathematical models and micrographs, animations offer a platform for scientists to experiment with various possibilities for movement. This in turn allows them to refine their hypotheses and devise new ones. These animations are integral to research practice. Such integration of animation into scientific research is what Jason Sharpe and his colleagues envision as well. But for them, and in keeping with the goals of computational biology more generally, computer animation holds out the promise of aligning human vision with the mathematical complexity of the biochemical world. Here too there is an explicit attempt to align computer animation with the dominant goals of a research practice—in their case, computational biology. This seems like a far cry from the attempts described above to make biology entertaining for non-specialists. Rather, this media is made specifically *for specialists.*

For this reason, it seems perfectly reasonable, and maybe even preferable, to argue that molecular animation encompasses such a broad and diverse set of practices that it hardly makes sense to call it the same thing for both functions. If nothing else, the designation leads to needless confusion. If we take this route, then it makes perfect sense to locate certain animation practices within a history of cinematic viewing and others within a history of visual epistemology at the intersection of experimental and systems biology. What's more, this may help to resolve the anxiety many scientists feel about using animated images in their research practices. While there are uses of molecular animation that "glamorize" and "anthropomorphize" the molecular world, there are other uses of the technology that can be optimized for biological research and experimentation. This is why Iwasa takes great care to discuss how to render molecules based on the confidence a scientist has in experimental data; and it is also why Gaël McGill and Campbell Strong developed Molecular Maya: to ensure that the geometries of the PDB files are preserved in the animation.

Despite the elegance of this solution, it overlooks some important details. For one thing, there is no clean separation between molecular animations intended for science audiences on the one hand and popular audiences on the other. In practice, this distinction is not neat and tidy. For instance, it is not as if Iwasa's animations are only minimalist models—downplaying what she does not have the data to support—that appeal exclusively to experimental scientists. Iwasa's animations are valuable to public audiences as well. Her TED talk titled "How Animations Can Help Scientists Test Hypotheses" has over 911,000 views (as of April 2020). There, she speaks about *Clathrin Mediated Endocytosis* (2012), an animation she made with Harvard professor of cell biology Tomas Kirchhausen, which has many of the same aesthetic and dramatic effects, including nondiegetic sound, that *The Inner Life of the Cell* and *Protein Packing* have.

Iwasa insists that this animation is not just good for communicating information to a broad audience but also helps researchers formulate hypotheses about molecular processes (similar to her work with Reck-Peterson). This animation, even with its stylization, is intended to be *a part* of the experimental apparatus. What this example, and there are countless others, illustrates is how there is no clean division between molecular animations built for mass audiences and those built for the scientific community. Animations meant for scientists are publicly consumed, and publicly consumed animations are also used for experimentation.

There is an even more serious problem with the idea that molecular animation practices are too heterogeneous to put under the same heading. This explanation cannot account for the strange fact that this computer animation practice seems to fill a similar role in these very

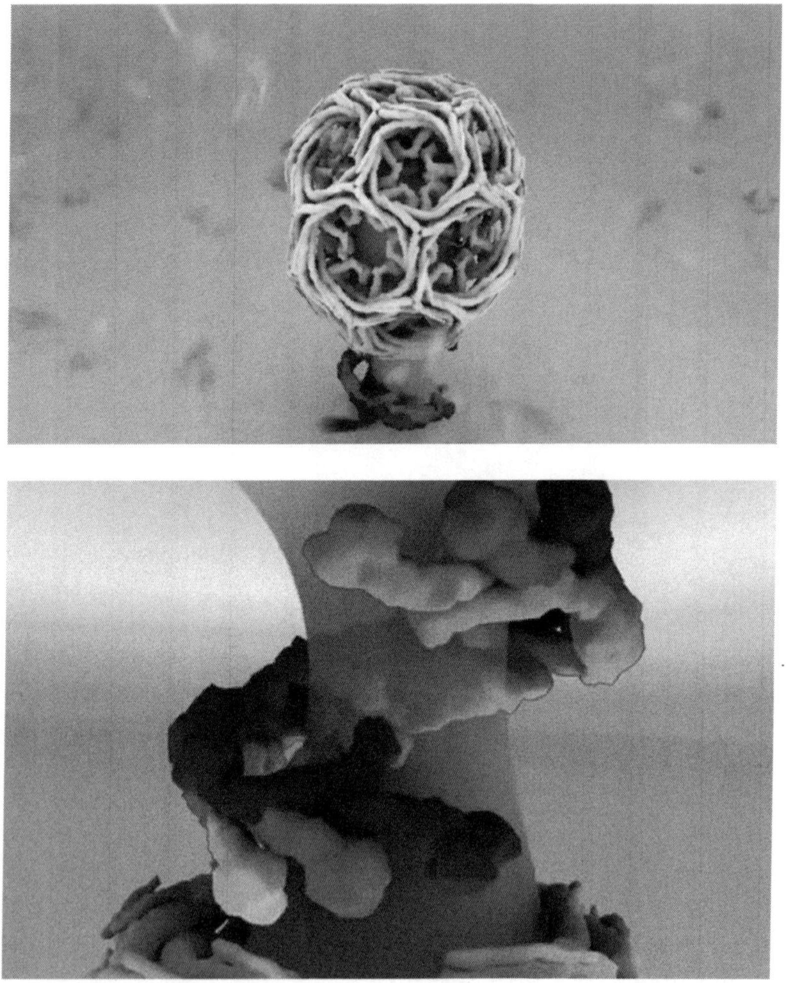

Figure 28. *Clathrin Mediated Endocytosis.* Directed by Janet Iwasa and Tomas Kirchhausen, 2012.

different visual and epistemic cultures. This counterintuitive claim—especially given that I have spent so much time showing the heterogeneity of these practices—should have become slowly apparent as I uncovered the distinct genealogies for the technology: a similar mode of visual perception is not only propagated but also reinforced across science and entertainment.

To appreciate this claim, recall the steps we took in chapter 1 to demonstrate how short animations such as *Inner Life of the Cell, Protein Packing,* and *X Inactivation and Epigenetics* operate according to the same perceptual parameters as other forms of popular screen-based media. We examined these animations in the context of the history of cinema and animation, and especially in terms of Manovich's taxonomy of new media objects and software metamediation. We showed how molecular animations seem to fit neatly into these software taxonomies and how there is little to distinguish digital entertainment media from the computer animations of cells and molecules. We also examined why Thomas LaMarre's notion of cinematism or hypercinematism (indebted to both Paul Virilio and Félix Guattari) was better equipped to handle the tendency of the moving image expressed in a molecular animation than Manovich's thesis. Taken together, these works illustrate how molecular animations are expressions of a logic or a machinic tendency of the image shared by many other forms of screen-based popular media. This is a tendency of the image that dislocates vision from the spatial and temporal coordinates of a situated observer.[2] Visual perception is abstracted from the restrictions and limitations of the human eye and reassembles itself on a plane of infinitely recombinable data. Such a reorganization of the visual promotes a logic where there is no truly exogenous force, since every external object, system, or structure can (in principle) be made visible. According to this logic, the endless proliferation of processes, objects, and systems does not escape the visible. Visual perception is immanent to the various and complex permutations of the informatic universe. Molecular animations seem perfectly comfortable in a genealogy of visual media shared by Hollywood blockbusters and a wealth of other popular media practices.

In chapters 2 and 3 we set out to chart a different genealogy of molecular animation. The claim in these chapters is that the technology is also an important outgrowth of visual epistemology in the biosciences. At stake in this discussion is a dramatically different conception of molecular animation: one that underscores how the medium is a part of a research apparatus that fills a gap within the visual economy of the life sciences. To make this point required a much lengthier, and in some ways more complex, argument. This is primarily because this genealogy of media is not a well-worn path in the history of the biosciences.

Chapter 2 endeavored to show the importance of visual technologies to the history of knowledge practices in the life sciences and how experimental biology is left with certain blind spots in its field of vision. This is in large part because overall systems of interaction are not visible. In chapter 3, I dove into a lengthy discussion of the struggles mathematical and systems biology faced throughout the twentieth century and how they eventually won the favor of many biologists due to the inundation of data in the post-omics era. With systems modeling, entire systems of interaction can be conceptualized, but unfortunately these models lack the experimental and observational verification of experimental biology. With molecular animation we witness a profound attempt to build experimental aspects back into systems theory—visualization is no longer limited to the individual parts and interactions of a system. Now, entire systems come into view. Mathematical models are given a phenomenological depth they never had before. Molecular animation designs a dynamic visual field for large-scale mathematical models of molecular systems. In this way, many of the affordances of experimental knowledge practices are infused into the computational biosciences.

But on the other hand, molecular animation also provides the means for integrating those components of the overall system that are "unknown," and thus do not yet count among the items of knowledge. This is precisely what Iwasa is able to do with the Reck-Peterson group. Molecular animation does not overlook components of the molecular world that have not been tested and that have been the preoccupation of experimental biologists. Instead, it performs experiments with these unknowns and can (ideally) integrate them into the total field of visual knowledge (dynein walks like a "drunken sailor"). Thus, if one of the main challenges for systems-level thinking in biology is bridging the gap between experimental knowledge and systems-level analysis, then molecular animation proposes a strategy for dealing with this rift in knowledge production: it has invented a powerful means for incorporating exteriority.

What this means, as I alluded to at the end of the previous chapter, is that these two case studies (Sharpe et al. and Iwasa) are not opposed or in conflict with one another, even though they seem to have different approaches to the use of animation in biology. Instead, they are deeply complementary. Of course, there is no single animation, just as there is no single shot of a molecule, that can do all of the experimental and systems-level heavy lifting that I am proposing. What I am suggesting is that when these practices are taken together, they express different ends (systems and experimental) of a visual epistemic domain in the biosciences. And this domain is what I have been calling "molecular animation."[3] This is a visual epistemic field that comes into view when the practices that compose it are conceived in terms of a genealogy that uncovers how heterogeneous actors, practices, and technologies coalesce over time to contribute to wider tendencies of seeing and knowing the living world. And this domain of molecular animation is defined by practices and materialities that are meaningful to one another because they contribute to the visualization of temporally unfolding systems while also integrating new and unknown components into its visual knowledge matrix. In short, the visual logic that underwrites these practices is primed to fill in blind spots in the history of bioscientific knowledge: it connects parts to the whole and leaves no system or individual part outside of what is possible to visually capture.

But in articulating this diverse and complex history of visual epistemology in chapters 2 and 3, we have stumbled upon something rather astonishing—namely, that molecular animation seems to be firmly situated in both popular visual culture (chapter 1) and visual epistemology in the life sciences (chapter 3), because they value a similar tendency or logic of the image. Whether we call it (hyper-)cinematism (Virilio, LaMarre) or something else seems to matter little at this point (we will clarify this in later chapters). What matters is that this is a logic that is valuable to the cultures of entertainment and the biosciences because it fosters flexible modes of perception capable of capturing the endless proliferation of objects, systems, and processes at any temporal and spatial scale. There are no barriers to visual perception in this shared tendency since every new element, every new dimension or scale of an object, presents a new opportunity for visual capture.

Part I of this book therefore attempted to demonstrate that what matters to these seemingly disparate visual regimes is this same visual logic. And molecular animation is a point where this intersection between science and entertainment becomes apparent. In other words, despite the diverse cultures and materialities that molecular animation traverses, this logic remains constant across them. Much as Foucault (1994) excavates the *epistemes* distributed across disparate domains of knowledge production, here too we have uncovered a visual economy traversing different regimes of seeing and knowing. Molecular animation has therefore opened up a much deeper set of questions about how two different domains of seeing and knowing the world are woven together through their shared value of a visual logic.

Before we move on to theorize computer animation's unique ability to yoke the visual cultures of science and entertainment together, it is worth mentioning that scientific, artistic, and popular ways of knowing the world have a long and complex history of entanglement. Molecular animation is by no means the first and it certainly will not be the last medium distributed across these heterogeneous domains. Jonathan Crary's (1990) work, for example, addresses many of these convergences in nineteenth-century optical technologies (see chapter 6). Likewise, Lisa Cartwright (1995), Hannah Landecker (2007), and Oliver Gaycken (2015) examine how early twentieth-century biologists, such as Jean Comandon and Alexis Carrel, experiment with cinematographic techniques in order to understand the temporal dimensions of cellular life (see chapter 7). José van Dijk (2005), for her part, situates medical imaging technologies—X-ray, endoscopy, ultrasound, and so on—at the intersection of scientific and popular fantasies of transparency. And Patricia Pisters (2012) has gone to great lengths to frame contemporary Hollywood cinema in terms of neuroscience and neuroculture. Elsewhere, literary scholars have unearthed the myriad and complex relations between scientific and literary ways of knowing (Meyer 2018). In this context, Robert Mitchell examines how Romantic-era authors incorporate the experimental practices of the life sciences in their conceptualizations of life. And in mid-twentieth-century design research, Bauhaus-era designers, such as László Moholy-Nagy and György Kepes, experiment at the intersections of cybernetics, biology, architecture, and advertising in the graphic arts (Martin 2003; Halpern 2014). These examples do not begin to account for the range of important research on the convergences of science, art, and popular culture now populating the rapidly growing fields of literature and science, ArtScience, as well as media, science, and design studies.

The intersection of science and entertainment that I am proposing sits neatly alongside much of this work. Molecular animation is a kind of repository—not unlike other technical media—for converging knowledges, cultures, and meanings. And in later chapters I make the connection to earlier technical convergences explicit, especially to microcinematography and the graphic arts. However, there are also important features that distinguish molecular animation. Chief among them is that molecular animation is a contemporary media practice. Where many accounts of convergent knowledge practices are from previous eras, molecular animation weds the diverse practices of scientists and entertainment consumers in the twenty-first century. What's more, this wedding of cultures is different from the mutual influence of cultures. Molecular animation is both science and entertainment. It is a site where the values of science and entertainment collide. And yet, this study does not take this collision for granted. As I discussed in the Introduction and throughout Part I, there are real divergences in the historical and cultural formation of popular and scientific media, and these differences should not be discounted or elided in the service of unifying histories. These divergences matter. And this could not be more evident than in molecular animation. For even if scholars regard science and entertainment media to be deeply entangled and even a part of the same episteme, then molecular animation does not broadcast this coincidence. Just the opposite: it conceals it. As a scientific tool, molecular animation conceals its identity as a medium for popular entertainment. And as an entertainment and educational medium, molecular animation obscures its use for scientific research (see Bolinsky's online response to his TED Talk). Molecular animation is the site of a covert alliance.

This means that we have to account for molecular animation's diversity and uniformity, the fact that it combines a multiplicity of things that do not lose their diversity when they are one thing. But this leaves us with a number of questions concerning what molecular animation is and how it came into being. In particular, how is it possible for a heterogeneity of things to converge into a single thing—namely, a unified domain of seeing and knowing? And then does molecular animation retain its diversity (it is science and entertainment) when it is also one thing? In other words, molecular animation confronts us with certain time-honored metaphysical questions: Is it one thing, many things, or both? If the first part of this book revealed how molecular animation is "many" things—this is to say, composed of a heterogeneity of materials, meanings, histories, and practices—then it also showed how this multiplicity becomes some kind of "one." And what's more, the many seem to become one through shared "values," although it is not at all clear how value is distributed across this heterogeneity. So in the end, we are left puzzling over what kind of visual medium molecular animation is and whether there are conceptual tools that could help clarify its unique composition of elements. *How does the many become one through shared values?*

It is in order to respond to this question that I turn to Whitehead's conceptual scheme. As I go on to show, it is overcoming certain modern preconceptions about the relation between the many and the one that serves as a foundation for Whitehead's entire metaphysics. It is by overcoming these preconceptions that we come to appreciate how diversity is woven together through valuations, and this is the basis for the construction of social orders that are ecologically embedded within other forms of social order. I use this conception to anchor what I am calling an "eco-social theory of media" that is at once capable of responding to (portions of) our problem and is also in conversation with the latest developments in media materialism and media ecology. What this means is that molecular animation poses a metaphysical problem (about the many, the one, and shared values) that not only requires Whitehead's philosophical scheme to unpack (since it directly engages these questions) but also challenges us to see his theory as the basis for a conception of media governance (via Foucault) adequate to theorizing the design of the molecular animation apparatus. But this will come.

MEDIA ECOLOGY AND WHITEHEAD

Before we dive into Whitehead's theories of social order and value, I want to situate my use of them within a much wider framework of media ecology and materialism. Although media ecology and media materialism have become popular in the last few decades, their meanings and uses are notoriously difficult to pin down. For instance, "media ecology" was originally coined by the media commentator Neil Postman in the late 1960s. However, as Matthew Fuller (2005) notes, it was conceptually inspired by the work of Marshall McLuhan, Harold Innis, Jacques Ellul, and Walter Ong. Originally, media ecology referred to a conception of media "environmentalism" where "there has passed, or that there will be reached, a state of equilibrium: that there is a resilient and harmonic balance to be achieved with some ingenious and beneficent mix of media" (Fuller 2005, 4). Fuller points out that since its inception, numerous other strains of media ecology have been forged as well: for instance, it gained traction among literary and cultural theorists such as Friedrich Kittler and N. Katherine Hayles, whose poststructural inclinations tended to counter the humanist and eco-theological persuasion of media environmentalism (5).

More recently, media theorists such as Jussi Parikka, Fuller, and others have drawn on the ecological thinking of Guattari and Gregory Bateson, as well as the systems thinking of Humberto Maturana and Francisco Varela and theories of individuation and machinic organization indebted to the works of Gilbert Simondon, Deleuze, and Guattari. LaMarre's *Anime Machine* is a good example of a media theory that draws on Guattari's notion of the abstract machine and Simondon's notion of individuation to theorize the material (and arguably ecological) relations of the animation apparatus. In this regard, LaMarre is careful to situate his work on animation in terms of science and technology studies instead of art history, mainly because he wants to focus on the (im)material relations of the apparatus (more specifically, the Guattarian "machine" in the apparatus) instead of an analysis of the animated image that is overdetermined by formal and aesthetic analysis (LaMarre 2009, xxvi). Parikka, for his part, expands our engagement with the material and ecological connectivity of media devices and apparatuses in his research on media archeology. In numerous works, he offers impressive archeologies of our media in order to excavate histories that are entangled with a wide variety of material systems—for example, biological, financial, and geological systems (Parikka 2010, 2012b, 2015).

The rise of speculative realism, object-oriented ontology, new materialism, and various strands of posthumanism have also ramified materialist and ecological approaches to media. In many ways Quentin Meillassoux's (2008) critique of correlationism served as a rallying cry for media theorists who sought to overthrow the hegemony of human interpretive frameworks (cultural, ideological, semiotic, psychoanalytic, etc.)

in order to theorize the nonhuman forces operative in our media: from the material relations of software and hardware to the ubiquity of planetary computing, digital infrastructures, and global finance networks. Building on this post- and nonhuman momentum, media studies has also begun to expand what it considers to be media (echoing McLuhan in certain ways) to focus on broader practices of mediation (Peters 2015; Galloway, Thacker, and Wark 2013). This has opened the door to a wide range of objects and processes counting as "media" and new and exciting cartographies that usher in notions of animal/biomedia and geological media. Parikka has pushed this trend in new directions, challenging us to think of contemporary computational media practices in terms of the "geophysical elements that give us digital culture," the vast timescales of the Earth, and the devastating effects that our production and consumption of these media have had on living and nonliving systems (Parikka 2015, 139). Likewise, in his conception of elemental media, John Durham Peters invites us to think of media as "enabling environments," as "life supports," which in turn allows us to conceive of "water, fire, sky, earth, and ether" as media. This makes a very different conception of environmental media possible. And in a different, though related, conceptual register, Sean Cubitt (2016) and a host of other media theorists, such as Jennifer Gabrys (2016, 2011) and Elizabeth Grossman (2006), have examined the troubling connections between the global production and consumption of digital media and environmental degradation and push us to reexamine what we mean by media and communication in our geological era.

Amid these diverse conversations in media theory, Whitehead's philosophy is increasingly becoming a more important touchstone. It is no longer uncommon to see Whitehead cited alongside Deleuze, Deleuze and Guattari, and Simondon in discussions of media, materiality, ecology, and process (Sha 2013; LaMarre 2013). And while it is less common to see a deep and careful engagement with how Whitehead's philosophy could inform media theory and practices, there are some important exceptions. Andrew Murphie has long invoked Whitehead in his work on vital technics and matter. Likewise, Whitehead is the philosophical centerpiece of Luciana Parisi's (2013) work on algorithmic architecture and computational thinking. Steve Goodman also draws explicitly on Whitehead throughout his influential book *Sonic Warfare: Sound, Affect, and the Ecology of Fear* (2009), and so does Adrian Ivakhiv (2013) in his development of an eco-philosophy of cinematic images. In a much different intellectual context, Sha Xin Wei (2013) mobilizes Whitehead's account of pan-experiential metaphysics to "re-enchant" the material world through responsive media art and research. And perhaps most recently, Mark Hansen (2015) provides an ecological and materialist treatment of "mobile and ubiquitous media" using Whitehead's philosophy in order to revive phenomenological conceptions of media.

Although this list is not exhaustive, it gives a rough overview of some notable attempts to connect Whitehead to media theory. My hunch is that what most media theorists find generative about Whitehead's work is that it offers a non-anthropocentric and processualist view of the material world that yields a fresh take on media ontology. And what's more, because Whitehead's work is often invoked alongside the various speculative and new material turns in continental and cultural thought (it's a precursor of sorts), media theories that draw on Whitehead tend to be a part of the latest conversations in contemporary philosophy/theory (Gaskill and Nocek 2014).

Despite this interest in Whitehead, there is relatively little attention paid in media theory to his work on social order ("societies") and even less attention paid to his theory of value. To Whitehead specialists, this may come as a surprise, given how central societies are to his metaphysics. Anything that achieves unity over time and endures is a society (PR, 34). Media systems, biological cells, molecules, subjects, and nation-states are all societies for Whitehead, and they all have their specific "manner" of being social (Stengers 2011b; Halewood 2011; Debaise 2017). This is why Isabelle Stengers writes that "for Whitehead, it is always societies that we study. Everything is sociology . . ." (2011b, 325). For this reason, prominent readers of Whitehead have focused on the importance of the social in his thought, as well as how "every society requires a wider social environment" (PR 90). Hence, "eco-sociality" is inherent to Whitehead's metaphysics. Value (or valuation), on the other hand, is also deeply connected to social existence in Whitehead's universe, but it seems to be even less interesting to media theorists. This may be less surprising given that Whitehead's theory of value is not straightforward and is rife with controversy (Schindler 1983; Henning 2005; Halewood 2011). Nevertheless, value creation is no less central to Whitehead's thinking: “Everything,” he explains in *Modes of Thought,* “has some value for itself, for others, and for the whole. This characterizes the meaning of actuality. . . . Existence, in its own nature, is the upholding of value intensity” (MT 111).

There may be any number of reasons why these concepts have not been widely deployed in media theory. The answer could be as simple as they do not seem especially relevant to our current understanding of media and mediation. With theorists pleading for ever more “deterritorialized” notions of media rooted in nonhuman and even geologic scales of space and time (Parikka 2012a, 2015), or even “excommunicative” theories of media that dispense with the presumption of communication altogether (Galloway, Thacker, and Wark 2013), Whitehead’s theory of social order might seem especially conservative. Likewise, his concept of value is invariably bound up with his theology—which has proven to be hard to stomach for many contemporary scholars—and so may not be a fruitful conceptual resource for media theory. Whatever the reason, in the remainder of this chapter I examine how Whitehead’s notions of valuation and social order are developed out of his reframing of the problem of the “many” and the “one,” and this gives us an especially rich theoretical frame for understanding what interests us about molecular animation. In particular, molecular animation is a media form whose historical and conceptual genealogies are diverse and wide-ranging, and so are the physical, semiotic, and epistemological regimes that make it possible. It is for these reasons that the convergence of such a heterogeneous ensemble of elements in a shared visual logic proves to be both a demanding and pressing concern for theorists and philosophers of scientific media.

Equally demanding is using Whitehead’s conceptual scheme to understand the metaphysics of media that underlies molecular animation. If there is an upside to the genealogical and philosophical heavy lifting that will ensue, then it is this: Whitehead’s notion of the social will give us the theoretical vocabulary we need to address how myriad agents, processes, and materials, which traverse multiple institutional, epistemic, and historical contexts, are able to cohere in a set of technical practices through mutual valuations of a “defining characteristic”—and in this case, the shared valuation of a visual logic. To address this requires that I situate Whitehead’s conceptualization of society and valuation in terms of his larger speculative project. The hope is that this excursion into Whitehead's work will not only demystify some of his unique and difficult vocabulary but will also shed important light on the centrality of social order and valuation to his philosophical project more generally.

ACTUAL ENTITIES AND ETERNAL VALUATIONS

As many Whitehead commentators have noted, *Process and Reality* is perhaps Whitehead's most ambitious work. He sets himself the demanding task of constructing a speculative philosophy that would be able "to frame a coherent, logical, and necessary system of general ideas in terms of which every element of our experience can be interpreted" (PR, 3). Whitehead endeavors to construct a set of notions capable of expressing *all* aspects of experience, without bifurcating reality and reducing or explaining away certain aspects (subjective feeling) in favor of others (objective facts). As he puts it in *The Concept of Nature,* he wishes to resist the "bifurcation of nature" and put everything "in the same boat, to sink or swim together" (CN, 148).

As noted in the Introduction, Whitehead was a mathematician, which means that the concepts he deploys are never detached from the problems they attempt to solve. And although there is no single problem that animates Whitehead's metaphysical works, each articulation of the problem challenges us to guard against the "accidental error of mistaking the abstract for the concrete" (SMW, 51). We cannot "[allow] certain knowledge-claims to step beyond their relevant domains and thereby eliminate the real values and achievements available in other modes of experience" (Gaskill and Nocek 2014, 9). Whitehead therefore aimed to cultivate a habit of thought that was entirely general and that resisted the temptation to bifurcate reality into essential and inessential features (PR, 3). And to the extent that his notions promote this habit, they are successful. But this is also why Whitehead's speculative philosophy is pragmatic in orientation: a concept is verified based on what it does, on how it transforms our habits of thought, rather than what it represents. For Whitehead, metaphysics is a speculative pragmatics (Stengers 2011b).[4]

It is with this in mind that we can begin to appreciate the work that Whitehead's metaphysical abstractions are supposed to do. As I have written elsewhere (Nocek 2014a; Gaskill and Nocek 2014), Whitehead begins *Process and Reality* by claiming that creativity is the ultimate generic principle—it is the "Category of the Ultimate" (PR, 21). With creativity Whitehead does not wish to introduce a transcendent metaphysical principle. Rather, he fabricates a productive constraint for cultivating a fully general habit of mind. Creativity is the ultimate "matter of fact" that undoes one of the longest-standing dualisms in the West: between the "many" and the "one." Creativity "is the ultimate principle by which the many, which is the universe disjunctively, become the one actual occasion, which is the universe conjunctively" (21). What creativity explains is how every one is the coming together of the disjunctive many into a novel one, which then increases the many by one. Creativity is the principle that accounts for how every one, no matter how mundane, repetitive, or insignificant, is a novel coming together of the many that adds something new to the world (the disjunctive many). Once this Ultimate is established, everything else becomes a "creature of creativity"—all other notions presuppose it—that is, specify the processual unfolding of the many and the one.

Actual occasions or actual entities are the primary units of the creative process. They specify the instances of the many becoming one of the disjunctive many. Actual entities are what they are by virtue of their integration of the many, or the entire history of actual entities, into their own constitution. With this notion, Whitehead casts aside the modern habit of thought that conditions us to think of individual substances as primary and disconnected from other entities. Rather, and crucially, entities actually enter into the constitution of each other in order to *become* what they are. Each entity *just is* an integration of the disjunctive many, which creates a real continuity among occasions of experience. This is why Whitehead explains that actual entities are the only reasons: "There is no going behind actual entities to find anything more real. . . . Actual entities are the only reasons." This is what Whitehead calls the "ontological principle" (19).

The trouble is that an actual entity doesn't simply reproduce the totality of the past exactly as it was. Each entity must "decide" how it will integrate each past perspective as datum for its own becoming a subject.[5] The name Whitehead gives this process is "prehension," which essentially describes how an entity incorporates elements of the past into its own constitution through unique and determinate *feelings* of it. "Actual entities," he continues, "involve each other by reason of their prehensions of each other" (20). With this idea, Whitehead accounts for how a diversity of things over "there" can be constitutive of a vividly felt "here": "Realisation," he explains, "is a gathering of things into the unity of a prehension. . . . What is thereby realized is the prehension and not the things" (SMW, 69). In this account, modern distinctions between "subject" and "object" are undermined, since a subject *just is* its experience of other objects. These objects retain their connection to coming from over "there," but that "there" is felt and enjoyed privately as being relevant for the constitution of a concrete, particular, and felt "here." Each occasion then is the *concrescence* of the distinctive and determinate emotional bonds that it forges with the manifold of prehended datum. The result is a unique and fully determinate *satisfaction* that is *superjected* to become data for future prehensions: "The creative process is rhythmic: it swings from the publicity of many things to the individual privacy; and it swings back from the private individual to the publicity of the objectified individual" (151).[6]

But to specify how this subject emerges from the world as a determinate concrescence of a manifold of prehensions, Whitehead needs to show how a subjective form—or *how* a subject feels its world—comes into being in the first place. It is at this point that Whitehead breaks down an occasion's concrescence into (nontemporal) phases or stages, which vividly express an occasion's integration of the disjunctive and objective many into a privately felt satisfaction of the universe (a unique one) that is then capable of being prehended by future occasions. Whitehead's analysis of these phases is baffling in complexity and detail, so I cannot describe them in full here. But it is necessary that he provide an exhaustive account of the occasion's genesis if he aims at full generality. To put it schematically, the initial phase of concrescence describes the process by which a concrescing occasion feels a multiplicity of already satisfied occasions (superjected occasions) growing into datum for its own unique satisfaction. These are primary, physical feelings of the sheer multiplicity of past perspectives (a multiplicity of actual worlds) entering into the subject's own constitution. These physical feelings are then supplemented by more complex "conceptual prehensions." What distinguishes primary feelings from subsequent, conceptual feelings is that in the former the concrescing occasion is feeling the energy transfer from past to present via efficient causality, where in latter, conceptual phases, the entity prehends an "eternal object" (PR, 85). Thus, in a conceptual prehension, the occasion has an eternal object as its datum instead of an actual entity.

Eternal objects are among Whitehead's more challenging notions and are one of the Eight Categories of Existence. He refers to them as "Pure Potentials for the Specific Determination of Fact, *or* Forms of Definiteness," and then remarks that, along with actual entities, they "stand out with a certain extreme finality" (PR, 22). The trouble with eternal objects is that they are vaguely reminiscent of Platonic forms, an association that Whitehead does not totally dispense with (PR, 291).[7] Eternal objects are *not* Platonic forms, however: they are pure potentials for *how* an occasion might be objectified in another (149). They are "adverbial qualifiers" that determine how one thing could enter into another—hotly, coldly, warmly, fiercely, and so on.[8] Nevertheless, these objects are *pure* potentials in the sense that they do not depend upon any specific actuality or epoch for their existence. Instead, they are said to "transcend" the occasions that actualize or *ingress* them, since potentiality is always in excess of what is realized. Eternal objects in this sense are said to be "objective," since they do not depend upon subjects for their existence, and also eternal, since they are patterns not effected by time (a specific epoch, for example). In itself, each eternal object has a "unique character" that is also inextricably related to all other eternal objects (Kraus 1998, 32).[9] This does not mean that eternal objects are completely independent of the unfolding world. Without the actualities into which eternal objects are ingressed they are meaningless, since they *just are* potentials for actualities. There is no separate sphere or realm for these potentials located outside of actual entities: "Apart from things that are actual, there is nothing—nothing either in fact or in efficacy" (PR, 40).[10] Likewise, all actual entities require the forms of definiteness that eternal objects provide. Actual entities and eternal objects require each other (they are in mutual presupposition) in order to realize the creative advance of a world that does not bifurcate.

Conceptual feelings are not passive modes of contemplation: they describe the process by which an adverbial qualification is admitted into the concrescing occasion. There is nothing haphazard about these qualifications being admitted into experience either: an eternal object's ingression occurs through *valuation.* For this reason, valuation has a central role in Whitehead's metaphysical description of an occasion's becoming an individual: it describes the process by which an occasion of experience becomes the form of definiteness that it is. Here's Whitehead: "By reason of the actuality of this primordial valuation of pure potentials, each eternal object has a definite, effective relevance to each concrescent process" (40). Eternal objects are valued as important or unimportant (valued up or down) elements for the concrescing individual as set forth by the *subjective aim* (241). This aim, which is not to be confused with the subjective form, provides the occasion's own concrescence directionality (its final cause). Whitehead defines the subjective aim as a "lure" for further integration, a "lure for feeling" that is determined by God's primordial envisagement and that guides the self-creating subject to make valuations of the relevance of eternal objects that will lead to the most intensely contrastive satisfactions for the subject (I return to the notion of intensive contrast below and to God and subjective aim in the next chapter) (47).

Despite the centrality of value or valuation to Whitehead's conception of concrescence, there is nothing easy or straightforward about it. Commentators have struggled over this aspect of his work for decades. For instance, David L. Schindler argues that there are serious inconsistencies in Whitehead's theory of value. He pays special attention to Whitehead's assertion that "value is inherent in actuality itself. To be an actual entity is to have self-interest. This self-interest is a feeling of self-valuation" (RM, 87). Schindler then zeros in on this notion of self-interest and asks: "If the value of actuality lies in actuality's character as subject, . . . what warrants my assigning value to others, the data, that is, the *objects* in relation to which I (or any actual entity) constitute myself as subject?" (1983, 121). In short, he accuses Whitehead of ontological subjectivism and of constructing the kind of universe that he wished to undermine: one filled with value-seeking, atomized subjects (Henning 2005). Michael Halewood forcefully responds to this charge and argues that Schindler's critique stems from a common misreading of Whitehead's notion of value: that it is based on a notion of the ethical subject inherited from the nineteenth century, which already has definite content. Furthermore, Halewood observes that Whitehead rarely uses the term "value" in all of *Process and Reality* and uses "valuation" instead (see above). According to Halewood, this changes the emphasis entirely: from something already settled and determined to an activity of the subject that is always under construction—the subject's becoming is a valuating activity. This is why Whitehead explains that "the subjective form of a conceptual feeling is valuation" (PR, 248). Unfortunately, this still does not tell us much about how valuation occurs and how exactly it avoids ontological subjectivism.

Brian Henning (2005) has done important work in this area and demonstrates how our understanding of value or valuation in Whitehead's philosophy depends upon how we interpret the division between subject and superject. According to "ecstatic" readers of Whitehead (Jorge Luis Nobo, Nancy Frankenberry, Elizabeth Kraus, and Judith A. Jones), there is no ontological distinction between subject and superject, such that the present subject is the sole creative agent in the universe and the past (qua superjected occasion) is dead, inert, and inactive (Ivor Leclerc [1958] and William Christian [1959]). Rather, the causal efficacy of the past datum requires that past satisfactions are actually in the present, not as vague reminiscences or representations, but as real constitutive elements that make up the *intensive contrasts* that lure the present achievement into its final satisfaction. Hence, each present satisfaction reaches back into the past and feeds forward into the future to establish the very conditions for future becoming. Actuality extends into the past and into the future, ensuring that no bifurcation exists between the subject and the superject (Henning 2005, 460–65). The ecstatic interpretation of Whitehead dispels the idea that an occasion's valuation of eternal objects is based upon a subjective aim that would lure the occasion to make ingressions based on self-interest. Rather, its valuations are grounded in bringing about *relevant contrasts* among the valuations made by past occasions that are now physically felt and that may be radically discordant and unharmonious (see my discussion in chapter 5). Conceptual feelings of eternal objects are supplemental feelings valued according to their capacity to structure data in ways that bring about maximal intensity of satisfaction relevant to the concrescing occasion and to future satisfactions. Thus the occasion's valuations are made with an eye to ensuring intensity of contrast "for itself, for others, and for the whole" (MT, 111).[11]

But if Whitehead insists upon the importance of a value-filled universe of interconnected entities, then he also needs to account for the specific kinds of bonds established among these entities, and moreover how they are connected to the world we ordinarily experience. Whitehead is not trying to replace our ordinary experience of the world—the various objects, histories, and technologies that we have—with a more accurate or true metaphysical description. "There remains the final reflection," Whitehead notes at the end of the preface to *Process and Reality*, "how shallow, puny, and imperfect are efforts to sound the depths in the nature of things. In philosophical discussion, the merest hint of dogmatic certainty as to finality of statement is an exhibition of folly" (PR, xiv). Much more challenging than providing an account of *what* is really real via a new set of ultimate entities is the demanding task he sets for himself of transforming *how* we experience the world already given. This includes "experience drunk and experience sober, experience sleeping and experience waking" (AI, 226); all must be put "in the same boat, to sink or swim together" (CN, 148). In order for actual occasions and eternal objects to lure us into encountering a nonbifurcated world, Whitehead must show how they are connected to the objects of our experience, to the world we normally encounter, and ultimately to our experience of scientific visual culture. Otherwise, his project fails because it has explained away certain experiences (PR, 19; see also Gaskill and Nocek 2014, 20). Whitehead answers this call by characterizing the specific kind of "social" bonds that occasions share. Social bonds are what make up our world of everyday objects and experiences.

SOCIETIES AND VALUATIONS

Whitehead has shown us that "actual occasions involve each other by reason of their prehensions of each other," which means that there are "real facts of togetherness of actual entities," and these facts are which he calls *nexūs* (PR, 20). *Nexūs* of occasions are the forms of ordered togetherness that an occasion feels among the entities in its actual environment. And insofar as an occasion is an outgrowth of its environment, then the degrees of order and disorder within *nexūs* are the primary ingredients for the datum that contribute to an occasion's intensity of satisfaction. So, for instance, a nexus may be highly uncoordinated, which means that there is excessive "width" in the data, and so it does not promote feelings that mutually reinforce each other. In such instances, the satisfaction is trivial: "Triviality," according to Whitehead, "arises from lack of coordination in the factors of the datum, so that no feeling arising from one factor is reinforced by any feeling arising from another factor" (111).

A molecular animation that combines too many elements, or too many disjointed factors, to deliver a message in which the diversity of composition is ultimately complementary and reinforcing offers a trivial satisfaction. The trouble is that there is no ultimate gauge as to the triviality of a satisfaction; it is environment-specific. Take the molecular animations from the 2012 BBC documentary *Our Secret Universe: The Secret Life of the Cell* (see Figure 17). If the organization of elements in this animation (discussed in chapter 1) were used in the context of experimental research on motor proteins or some other protein domain (as Iwasa does with the Reck-Peterson group), then they would likely deliver trivial satisfactions—a lack of coherence for the purposes of research. On the other hand, Iwasa's animation of dynein (see Figure 27) would most likely lack the diversity and contrast that ensures maximum satisfaction for the consuming public. In such cases, the environment displays too little diversity, which means that there is a "lack of differentiation among the component objectifications" (112), and so there is an "intensive narrowness in the prehension of the whole nexus, by reason of the common character, combined with vagueness, which is the irrelevance of the differences between the definite actual entities of the nexus" (111).

The notion of nexus grounds Whitehead's extensive discussion of the forms of order that we encounter in everyday experience: social order. Social order explains those forms of togetherness that endure—from computer animations to rocks, trees, bacteria, human subjectivity, social and political systems, and so on. Societies do not introduce a new division in Whitehead's metaphysical scheme but rather make clear the specific dosages of order and disorder that account for what humans normally experience. However, what distinguishes this (namely, a social) interconnection of entities from others is that it shares a "common element of form" or "defining characteristic," also known as a complex eternal object (33, 90). In this way *nexūs* of occasions are said to be social when the same eternal object is positively valued by occasions in their mutual prehension of each other. This shared conceptual feeling is what turns the diversity of actual experiences among members of the nexus into relevant contrast that contributes to an intensity of satisfaction that is mutually reinforcing. Such a positive transmission of an eternal object throughout the nexus is not imposed from outside, however. Whitehead explains that

> the reproduction of the common form throughout the nexus is due to the genetic relations of the members of the nexus among each other, and to the additional fact that genetic relations include feelings of the common form. Thus the defining characteristic is inherited throughout the nexus, each member deriving it from those other members of the nexus which are antecedent to its own concrescence. (34)

The defining characteristic of the social nexus of occasions arises immanently from out of the mutual prehensions of its members. “A society,” Whitehead continues, “does not in any sense create the complex of eternal objects which constitutes its defining characteristic. It only elicits that complex into importance for its members, and secures the reproduction of its membership” (92).

What this means is that an eternal object is a defining characteristic of a nexus only insofar as the member occasions share positive valuations of it. The social order’s endurance therefore depends upon the complex eternal object continuing to be valued as a common element of form among member occasions. What this also means is that social order is finite. Its endurance is contingent upon whether the specific valuations required for it to be a society continue to be made. This depends, however, on whether the wider social order of which it is a part continues to support the maintenance of these valuations. For instance, molecular animation societies are composed of diverse computational, cultural, and semiotic elements (each of them societies in their own right) that come together because they share a positive valuation of some quality of organization; and this quality allows for a specific coagulation of elements to be identifiable *as* a molecular animation rather than some other computer animation or 3D scientific visualization society. But this society endures qua society of computer-animated molecules not only insofar as the elements within it continue to incorporate a certain form of organization but also insofar as the wider environments—institutional, historical, epistemic, infrastructural, technical, economic, etc.—support this incorporation. In short, molecular animation has to be beneficial to the wider social orders that support it.

Just as there are no isolated occasions, and every occasion is interfused with every other one, so too there are no isolated societies: each society permeates all the others. Thus, in order for a society to endure, there must be favorable conditions from the more generalized social orders. I quote Whitehead at length:

> There is no society in isolation. Every society must be considered with its background of a wider environment of actual entities, which also contribute their objectifications to which the members of the society must conform. Thus the given contributions of the environment must at least be permissive of the society. Also, in proportion to its importance, this background must contribute those general characters which the more special character of the society presupposes for its members. But this means that the environment, together with the society in question, must form a larger society in respect to some more general characters than those defining the society from which we started. Thus we arrive at the principle that every society requires a social background, of which it is itself a part. In reference to any given society the world of actual entities is to be conceived as forming a background in layers of social order, the defining characteristics becoming wider and more general as we widen the background. (90)

This lengthy passage helps to secure Whitehead's legacy as an ecological thinker: there are no isolated systems or idealized notions of Nature in his social ecology (anticipating much contemporary eco-theory); all physical, chemical, biological, semiotic, political, and economic orders interpenetrate each other in ecological interdependency.[12] So, for instance, two specialized societies, with different histories and aims, might find themselves supported by the same wider social order. This wider order is not a container for the narrower, more specialized ones (we are not dealing with nesting dolls). Rather, the more specialized societies are also a part of the more generalized one, and require the latter to carry out its more specialized tasks; hence, there is genuine ecological interpenetration of social orders. This is why Whitehead says that "we arrive at the principle that every society requires a social background, of which it is itself a part."

Elizabeth Kraus's example of biological cells is helpful: "[A] cell in an organ," she explains, "has specialized characteristics in virtue of its inclusion in an organ, characteristics not shared by cells in other organs. But in addition to being a liver cell, for instance, it is likewise a bodily cell and shares certain general characteristics with all other bodily cells. Without these more generalized characteristics, it could not support its specialized ones. Its social environment is the liver *and* the body" (1998, 68n6). But of course if the bodily environment decays, then there are no longer favorable conditions for the liver cell and other bodily cells. "The favorable background of a larger environment," Whitehead explains, "either itself decays, or ceases to favor the persistence of the society after some stage of growth: the society then ceases to reproduce its members, and finally after a stage of decay passes out of existence" (PR, 91). Something similar could be said about molecular animation societies: each film, clip, image, line of code, and so on has a specialized function with a unique set of characteristics that distinguishes it from other elements. Each of these elements coexists with and depends upon the existence of the larger society of computer molecular animation, and even larger societies of computer animation, molecular visualization, and so on. The erosion or transformation of any one of these wider environments affects the more specialized ones. A society's continued existence is therefore dependent upon whether or not more generalized social orders offer suitable conditions for its maintenance. In this way Whitehead exposes the deep ecological fragility that pervades all social organization and the valuations upon which it depends.

TOWARD AN ECO-SOCIOLOGY OF MEDIA

I would like to keep Whitehead's "eco-social theory" in mind as we find our way back to the question that animated our detour through his philosophy—namely, can we devise a conception of media capable of explaining how molecular animation is a heterogeneous "many" that becomes "one" through shared "values"? Before we address this question head-on, I want to rehearse one of the attendant risks of using

Whitehead's work to understand any phenomena: that we leverage his metaphysics to explain how the world *really* operates outside of the questions that we pose to it. It is tempting to treat Whitehead's notions as generic metaphysical principles against which all previous and future conceptions should be measured. This temptation is especially strong today when much of Whitehead's revival has been associated (for better or worse) with the booming industry of speculative metaphysics, realism, and materialism in continental thought.[13]

But as Stengers reminds us, Whitehead does not offer us a new "'enlightened,' scientifically grounded conception of the world" (2014, 44). He does not devise generic notions that gain "access to some concrete truth hidden by our specialized abstractions. If no experience is devoid of interpretation, then what is prohibited from the start is that we should retain some nostalgic memory of what we previously believed we genuinely knew about nature, and entertain the possibility of a more authentic experience" (47–48). I want to heed Stengers's advice about reading Whitehead's *Process and Reality*: we need to "avoid a certain conceptual trap—namely, the idea that the proposition 'actual occasions exist' is something Whitehead wanted his reader to accept as a matter of fact, as physicists have done with such propositions about atoms." Instead, Stengers challenges us to feel "the kind of efficacy that these fantastic metaphysical existents that he named 'actual occasions' can produce, in terms of their being speculative abstractions" (57).

As already noted in the Introduction, Whitehead deepens this insight when he insists that the efficacy of a speculative abstraction is determined in large part by its applicability. He explains that "the success of the imaginative experiment is always to be tested by the applicability of its results beyond the restricted locus from which it originated" (PR, 5).[14] Success depends on whether the imagination is able to find notions that are capable of being exemplified beyond the narrow field of experience from which it began: "In default of such extended application, a generalization started from physics, for example, remains merely an alternative expression of notions applicable to physics" (5). In this same spirit, we must ask ourselves whether Whitehead's speculative notions help to understand the coordination of experiences and practices beyond the narrow field from which we began our thinking—namely, molecular animation practices. And in what direction(s) do they lead us? It is with these questions in mind that I want to address what Whitehead's theories of eco-sociology and valuation do to our understanding of scientific media.

The first takeaway from Whitehead is that the beings we encounter in the world are "eco-social beings." What this means, as I have already alluded to, is that a media system, like molecular animation, is an intensively integrated many forged at the intersection of technologies, software systems, practices, histories, aims, semiotic systems, and so on. This vast heterogeneity of elements is able to integrate across differences through the evaluative feeling of a common element of form. While this form is irreducible to the diverse components in the system, these elements nevertheless require it in order to experience their diversity as more than *mere* diversity, but as mutually reinforcing contrasts that lead to especially intense satisfactions. This form is not imposed from outside, or "hylomorphically," but it is through the elements' own immanent production of the conditions for their own organization (I return to this notion of "imposition" of social form in chapter 5).[15]

This social form is not unlike what Guattari calls an abstract machine, although with some important differences. As we discussed in chapter 1 in relation to LaMarre's work on the animation and cinematic machine, Guattari reverses the priority we usually give to technology over the machine and insists that we need to find the machine on which a technology depends (Guattari 1995, 33–35, 43). This means that we need to search for the diagrammatic form or machine that integrates the materials across their differences. Any technical ensemble is composed of an incalculable number of materials that have no inherent or natural relation to one another. The practice of molecular animation is composed of algorithms, software packages, hardware, datasets, wet and dry biology, human imagination and reason, institutional settings, epistemic norms, and so on. And none of these elements naturally belong together. Maya's scripting language MEL has no inherent relation to the PDB (Protein Data Bank) files from which its content is derived (in fact, they don't communicate very well), and so plug-ins such as mMaya were designed to reduce the friction between these systems. The coordinated relations these materials and infrastructures come to have in a technical system is not pregiven in any of the materials. In other words, the fact that these materials integrate across their differences cannot be explained by the materials themselves. Rather, we require an account of the diagrammatic form that makes the integration possible. And for LaMarre, it is this Guattarian framework that helps explain how a media apparatus could integrate materials to actualize different diagrammatic tendencies of the moving image—animetism or cinematism (2009, 32).

But what Whitehead helps us appreciate, and in conversation with Guattari, is that in order for a media system (or a machine) to reproduce itself over time, evaluations have to be made. The production of a form needs to be positively *valued* (it is in this sense that technical media are never neutral). But these shared valuations do not occur in isolation of the wider environment in which the social system is situated; in fact, they are meaningless without it. The reason that a specific organization of materials endures over time is that positive valuations of this form continue to be important to the wider social environment that supports and encourages its existence. There are vast eco-social infrastructures that support and maintain its production. Hence, there is no media system in isolation. But this does not simply mean that each media system is embedded in a wider environment of media relations. It also means, and much more radically, that each media system has a determinate and internally felt relation to every other social system. It is in this sense that Whiteheadian eco-social media is in conversation with much of the work in media ecology and media materialism, especially those theorists who insist on situating our technical systems in terms of vast scales of temporal and spatial organization. Thus geological and biochemical universes, along with psychological, political, and economic universes, have to be thought in relation to the support and maintenance of digital media systems (see also Goffey and Fuller 2012). This means that a media society is able to reproduce itself only insofar as its wider environment—which, in principle, extends to the widest social order, the cosmic epoch—continues to deem the replication of its valuations important to its own reproduction and integrity. Whitehead therefore gives us the conceptual vocabulary to address how the dissemination of media assemblages is inseparable from the replication of values throughout the human and nonhuman world.

What we call molecular animation, then, is a highly complex set of relationships established among heterogeneous elements that contribute to producing a visual regime with a particular logic or a certain pattern of relations. The coordination of elements is "social" in character inasmuch as there is integration across vast differences in materiality and functionality for the purpose of expressing a specific pattern. And it is this pattern or logic to the visual that functions like a "common element of form." I want to stress here that it is the pattern or logic of the visual (and not the visual itself) that exemplifies Whitehead's common form.[16] What stands out about molecular animation is how diverse the elements in its social assemblage are. Not only do software systems, databases, and programming languages have to be integrated into a social ensemble but so do scientific and popular ways of knowing, feeling, and experiencing the world. According to Whitehead, these elements all make positive valuations of the same form. But if molecular animation is a social formation, then its emergence and continued existence depends upon the wider environmental conditions that promote the many convergences that take place within it. In order for heterogeneous visual cultures to have shared valuations and be drawn together in the same social order, the wider environment has to provide favorable conditions for these valuations to exist. What this means is that the right environmental conditions have to obtain in order for scientific and popular media practices to share valuations of a visual logic. We therefore need to understand what eco-social mechanisms make it possible for molecular animation to exist.

The remaining chapters of this book examine these wider environmental conditions. They explore how this favorable environment did not emerge all at once. It emerged in fits and starts, sometimes with great speed, and sometimes more slowly, and even imperceptibly. And although Whitehead has an arsenal of speculative abstractions that permit us to speak with full generality about the world, and in doing so, guard against the temptation to reduce or explain away certain items of experience, they need to be reframed in some instances in order to address how the eco-social world unfolds with historical specificity. The specific histories, practices, asymmetries, and divergences that we experience in our world cannot always be expressed with his generic descriptions. But it is precisely these specificities that account for how social assemblages, like molecular animation, have come into being and cohere over time. If Whitehead gives us speculative tools to address the eco-social metaphysics of media, then we need a complementary framework that allows us to explain how specific environments make certain materials enter into social ensembles for finite periods of time. Or, in other words, what are the wider social, political, and economic systems of power driving the convergence of materials and cultures in molecular animation?

5

GOVERNING THE SOCIAL

> Freedom may indeed appear as the condition for the exercise of power.
>
> —Michel Foucault, "The Subject and Power"

In chapter 4 we started to build a conceptual framework for molecular animation. It became clear that we needed a way to address how heterogeneous visual cultures converge through mutual valuations in a media practice. It seemed to us that we required a theory that could at once accommodate the specific histories and practices bound up in molecular animation and could place them in a wider ecological frame to account for the way in which these divergences collide and even mutually reinforce one another. For this reason, the speculative philosophy of Whitehead became an important touchstone, not simply because of its newly acquired relevance in media ecology and materialism but because his work on social order and valuation helps us to unpack the ways in which (1) the reproduction of an ordered system brings a multiplicity of perspectives together for a finite period of time through shared valuations, and (2) how every social order requires a vast ecological infrastructure to ensure its endurance over time. As Whitehead explains, "There is no society in isolation," so that "in reference to any given society the world of actual entities is to be conceived as forming a background in layers of social order, the defining characteristics becoming wider and more general as we widen the background" (PR, 90). With this theoretical framework in place, my hypothesis is that we are able to offer the blueprint for an eco-social metaphysics of media that accounts for how widely divergent systems are capable of converging on a set of values for a finite period of time.

But as I alluded to at the end of the previous chapter, Whitehead seems to only get us part of the way there. Whitehead characterizes societies in generic and abstract terms. While societies have different classifications—they can be enduring, structured, corpuscular, or living (34–35)—Whitehead shies away from concrete and empirical descriptions of them (of molecular, human, or media societies, for instance) and opts for generic accounts of sociality that are designed to capture the numerous possibilities for the real forms of togetherness that emerge, endure, and pass away in our universe. But what these descriptions do not offer are specific, empirical content by which we could verify or correct his general propositions. Although Whitehead proposes the form our thinking should take, and more specifically, has helped us to conceive the possibility of converging popular and scientific social orders through valuation, he seems less able to account for the conditions under which this might occur. What the driver of this eco-social convergence might be, what historical, institutional, and political–economic mechanisms might be at work in its formation, are left for us to fill in.[1]

This is not inconsistent with Whitehead's general method and moreover should not be taken as an oversight in his work. The speculative propositions that Whitehead constructs through imaginative reason aim at generality and are neither true nor false in themselves; they are lures for feeling the world differently. But his imaginative rationalizations *do* require empirical verification. Such empirical work obliges us to ask: "Have we resisted the temptation to exclude, reduce, or explain away any item of experience?" To the extent that we have failed to resist this temptation, our propositional scheme fails and requires modification. Such verification and modification are largely what I will be undertaking in Part III. In particular, I will be inquiring into whether our speculative notions help us to understand the specific historical, political, and economic conditions that made it possible for the alliance of popular visual culture and the scientific epistemic practices in molecular animation.

But in order for our flight of the imagination to land in experience and make a significant contribution to our understanding of this alliance, we need to have a framework for engaging how media history unfolds.

And more specifically, we need to address how it unfolds in relation to questions of technology and political economic power. These are the wider environments that envelop the visual cultures of science and entertainment and underwrite the eco-social convergences we witness in molecular animation. If Whitehead gives us a speculative frame to envision molecular animation as a social ensemble that positively values flexible modes of perception equipped to capture objects and processes at any possible spatial and temporal scale, then we also need to know how this media valuation took shape. Whitehead's thought therefore needs to be primed to address the political and social realities of the human and nonhuman technical world in order to return to molecular animation with fresh eyes in Part III.

And yet the million-dollar question is: How does Whitehead prepare us to talk about the historical formation of political and economic realities with any specificity, and especially those realities that bear on our media culture, when he rarely, if ever, addresses politics or the political? Of course, there are a host of contemporary theoretical frameworks, especially those issuing from new materialism, neovitalist metaphysics, and object-oriented ontology that are depoliticized, or dubbed "prepolitical," but nevertheless weigh in on politics and notions of the political. This chapter is not denying that there are conceptual gains to be made in this kind of work—witness, for instance, how preindividual Being in Gilbert Simondon has been read politically by thinkers such as Alberto Toscano, Roberto Esposito, and others (Toscano 2012; Esposito 2004). However, we cannot overlook that there are tremendous risks involved in marshaling largely depoliticized ontologies for political ends, which is something that Alexander Galloway and a host of others have made crystal clear.[2] The aim here is not to leverage Whitehead's propositional scheme to reinvent political ontology. Instead, what is at stake is understanding whether and to what extent Whitehead's philosophical conceptions can be reconceived in such a way that the unfolding of political and economic power within human technical societies becomes vivid in a new way for us. We will then be in a position to weigh in on visual media cultures in the present. Ultimately, this chapter aims to draw out those aspects of Whitehead's philosophical system that are amenable to such retooling and allow us to speak about the formation of the present with both speculative generality and genealogical specificity.

Unfortunately, the path for this inquiry is not laid out for us in advance. This is because we know very little about how Whitehead's work actually maps onto political and economic history. Whitehead rarely, if ever, addresses political economics, or how he might leverage his metaphysical work to address the political and social realities of his day. For this reason, his philosophy, like the work of many other speculative metaphysicians, often feels disconnected from the power asymmetries operative in human societies. Unlike the work of some of his contemporaries, John Dewey and George Santayana, for instance, Whitehead's philosophy tends to abstract from the messiness of political struggle. One of his more notable attempts to discuss political societies comes in his Barbour-Page Lectures, *Symbolism: Its Meaning and Effect* (1927). But there too he only addresses these societies in the broadest sense, and mostly concerns himself with the ontology of human and cultural symbols in the context of the two primary modes of perception, presentational immediacy and causal efficacy (S 23, 43–44). He leaves us to wonder what he might have to say about the various crises of liberal economics rocking the Western world during the interwar period (Dardot and Laval 2017, 37).

Nevertheless, there are some important moments in Whitehead's oeuvre that call to mind much larger questions about the constitution of modern political societies. As I have written elsewhere, there are a few exceptional passages from *Science and the Modern World* where Whitehead meditates on the problem of the modern professional (Nocek 2017). In particular, he criticizes the nineteenth-century "discovery of the method of training professionals, who specialize in particular regions of thought and thereby progressively add to the sum of knowledge within their respective limitation of subject" (SMW, 196). What worried Whitehead was the crippling effect that modern professionalism has on thought: professionals "live in contemplating a given set of abstractions. The groove prevents straying across country, and the abstraction abstracts from something to which no further attention is paid" (197). When this happens, thinking loses its hold on the complex environmental coordination required for a specific set of abstractions to exist.

But once again, we are left to wonder about the nature of the wider social environments that make it possible for professionalized abstractions to exist. Unfortunately, Whitehead does not give us much to work with other than his diagnosis of the state of scientific education in the world of modern *science*. The social environments of science, or in other words, those wider modes of togetherness that account for how modern scientific professionalism came into being, are not spelled out for us. But it is these wider social spheres that are integral to our understanding of the "environmental coordination" of facts that Whitehead famously champions (SMW). Such an ecological framing would no doubt have to take account of the fact that the production of "minds in a groove" happens against the backdrop of the pervasiveness of liberal ideals of competitive individualism. For instance, Michael Polanyi, an influential Hungarian–British scientist and philosopher, argued that true science only occurs in a (classically) liberal society where the values of individualism and competition promoted in the free market are implemented in the domains of science. Only under such conditions is scientific knowledge free from state authority; this is "the decisive reason," Polanyi writes, "for individualism in the cultivation of science" (1951, 89). Polanyi spent nearly thirty years in correspondence with the economist Friedrich Hayek, who adopted many of his ideas, including the notion of "spontaneous order" (Jacobs 2000).

And yet, how Whitehead's speculative system would weigh in on this, or any other political and economic reality in the early twentieth century, remains largely a mystery. As the scientist Joseph Needham laments, "One looks in vain in Whitehead's writings for some clear lead among the social tendencies of our time" (1941, 267). This doesn't mean that others have not tried to unpack the social and political implications of his thought. Phillip C. Morris's *Process Philosophy and Political Ideology: The Social and Political Thought of Alfred North Whitehead and Charles Hartshorne* (1991) is exemplary for attempting to cash out the political stakes of Whitehead's (and Hartshorne's) work. For Morris, however, this means uncovering the social and political environments in which Whitehead's metaphysical views were formulated with the hope of reconstructing his personal beliefs. Autobiographical details are of seminal importance, and in particular Whitehead's assertion that "my political opinions were, and are, on the Liberal side, as against the Conservatives. I am now writing in terms of English party divisions" (SP, 14). This does not add up to much, though, especially when we are trying to say something substantial about how Whitehead's metaphysics might relate to contemporary capitalism(s). As Steven Shaviro remarks, "Whitehead himself does not offer us any concepts of directly political import. He has nothing in particular to say about capitalism; his politics, though moderately left of center, were certainly not Marxist" (2009, 157).

But this has not stopped theorists from speculating about what a Whiteheadian critique of capital would look like. While Shaviro is cautious not to overstate the relevance of Whitehead's ideas (as it would reinstate the fallacy of misplaced concreteness), he nevertheless insists that "Whitehead's aestheticism is radical enough that it nudges and cajoles us away from the complacencies and satisfactions of commodity culture" (158). But perhaps more explicitly, Michael Halewood puts Whitehead into conversation with foundational sociological thinkers—for example, Max Weber, Émile Durkheim, and Karl Marx—in order to rethink the very meaning of the social for sociological theory. In the final chapter of *A. N. Whitehead and Social Theory: Tracing a Culture of Thought,* Halewood exposes deep affinities between Whitehead and Marx on the question of abstraction and even shows how commodity fetishism is a special case of the fallacy of misplaced concreteness (2011, 159). Whitehead's thought is therefore at the very center of a Marxian critique of capital for Halewood. Likewise, in an earlier work, Anne Fairchild Pomeroy recognizes the potential for a Marx and Whitehead alliance in *Marx and Whitehead: Process, Dialectics, and the Critique of Capitalism* (2004). Crucial for her argument is that Whitehead helps us to understand the metaphysical underpinnings of Marx and capitalism generally. She leverages this connection to show the metaphysical inadequacies of the capitalist system: a system that values the production of private property cannot grasp the processual interrelatedness of the world. "Capitalism is a disaster for processive reality," Pomeroy insists (2004, 164).

These and related works help to advance the relevancy of Whitehead's thought for political and social theory. My sense, however, is that much of this research is still especially abstract, committed to rethinking political and social ontologies at the most general level, with little to no connection to specific histories and practices. This is intended to be less of a criticism than it is meant to signal my departure from this orientation in thought, as well as my commitment to following Whitehead's injunction to verify the efficacy of imaginative abstractions empirically.

Keeping this in mind, one of the central hypotheses of this chapter is that Foucault, instead of Marx, is better equipped to tease out the political efficacy of Whitehead's metaphysics. There is a sense in which Foucault is an obvious partner in this conversation. For who else is more qualified to theorize the relations of power, domination, and suffering that have shaped our present than Foucault? But there is another sense in which this reading might seem forced and fly in the face of my previous admonitions, yielding little more than abstract comparisons. There is little, if anything, in Whitehead's oeuvre that jumps out as especially "Foucauldian." There may be some superficial similarities—for example, their critique of substance ontologies—but there has yet to be an especially compelling reason to bring Whitehead and Foucault into conversation. And there may be good reason for this. On the surface, they couldn't be more different: where Foucault sought to bring historical specificity to his analyses of institutions and power relations, Whitehead, by contrast, endeavored to construct a metaphysical system that aspired to full generality. These are profound differences that should not be overlooked.

What I propose does not aim to reduce or explain away the differences between their projects. In fact, I insist that their differences are just as important as their similarities and are what establish essential contrasts between the two. To make this case, this chapter looks to Foucault's late work on power and governmentality and stresses how there are a number of important, if surprising, resonances between Foucault's general characterizations of power and government, or the "conduct of conduct," and Whitehead's conception of the "imposition" of a defining characteristic in a society of occasions. I draw out these connections in order to conceive of a notion of "infectious social power" that resonates with both Foucault and Whitehead. Using this framework, I also discuss how Whitehead helps Foucault to reconceive the meaning of immanent resistance to social infection and how it unfolds through engaging with Whitehead's conception of propositional feelings. In these ways, it's possible to see how the many divergences between Whitehead and Foucault are actually reinforcing and enhancing instead of negating and detracting. I propose that Whitehead and Foucault are in *contrast* and not in conflict with one another, and this relation expresses what Whitehead says about aesthetic relations more generally. My thesis is that an aesthetic reading of the two opens the door to thinking about how governmental control is operative in the social and ecological orders that Whitehead describes. This will provide the requisite conceptual framing for a more specific engagement with how the governmental control of visual perception operates across the social–ecological registers of entertainment and science to generate the conditions for molecular animation to emerge as a dispositif in the contemporary.

ON POWER AND GOVERNMENT

Foucault's work is notoriously difficult to characterize. This is partly because it demands so much of the reader. Brief forays into Foucault risk oversimplifying his conceptual scheme and fail to appreciate the periodization of his ideas, how they develop in response to problems identified in specific eras of his thinking, and how his thought is constantly in dialogue with other movements and thinkers (structuralism, poststructuralism, psychoanalysis, Marxism, Deleuze, and so on).[3] For these reasons, providing an overview of power and government in Foucault's work seems like a task best reserved for a much longer and detailed study than I can provide here. Keeping all of this in mind, Foucault nonetheless offers a general and instructive characterization of power and government in his late essay from 1982, "The Subject and Power."[4] The essay gives us a unique opportunity not only to weigh in on many of the developments in his later work but also to appreciate their overlaps with Whitehead's thought. This chapter will be less concerned with specific mutations in power (which are discussed at length in chapter 6) and more interested in his philosophy of power and government generally.

In the second part of his essay "How Is Power Exercised?" (the original French portion of the essay), Foucault writes that "what characterizes power is that it brings into play relations between individuals (or between groups). For let us not deceive ourselves, if we speak of the structures or the mechanisms of power, it is only insofar as we suppose that certain persons exercise power over others" (1982, 217). In saying this, Foucault is careful to emphasize that power is not an entity or object in itself that would be capable of bringing these relations about. Power itself does not exist. "That something called Power," he stresses, "with or without a capital letter, which is assumed to exist universally in a concentrated or diffused form, does not exist. Power exists only when it is put into action, even if, of course, it is integrated into a disparate field of possibilities brought to bear upon permanent structures" (219). Foucault elaborates on what he wrote six years earlier in *The History of Sexuality, Vol. 1*—namely, that power is "not something that one holds onto or allows to slip away" (1978, 94)—by asserting that power is nothing by itself. Power *relations* exist, power itself does not.

Foucault then sketches the "actions" proper to power relationships. He refines his (above) characterization somewhat by explaining how a power relationship is not merely a relationship between entities (individual, collectives, etc.). Instead, it is characterized by the "way in which certain actions modify others," which is to say, a power relationship is a "mode of action which does not act directly and immediately on others. Instead it acts upon their actions: an action upon an action, on existing actions or on those which may arise in the present or the future" (1982, 219, 220). Foucault stresses that this "modification" of action by action is not equivalent to violence or consent (though it certainly does not preclude either one, and it very frequently makes use of them) and is perhaps best understood in terms of the equivocal nature of the term "conduct." Playing with the dual meaning of the French infinitive "to conduct" *(conduire* and *se conduire),* Foucault captures the sense in which "to conduct" means both to lead or to direct as well as to behave, as in "a way of behaving within a more or less open field of possibilities" (220–21). He exploits this duality to articulate the way in which power is less about adversarial relations or acceptance and more about "guiding the possibility of conduct and putting in order the possible outcome" (221).

It is from this perspective that the meaning of government in Foucault's writings on "governmentality" comes into sharper focus. For what characterizes the exercise of power, at its very core, is not violence, brutality, or consent, although it may involve any of these; it is really a matter of "government."[5] And by "government," Foucault is not referring to the narrow conception of government that we use today—modern states and their institutions—but to the much broader conception of government employed in the sixteenth century (and up until about the eighteenth century; see Foucault 2007, 121–23). In this earlier conception, to govern is to organize and structure the possible actions of self and others. Government therefore designates

> the way in which the conduct of individuals or of groups might be directed: the government of children, of souls, of communities, of families, of the sick. It did not only cover the legitimately constituted forms of political or economic subjection, but also modes of action, more or less considered and calculated, which were destined to act upon the possibilities of action of other people. (Foucault 1982, 221)

If power relations are best characterized by the way "certain actions modify others," then this modification is centrally concerned with the strategic government of action, understood as the "conduct of conduct," which covers everything from the governing of others to the governing of self. And in Foucault's 1978 and 1979 lectures at the Collège de France, *The Birth of Biopolitics,* he offers a genealogy of governmentality from its classical origins through the science of the police to its liberal and neoliberal forms in the twentieth century—or, in other words, to those points where the state became "governmentalized." What is central to Foucault's understanding of governmentality, however, inasmuch as it is fundamental to every act of governing conduct, is *freedom.* As Foucault explains:

> Power is exercised only over free subjects, and only insofar as they are free. By this we mean individual or collective subjects who are faced with a field of possibilities in which several ways of behaving, several reactions and diverse comportments may be realized. Where the determining factors saturate the whole there is no relationship of power; slavery is not a power relationship when man is in chains. . . . Consequently there is no face to face confrontation of power and freedom which is mutually exclusive (freedom disappears everywhere power is exercised), but a much more complicated interplay. In this game freedom may well appear as the condition for the exercise of power. (221)

The point here is that Foucault disarticulates power and government from any conception that would totalize it, that would overwhelm the entire field of action and refuse any and all potentials for resistance to it. Power always presupposes that there are alternative modes of conduct, which is to say, that there are other modes of governing oneself and others within power's current mode of articulation. For Foucault, the government of self and others assumes the existence of struggle. "At the very heart of the power relationship, and constantly provoking it," Foucault insists, "are the recalcitrance of the will and the intransigence of freedom. Rather than speaking of an essential freedom, it would be better to speak of an 'agonism'—of a relationship which is at the same time reciprocal incitation and struggle" (221–22). And in his 1977–78 lectures, *Security, Territory, Population,* Foucault speaks about this struggle at the very heart of government in terms of "counter-conduct" (2007, 200–3). The term is less than perfect, he admits, but it nonetheless "has the sole advantage of allowing reference to the active sense of the word 'conduct.'" "'Counter-conduct'" therefore captures "the sense of struggle against the processes implemented for conducting others" and avoids the pitfalls of "'misconduct *(inconduite),*' which only refers to the passive sense of the word, of behavior: not conducting oneself properly" (201).

But in the 1982 essay, Foucault deepens his conception of freedom at the heart of the power struggle, not by reference to counter-conduct, but by reference to strategy. He notes that there are currently three ways that strategy is employed: (1) as the rational means used to obtain an end—"it is a question of rationality functioning to arrive at an objective"; (2) as the "manner in which a partner in a certain game acts with regard to what he thinks should be the action of the others and what he considers the others think to be his own; it is the way in which one seeks to have the advantage over others" (224–25); and (3) as what characterizes the "procedures used in a situation of confrontation to deprive the opponent of his means of combat and to reduce him to giving up the struggle; it is a question therefore of the means destined to obtain victory" (224–25). Foucault draws on these three senses of strategy to point out that "one can therefore interpret the mechanisms brought into play in power relations in terms of strategies," which is to say that power concerns rational mechanisms that seek to have the advantage over others by depriving them of the ability to engage in struggle. But of equal importance is the fact that these power relations also contain within them "confrontation strategies" (225). This is why Foucault insists that "every governmental power relationship implies, at least *in potentia,* a strategy of struggle, in which the two forces are not superimposed, do not lose their specific nature, or do not finally become confused. Each constitutes for the other a kind of permanent limit, a point of possible reversal" (225). Power relations therefore generate the very strategies that would make "escape" or "flight" from them possible (225).

POWER AS SOCIAL INFECTION

I want to use this rudimentary characterization of power and government in Foucault to recast certain aspects of Whitehead's work on social order. Before I do so, a few preparatory remarks are in order. One of the immediate challenges facing such a reframing of Whitehead is he never (or practically never) speaks about social relations in terms of strategies that would structure the field of action against the will or desire of those it includes. Whitehead is clear in *Process and Reality* that a social order is distinguished from other kinds of togetherness by the fact that the "common element of form" is "positively" felt by its members (PR, 61). This means that the form, which serves to structure the quality of relation among prehending subjects and their objects (superjects), is affirmed by each member subject, which means that it is not negatively prehended by them.

It's crucial to realize that Whitehead is not speaking about "positive feelings" for social relations that are conscious, willed, or desired by humans. Consciousness is introduced later, if at all (280). Instead, he is trying to come to terms with how the elements of our experience seem to share a distinguishing feature or form that serves to structure or organize the various modes of togetherness found in the world. Accordingly, we might say that Whitehead is working through the preconscious microphysics of how order arises from heterogeneous forces. The positivity associated with the prehension or feeling of a form should not therefore be confused with conscious consent. Rather, it concerns the admittance of some form of adverbial organization into a subject's relation to others. As Whitehead writes, "Conceptual experience does not in itself involve consciousness; its essence is valuation" (280). "Positive valuations" do, however, carry the additional valence that a (non- or pre-conscious) "decision" is made to admit this structuring element into the subject's feeling, which has been imposed from the other members of the society. "The defining characteristic," Whitehead explains, "is inherited throughout the nexus, each member deriving it from those other members of the nexus which are antecedent to its own concrescence" (61).

Given this characterization, there is an important sense in which Foucault and Whitehead are both trying to uncover how heterogeneous actions are coordinated to reproduce specific forms of order. And neither Foucault nor Whitehead is interested in relations of servitude or conscious consent (or at least these relations cannot be reduced to consent or servitude). Instead, they each, and in their own way, characterize the subtle ways in which actions are steered in one way instead of another. And what's more, this modification is not imposed on already constituted subjects and individuals, but on actions (Foucault) and the becoming of subjects (Whitehead).

But for Foucault, it is specific forms of *power* that modify actions, that govern conduct, and that structure the possibility of action (1982, 221). More precisely, it is the dispositif, often translated as "apparatus," that specifies the ways in which heterogeneous forces are drawn together or captured through governance to achieve a particular outcome. Although I characterize this notion further in Part III and the Postscript, suffice it to say that the dispositif is a device that makes it possible for a diversity of elements to integrate and cohere for a particular period of time and for a specific purpose. And this integration is, for Foucault, the result of the mechanisms of governmental reason structuring the possibilities for action within social and subjective fields. But as he writes in the *History of Sexuality, Vol. 1,* this structuring of action does not come from above or on high: "Power comes from below; that is, there is no binary and all-encompassing opposition between rulers and ruled at the root of power relations, and serving as a general matrix" (1978, 94). The guiding or governing of the possibility of conduct is not imposed from a domain external to or outside of the field of action in a society.

Although Whitehead never uses anything close to the language of government, power, or apparatus to characterize the subtle guidance of a subject's becoming, he is in agreement with Foucault on at least this much: that the steering or guidance of action must come from *within* a social order. And a part of this guidance has to do with the admittance of a common element of form into experience. Recall that for Whitehead, the element of form in a society gives a definite quality to the relations that obtain among entities. Similar to the dispositif, the form makes the integration of diversity possible for a finite period of time, but in so doing, it also defines the "rules" or "laws" that hold for that particular social integration. The "laws which dominate a social environment," Whitehead explains, "are the product of the defining characteristic of that society" (90–91). This characteristic, which is passed genetically among members of the group, determines the rules of membership, that is to say, what it takes to be included in the society. This characteristic also determines how one society is distinguished from another—from the imperceptible details among physical societies to the much more prominent differences between human social and cultural systems. "But the society is only efficient," Whitehead continues, "through its individual members. Thus in a society, the members can only exist by reason of the laws which dominate the society, and the laws only come into being by reason of the analogous characters of the members of the society" (91). For Whitehead, members of a society must uphold the laws or rules that define it qua society; but these rules of membership are at the same time produced and maintained as rules by the members that are determined by them. Hence, the element of form that structures behavior, that gives a particular shape and meaning to actions among entities, does not come from above, but from below.[6] Much like in Foucault, the rules structuring the integration of heterogeneities are not imposed from outside or above; they are generated within a society of actors (more on this below).

But in Foucault, in order to ensure that individuals conduct themselves in one way instead of another, the governance of action is carefully designed. What Foucault offers in his late work is an especially illustrative genealogy of the various techniques, especially "the meticulous, often minute" ones (1977, 139) deployed to govern the actions of others and self in the history of the West: from the Christian pastorate to the neoliberal state. These techniques give concrete expression to "the way in which the conduct of individuals or of groups" has been directed and, more specifically, to how "the government of children, of souls, of communities, of families, of the sick" has occurred both inside and outside of the state. Thus the apparatuses of capture operative within a society are the result of power trying to achieve particular ends (Lambert 2016a; 2016b, chapter 6). And as Jeffrey Nealon (2008) has suggested, the history of power can be told from the perspective of how it intensifies its control over the field of action by becoming more efficient, which is what makes more supple and suffuse forms of control possible.

It is at this point that we need to pay special attention to *the means by which* defining characteristics are passed among entities in a society. If Foucault conceives of power as ordering the relations between actions, then how do we understand this process in Whitehead? If we return to Whitehead's description of the reproduction of the "common element of form," note his emphasis on "imposition":

> A nexus enjoys "social order" where (i) there is a common element of form illustrated in the definiteness of each of its included actual entities, and (ii) this common element of form arises in each member of the nexus by reason of the conditions *imposed* upon it by its prehensions of some other members of the nexus, and (iii) these prehensions *impose* that condition of reproduction by reason of their inclusion of positive feelings of that common form. (PR, 34; emphasis added)

Isabelle Stengers underscores just this aspect of Whitehead's definition and explains that "eternal objects do not dictate the particular mode of their ingression into an actual entity" (2011b, 319). In other words, complex eternal objects (the common element of form) do not have the capacity to determine *how* they will be ingressed. Each member occasion feels the form in its own way, which is to say that each occasion "decides" how it will integrate this potentiality. But what's essential is that the eternal object itself does not have the capacity to *impose* itself on occasions. Rather, imposition comes from other occasions: "Eternal objects, pure potentials, have no power to impose their ingression. If the fact that each member of a society includes a common element of form," Stengers continues, "it is not imposed by a power associated with the form itself but by the other members of society" (319). The point Stengers drives home is that the "endurance" and "preservation" of a particular form of order is not a "right," but something attributable to the "fact" that a form has managed to be imposed by other members in such a way that occasions have succeeded in "infecting" one another (319).

It is not by accident that she speaks of "infection" when discussing modes of social togetherness in Whitehead. Elsewhere Stengers underscores the importance of the concept in Whitehead's discussion of the "endurance of things" in *Science and the Modern World.* "That which endures," Whitehead explains, "is limited, obstructive, intolerant, infecting its environment with its own aspect" (SMW, 94). It is important that Whitehead's analysis of infection occurs in the midst of a longer account of the preservation of value across occasions of experience. Endurance is not characteristic of the individual entity, but on the contrary, it "depends first of all on the dynamics of infection. All that succeeds in enduring," she writes, "has succeeded in infecting its environment in a way that is compatible with this endurance" (2011b, 157). Stability, preservation, order—these are symptoms of the successful infection of a form that reproduces itself within a group of occasions: each entity prehends how the other entities prehend it through this characteristic. "Infection," Stengers insists, "designates the way in which the modes of prehension are reflected for each other, and success implies the co-adaptation of values" (158). Endurance is made possible by ecological co-production that is a value-conferring activity among entities (158).

For example, in a molecular animation society, a visual logic structures the field of action and meaning for the materials, technologies, actors, histories, and infrastructures that constitute it as a social assemblage. One shot of a molecule in *The Inner Life of the Cell* becomes meaningful in relation to another shot of a molecule from this film or another one, as well as to software suites, databanks, scientists, and knowledge practices, through its inheritance of this common form. This is a form that promotes flexible modes of vision capable of capturing the endless proliferation of objects, systems, and processes at any spatial or temporal scale. It is what defines membership in the molecular animation society instead of another data visualization society. This form is the media practice's "defining characteristic." Each member inherits its relation to this characteristic differently. A line of code in MEL does not incorporate the form the way a scientist working to upload PDB files does; nor does it incorporate it in the way another line of code does. But each of these elements is mediated through this form, and it is what allows them to be members of *this* ensemble rather than *that* one (while simultaneously belonging to countless other social ensembles). What's more, this characteristic is imposed by each member of the society on one another, rather than by an external force. Each element is meaningful to all the others through this structuring element.

Software packages, databanks, scientists, and the consuming public structure their relations to one another by means of this characteristic, however differently, and however fleetingly; it mediates their relation in such a way that there is a "co-adaptation" of elements. While there are countless other ways for the components to prehend one another (perhaps in another data-visualization society), they have managed to actualize this potential for relation. The fact that this characteristic can propagate among such a wealth of heterogeneous actors, and in such diverse ways, is the sign of successful social infection. And although this account does not yet explain what brought this modality of social infection into being, it nonetheless recasts the determinant bonds shared among elements in terms of the immanent imposition and infection of form.

There is an important sense in which this characterization intersects Foucauldian power dynamics. This overlap is not lost on Stengers: "From this viewpoint," she writes, "Whitehead is rather close . . . to the thesis of Michel Foucault, according to which power is not primarily repressive but inciting, inspiring interests, questions, and knowledge" (2011b, 160). And so if a Whiteheadian theory of power can be imagined (and in a way that is eventually equipped to frame our understanding of molecular animation societies), then it does not operate under the constraints of a critique of "power as such," since "a nature in which nothing succeeded in infecting anything else would not be a nature" (160). All "social achievements," Stengers continues, are the result of successful infections, including the social desire for a "powerless society" (160). Thus "social power" in Whitehead operates by means of infection, and this infectious power, like in Foucault, does not come from on high or from above, but from entities inciting the very conditions for sense and meaning in one another.[7] When Foucault spoke of power and government in 1982 in terms of the "way in which certain actions modify others" in order to "guid[e] the possibility of conduct and [put] in order the possible outcome," I would venture that he had something close to the dynamics of social infection in mind. And although Stengers raises the question of Foucauldian power relations in the context of Whiteheadian social dynamics, she does not pursue it any further and treats it more like a passing observation than a matter of concern that deserves her full attention. She writes that "what interests [Whitehead], . . . what will become requisite for his conceptual construction, is not the analysis of formations of power, but that of the link between power and adventure" (161).

It seems to me that it is on the relation between "power" and "adventure" that the differences between Foucault and Whitehead begin to take shape. But it is these differences that are essential for us to unpack in order to conceive of productive forms of contrast and complementarity in their thought. For although Foucault is quite close to Whitehead in his general characterization of infectious social power as well as his refusal to denounce "power" and sing the praises of human "emancipation" (161), his project is firmly rooted in a genealogical description of the emergence of power dynamics in the present. And as such, his work does less to imagine alternatives to present configurations of power and governance, although it has nevertheless cleared the space to think it. As Foucault writes, "Freedom may indeed appear as the condition for the exercise of power."[8] The real sticking point with Whitehead, I think, concerns the way in which intervention into infectious social power is staged. For the Whiteheadian "adventure" that Stengers alludes to above concerns the careful fabrication of other possible adventures for thinking and feeling a world that has fallen prey to certain modern social infections ("minds in a groove"). This is at the very heart of Whitehead's speculative project. Thus even though Foucault conceives of "struggle" and "agonism" as central to the power relationship, he does not spell out what this entails and leaves us to ponder how the "care of the self" might respond to neoliberal human capital (Foucault 1988). For this reason, Foucault has often been critiqued for his inability to elaborate in a clear and systematic fashion what resistance within power looks like (Foucault 1991), or even why one should resist in the first place (Fraser 1981). Although there are hints and allusions to the nature of resistance throughout Foucault's oeuvre (Pickett 1996), and for this reason the critiques of his relation to political struggle are surely oversold (Nealon 2008), such efforts are still a far cry from offering a program for political resistance that would satisfy the demands of those who look to Foucault for political answers. Whether this is a limitation in Foucault's thought (which I doubt) or readers who search in vain for an emancipatory strategy that leads them outside the variegated textures of power (which is more likely) is not what is at stake here.

What is stake is that Whitehead might turn out to have something important to say about how to counter the forces of social infection, and this could be used to amplify our own understanding of power dynamics generally. Even if Whitehead's work does not speak to "power" and "resistance" as such, the way in which contrast and divergence function as propositional lures immanent to social infection could end up being an important ingredient for thinking political resistance when it is complemented with Foucault's work on power and governance. In this way, Whitehead's thought not only becomes amenable to thinking about the dynamics of social power (which is what we require in order to address the dynamics circulating in contemporary media societies), it may also become an important partner in conceiving strategies for resistance to them. Here's our wager, then: if Foucauldian genealogies of power and governance help us to retool Whiteheadian social relations in order to diagnose the conditions under which molecular animation media societies emerge in the present, then Whiteheadian propositional lures help us to imagine alternative futures for media societies designed through Foucauldian power relations. We will have to wait until Part III and the Postscript to see whether this bet pays off, but suffice it to say that the theoretical ingredients for this proposition are sketched in the remainder of this chapter.

PROPOSITIONAL NONCONFORMITY

To make sense of this provocation, we need to understand how alternatives to social infection are generated in Whitehead. That is, where do they come from and how and when are they admitted into feeling? In other words, insofar as there is anything stable or enduring in the world, then social infection has taken place and succeeded. But this does not mean that the infection was inevitable or predetermined. For each entity "decides" how it will feel its world. But then how is this decision made in each case, and how is it possible to entertain alternative courses of action?

To conceive this, note that for Whitehead social infection is formally neutral, as it is in Foucault, and is neither "good" nor "bad" in itself (Stengers 2011b, 163). In fact, the content of the social relation is not in itself what structures a subject's aim. The propagation of a social form is beneficial when it helps to intensify the satisfaction of the concrescing subject. "The evocation of societies," Whitehead reminds us, "is purely subsidiary" (PR, 105). Rather, "the subjective aim" of each entity, "whereby there is origination of conceptual feeling, is at intensity of feeling" (27). And this intensity, as we noted in the previous chapter, is determined by the contrasts that a concrescing entity is able to hold together in its final satisfaction. Intensive contrasts are conceived as positive relations among heterogeneities that are mutually reinforcing and enhancing instead of contradictory and inhibiting. As Whitehead explains, each occasion aims at "balanced complexity," where "'complexity' means the realization of contrasts, of contrasts of contrasts, and so on; and 'balance' means the absence of attenuations due to the elimination of contrasts which some elements in the pattern would introduce and other elements inhibit" (278). Balance, then, is "the adjustment of identities and diversities for the introduction of contrast with the avoidance of inhibitions by incompatibilities" (278).

The important point is that intensity of feeling, procured by the realization of balanced contrast, is what every occasion aims at. Intensity "lures" each entity to evaluate the possibilities (eternal objects) for prehending its wealth of datum into a definite unity of feeling. "The 'lure for feeling,'" Whitehead explains, "is the final cause guiding the concrescence of feelings. By this concrescence the multifold datum of the primary phase is gathered into the unity of the final satisfaction of feeling" (185). Eternal objects must therefore be valued up or down according to their relevance for the attainment of depth of satisfaction. According to Whitehead, "The valuation values up, or down, so as to determine the intensive importance accorded to the eternal object by the subjective form of the integral feeling" (241). And because intensive contrast is what is sought in each and every experience, "there is the urge," Whitehead continues, "towards the realization of the maximum number of eternal objects subject to the restraint that they must be under conditions of contrast" (278). This means that "no realized eternal object shall eliminate potential contrasts between other realized eternal objects. Such eliminations attenuate the intensities of feeling derivable from the ingressions of the various elements of the pattern" (278). Each successful infection is therefore accompanied by a wealth of other possibilities, and as such is a kind of achievement. It was selected from other potentials for definiteness. Each potential has to be positively or negatively prehended according to its deemed relevance to the final satisfaction. Entities must therefore decide how they will ingress each potential into their conceptual feeling of the world. And this freedom to decide is what makes each occasion a concrete instance of self-creation. "The *relevance* of an eternal object in its role of lure," Whitehead explains, "is a fact inherent in the data. . . . But the admission into, or rejection from, reality of conceptual feeling is the originative decision of the actual occasion. In this sense an actual occasion is *causa sui*" (86).

One conclusion to draw from this is that things could have been otherwise. The occasion decides on certain forms of definiteness as opposed to others, and because of this must weigh the relevance of potentials for feeling. In doing so, the entity excludes certain adventures of meaning and significance. But even the act of exclusion does not mean that the potential simply vanishes or evaporates: if "a negative prehension is the definite exclusion of that item from positive contribution to the subject's own real internal constitution," then "this doctrine involves the position that a negative prehension expresses a bond" such that "the negative prehension of an entity is a positive fact with its emotional subjective form . . . however trivial and faint" (41–42). The negation of a potential nevertheless makes a positive contribution to feeling insofar as it is a potential that has been excluded. This expresses the fact that "an actual entity has a perfectly definite bond with each item in the universe" (41).

In an especially evocative passage, Whitehead argues that excluded potentials occupy a "penumbra" of eternal objects that surrounds our actual world (185). The significance of this penumbral region is captured in his example of the Battle of Waterloo, which I quote in full:

> For example, consider the Battle of Waterloo. This battle resulted in the defeat of Napoleon, and in a constitution of our actual world grounded upon that defeat. But the abstract notions, expressing the possibilities of another course of history which would have followed upon his victory, are relevant to the facts which actually happened. We may not think it of practical importance that imaginative historians should dwell upon such hypothetical alternatives. But we confess their relevance in thinking about them at all, even to the extent of dismissing them. But some imaginative writers do

> not dismiss such ideas. Thus, in our actual world of today, there is a penumbra of eternal objects, constituted by relevance to the Battle of Waterloo. Some people do admit elements from this penumbral complex into effective feeling, and others wholly exclude them. Some are conscious of this internal decision of admission or rejection; for others the ideas float into their minds as day-dreams without consciousness of deliberate decision; for others, their emotional tone, of gratification or regret, of friendliness or hatred, is obscurely influenced by this penumbra of alternatives, without any conscious analysis of its content. (185)

In this passage, Whitehead articulates how potentials that were rejected or only vaguely relevant to events that unfolded at the Battle of Waterloo can nevertheless become important elements in the entertainment of those events in the present. These potentials exist as other possibilities for the world, possibilities in the "penumbra" that can enter into our imagination, instead of being dismissed and negatively prehended. These possibilities are prehended as worlds that might have been. They add richness and texture to the world and enliven it with contrasts that could enhance the depth of satisfaction. His invocation of "imaginative writers" is especially illustrative and calls to mind those for whom the world is not a place of settled affairs and exhausted potentials. This penumbra of potentiality is not simply relevant to critical historians and writers of fiction who speculate about what could have been, but it is immanent to every instance of actualization.

Deleuze's discussion of "possible worlds" in *The Logic of Sense* has interesting resonances with Whitehead's example. Embedded within his critique of the possible as an "a priori Other" in Michel Tournier's retelling of Robinson Crusoe in his novel *Friday, or the Limbo of the Pacific* (which essentially rehearses the critique of the possible that we find in his *Bergsonism* and *Difference and Repetition*), Deleuze alludes to another articulation of the possible (Deleuze 1990, 301–21; Bogue 2007). He gives the example of encountering a frightened face before grasping the object of the other person's fright. In this instance, "a frightened countenance is the expression of a frightening possible world, or of something frightening in the world." But "let it be understood," Deleuze then writes, "that the possible is not here an abstract category designating something which does not exist: the expressed possible world certainly exists, but it does not exist (actually) outside of that which expresses it" (1990, 307). The possible world is a real world enfolded within the actual world that expresses this other world as a possibility for it. A similar structure underlies Deleuze's writings on Proust and is also alluded to in his discussion of *Friday.* Immediately following his meditation on fright, Deleuze writes that "Proust says of the perceived Albertine that she encompasses or expresses the beach and the breaking of the waves. . . . In short, the Other, as structure, is *the expression of a possible world*; it is the expressed, grasped as not yet existing outside of that which expresses it" (308).

Deleuze's discussion of possible worlds is of course illuminating in the context of his own thought, especially as it expounds a logic of the possible that is not purely negative. As Ronald Bogue notes, a different genealogy of the possible is opened up in Deleuze and Deleuze and Guattari's thought, one that helps express the meaning of their difficult and largely unqualified assertion in *What Is Philosophy?* that art "does not actualize the virtual event but incorporates or embodies it: it gives it a body, a life, a universe. . . . These universes," are neither virtual nor actual; they are possibles, the possible as aesthetic category, . . . the existence of the possible" (1994, 177). Art, in this context, is given an intermediary status in Deleuze and Guattari's philosophy: it exists somewhere between the virtual and actual insofar as it gives a universe or a body to the virtual potentiality that insists within every actualization, without that world having to be actual. Virtuality gains a material body in art in the form of a possible instead of an actual world. But I would also add here that this articulation of the possible in Deleuze and Deleuze and Guattari illuminates aspects of the penumbra of potentiality that surrounds each instance of actualization in Whitehead. "To admit" these alternative possibilities "into effective feeling" means, on this reading, to open up alternative adventures of significance for entities in the actual world. It is to fabricate a possible world in Deleuze's sense, but one that does "not yet [exist] outside of that which expresses it." Thus the possible world only exists as enveloped within the actuality in question, and not outside or beyond it. It is not pure potentiality, then, but a possibility for expression of the actual world that is not, in fact, actual.

Whitehead explains that this penumbra of possibility for the Battle of Waterloo is felt as a "propositional prehension." Propositions are neither actual occasions nor eternal objects, but they are "a hybrid between pure potentialities and actualities" (186). For this reason, propositional prehensions "are not pure conceptual prehensions; for their implication of the particular nexus with the Battle of Waterloo is an essential factor" (185). Thus, a proposition, Whitehead continues, "is the potentiality of the actual world including a definite set of actual entities in a nexus of reactions involving the hypothetical ingression of a definite set of eternal objects" (186). Working against the bivalent logic of the proposition that he developed with Bertrand Russell in *The Principia Mathematica,* Whitehead contends that the proposition is neither true nor false in itself. Rather, a proposition concerns actual occasions, or logical subjects, put into a potential relation with eternal objects, or predicative patterns (186). A proposition is the "could be" of actual subjects in a pattern of relation. However, and crucially, the proposition's "own truth, or its own falsity, is no business of a proposition. That question concerns only a subject entertaining a propositional feeling with that proposition for its datum" (258). In other words, truth and falsity concerns whether the predictive pattern does or does not "conform" to the actual world of the prehending subject.

It is important to note that while both propositions and eternal objects have "potentiality" in common, eternal objects are "pure potentials," which means that they do not reference any actuality or epoch; and propositions are "impure" since they always involve a particular set of subjects (actual occasions), in a particular epoch, that are put into a potential relation with a predicative pattern (eternal objects). As Whitehead notes, "Every proposition presupposes those actual entities which are its logical subjects" (188), and then elsewhere insists that "every proposition involves its logical subjects . . . no actual entity can feel a proposition, if its actual world does not include the logical subjects of that proposition" (259). And in those cases when the logical subjects are not in the actual world of the prehending entity, then the "proposition does not exist for that actual entity," and so "the proposition itself awaits its logical subjects" (188).

What Whitehead is suggesting here, which resonates with Deleuze and Deleuze and Guattari's conception of the possible, is that the proposition is what brings potentialities of definiteness together with datum in the world of the feeling subject. Propositions generate hypothetical forms of significance (in the form of "could be") within the actual world of the feeler. For without reference to the world of the feeling subject, there is no proposition. The propositional form requires the world of the feeler just as much as it requires the predicative patterns that it places that world into. Analogous to the worlds enfolded in the "perceived Albertine" or in the frightened face, the proposition constructs a possible world of relevance for occasions of experience. *It could be that there is no object of fright at all and that the frightened face is a trick designed to deceive me.*

That there may be different meanings folded into the actual face is what the proposition offers a prehending subject for entertainment. To admit this proposition into experience is what Whitehead has in mind when he insists that "the primary mode of realization of a proposition in an actual entity is not by judgment, but by entertainment. A proposition is entertained when it is admitted into feeling. Horror, relief, purpose, are primarily feelings involving the entertainment of propositions" (188). But qua entertained, the proposition can also enter into experience in the "mode of derision." "The negation, as well as the affirmation," writes Stengers, "impl[ies] the entertainment or realization of a proposition" as an "ingredient of the entity's satisfaction" (2011b, 413). But such derision is still a form of entertainment and does not yet concern the conformity or nonconformity (truth or falsity) of the proposition to the actual world (PR, 258).

If the proposition admitted into feeling is conformal (or true), then "the reaction to the datum has simply resulted in the conformation of feeling to fact" (187). However, if the proposition entertained is nonconformal, then "the reaction to the datum has resulted in the synthesis of fact with the alternative potentiality of the complex predicate. A novelty has emerged into creation. . . . That member of the locus has introduced a new form into the actual world; or, at least, an old form in a new function" (187). With this bold assertion, Whitehead undermines the especially strong desire among logicians to conceive of "propositions as merely material for judgments. . . . In that purely logical aspect, non-conformal propositions are merely wrong, and therefore worse than useless" (187). Instead, and against the prevailing wisdom of these "over intellectualized philosophers," Whitehead maintains that false propositions "pave the way along which the world advances into novelty. Error is the price which we pay for progress" (187). This is what motivates his well-known assertion that "It is more important that a proposition be interesting than that it be true" (259).

What's at stake here is the admittance into feeling of a form of definiteness not already exemplified in the actual world of the prehending subject. Such nonconformal patterns of relation become ingredients in the occasion's feeling—and possibly add depth of contrast to its satisfaction—instead of being rejected as incompatible potentials (and thus negatively prehended). It is the entertainment of such nonconformal propositions that Whitehead has in mind when he speaks of those imaginative writers who entertain other courses of history. In these instances, alternatives are brought to bear on the actual world and introduce new possibilities for subjects feeling it. Analogously, the frightened face may not turn out to be a trick, but even entertaining the possibility that it is one adds new depth to the experience of the subject who encounters it.[9]

Ultimately, nonconformity may serve to reinforce or undermine dominant modes of social infection. Whitehead emphasizes just this point when he notes that "the novelty [introduced by a nonconformal proposition] may promote or destroy order; it may be good or bad. But it is new, a new type of individual and not merely a new intensity of individual feeling" (187). What this means is that insofar as "error" is what accounts for the path "along which the world advances into novelty," as well as "mental progress" and "imaginative freedom" (S, 19),[10] then this progress and freedom are indifferent to the endurance of social orders—human or otherwise. In some instances, error or nonconformity may be an ingredient that promotes social togetherness. As Whitehead quips in *Symbolism:* "Aesop's dog lost his meat, but he gained a step on the road towards a free imagination" (19). But in other instances, error may serve to erode social bonds. Mistaking my partner's frightened face for a trick may demonstrate mistrust and serve to weaken the relationship. Propositional nonconformity generates novelty, but it cares nothing for the endurance of social togetherness.

But then how does an entity decide that one potentiality is more relevant to the achievement of balanced contrast than another? Or in other words, how can an occasion anticipate which patterns of definiteness (whether conformal or nonconformal) are more relevant than others for achieving maximum intensity of satisfaction when there can be no perspective outside of actuality from which to judge? These questions cut to the heart of our concerns regarding social infection and power. For if Whitehead has managed to create a space for propositional feelings in his thought, and in such a way that a prehending subject can entertain possibilities for the actual world that may be nonconformal with it, then we need to account for the conditions under which these propositional feelings are deemed relevant for a concrescing subject, instead of dismissed and thus negatively prehended by it. It is precisely because subjects can admit patterns of definiteness that do not conform to the actual world, but are nevertheless valued as important ingredients to the achievement of depth of satisfaction, that Whitehead might offer insight into how to resist insidious forms of social infection.

ANTISOCIAL LURE

It is at this point that Whitehead's thought becomes difficult to follow for some readers. And by "difficult," I mean that it is God who intervenes to establish the relevance of all eternal objects to an occasion's concrescence. As Whitehead explains, "He is the actual entity in virtue of which the *entire* multiplicity of eternal objects obtains its graded relevance to each stage of concrescence. Apart from God, there could be no relevant novelty" (PR, 164). The aim at intensive contrast is derived from God's primordial envisagement of the relevance of eternal objects to that achievement. "The initial stage of its aim," Whitehead continues, "is an endowment which the subject inherits from the inevitable ordering of things, conceptually realized in the nature of God" (244). Hence, God is "the lure for feeling, the eternal urge of desire" that "[establishes] the initial phase of each subjective aim" (344), but "its completion, depends on the self-causation of the subject-superject" (244). So although God's primordial valuations are what establish an "order in the relevance of eternal objects to the process of creation" (345), each occasion of experience is nonetheless causa sui, the "autonomous master of its own concrescence into subject-superject" (245). The occasion therefore "derives from God its basic conceptual aim, relevant to its actual world, yet with indeterminations awaiting its own decisions" (224).

This characterization of the ordering of potentiality is not easy to stomach for some commentators. In *Religion in the Making* (1926), for example, God is conceived of as a "principle," an "ideal entity," and even a "formative element" (Stengers 2011b, 265; RM), and is always referred to (in SMW, RM, and PR) using the third person singular pronoun, "He." Such a Judeo-Christian conception of God can strike some readers as counterproductive and, in the end, does little more than hinder the otherwise adventurous spirit of Whitehead's thought. Even though the God concept is far more developed by the time Whitehead delivers his Gifford Lectures in 1927 and 1928—which are, and not incidentally, delivered to a very different audience[11]—there are still certain difficulties that need to be dealt with.

So, for instance, God's primordial nature, which features prominently in *Religion in the Making,* is then complemented, in *Process and Reality,* by his consequent nature. The consequent nature of God "shares with every new creation its actual world; and the concrescent creature is objectified in God as a novel element in God's objectification of the actual world" (PR, 345). In this aspect, God is "consequent upon the creative advance of the world" (345). This means that by the time of *Process and Reality,* God has become a full-fledged actual entity, endowed with conceptual (primordial) and physical (consequent) poles (and not defying the ontological principle). Hence, God is also a "creature of creativity" (88), who, like every other entity, "is something individual for its own sake; and thereby transcends the rest of actuality" and is simultaneously a "creature transcended by the creativity which it qualifies" (PR, 88). But here is the issue: If Whitehead reserves a nontheistic role for God—indeed, one that even turns God into a "creature"—then why continue to call this function "God," and why insist on granting this function the unfortunate privileges that are associated with patriarchy? These questions seem all the more pressing and difficult to answer when Whitehead himself never stopped denouncing the problems associated with religious thought in the West (RM; Stengers 2011b, 451).

There are no good answers to these questions, although numerous commentators have tried to address them and have had varying degrees of success.[12] One might also add that by the time of *Adventures of Ideas* and *Modes of Thought,* God drops out of Whitehead's conceptual vocabulary, which may serve to illustrate the fact that God was simply a sketch that was never completed.[13] Stengers, for her part, seems less interested than the others in demonstrating how God is incomplete, or how God is more tenable when it is put in the context of concepts borrowed from more secular thinkers. She doesn't even attempt to downplay God's significance in order to showcase other aspects of Whitehead's metaphysics. In fact, Stengers offers one of the most generous readings of Whitehead's God without eliding the many challenges that the notion presents.

Without rehearsing the complexities of Stengers's account of God, it's important to note that if God's consequent nature is associated with the "tender care that nothing be lost," and the attendant idea that God "saves the world as it passes into the immediacy of his own life" (346), then we must also take care to appreciate how God's "sharing" in an occasion's actual world (God "shares with every new creation its actual world" [345]) does not mean merging or coinciding with its *private* enjoyment. Rather, God feels the world according his own "circumstances" ("divine pragmatics" [Stengers]), which is what permits him to "[lose] nothing that can be saved" (346). If the world is "saved" according to these circumstances, then this means saving the world in a way that is wholly "impartial"—which is to say, in a way that saves the contradictions, discords, rejections, vagaries, and so on from being lost. Stengers writes that "divine feeling . . . 'saves' individual accomplishment in such a way that its value includes what has been rejected, eliminated, or vaguely felt, so the accomplishment may be what it is" (2011b, 487). The significance of this aspect of God's nature cannot be overemphasized. All those insignificant or even wholly rejected patterns of feeling, which nevertheless constitute the subject-superject's determinate bond with the universe, are not forgotten, lost, or relegated to the dustbin of irrelevancy. The Battle of Waterloo, for example, is not "saved" simply for what was deemed relevant to the events that unfolded there, but also for what was deemed vaguely significant, and for what was seen to be wholly irrelevant and rejected outright. This penumbra of potentiality is saved impartially and allows for alternative ways of feeling the event and the imaginative freedom to envision other courses of history.

The second point I want to make about Stengers's reading concerns God's primordial nature. It is important to realize that insofar as God "is the actual entity in virtue of which the *entire* multiplicity of eternal objects obtains its graded relevance to each stage of concrescence," then his envisagement is not to be confused with a tender care for human satisfaction (PR, 164). This is a point that Stengers drives home in her reading of primordial envisagement: God is not interested in what concerns humans most—namely, their preservation (2011b, 317). As Whitehead insists, "The primordial appetitions which jointly constitute God's purpose are seeking intensity, and not preservation. Because they are primordial, there is nothing to preserve. He, in his primordial nature, is unmoved by his love for this particular, or that particular. . . . In the foundations of his being, God is indifferent alike to preservation and to novelty. He cares not whether an immediate occasion be old or new. . . . His aim is for depth of satisfaction" (PR, 105). Whitehead unmoors the concept of God from all theological conceptions and "his" concern for ensuring the preservation of human social orders. God is not a social being; God is an "it," a function, and it is fundamentally indifferent to social formations. The latter's preservation or destruction matters little to its primordial envisagement. Depth of satisfaction is all that God weighs. As such, this is not a God to whom humans should direct their praises; this is a "posthuman God" who is indifferent to the damage he may inflict. "What his tenderness is directed to," writes Stengers, "is deeply indifferent to us" (2011b, 317). "God's purpose in the creative advance is the evocation of intensity. The evocation of societies is purely subsidiary to this absolute end" (PR, 105). On the one hand, God saves the world "impartially," inasmuch as it saves without regard for what humans or any other social entity thinks is worth saving and forgetting. And then on the other hand, God lures the world into depth of satisfaction "indifferently." In sum: this is an impartial and indifferent God. The primordial and consequent natures of the divine ensure that the cosmos preserves and lures itself in a way that is indifferent to human affairs.

But notice, if the God function is indifferent to social order, then this does not mean that it aims to destroy or dismantle it tout court. It means that it cares for no social order in particular, and only for those modes of togetherness that promote maximum intensity of contrast. For this reason, imaginative writers who generate alternative narratives for the Battle of Waterloo inject contrast into their stories that traditional historians do not. It might be bad for one's career as a historian (one social order), but it might also promote imaginative and aesthetic interest (in another social order). What's important is that social infection is never aimed at *for itself,* but only insofar as it heightens and intensifies the possibility for alternatives to be put into mutually compatible relations. Certain forms of infection, like those exemplified in a painting or a novel, may exhibit tremendous depth of contrast. And thus, when it comes to the forms of social infection propagated in today's systems of political and economic power, they are not problematic in themselves. Rather, they are only a problem to the extent that they fail to promote intense contrast of feeling. In fact, the propagation of social infections associated with advanced capitalism testifies to the fact that each experience is causa sui: an occasion is able to negatively prehend modes of feeling a situation that would promote intensity of satisfaction.

In Whitehead, the divine is the function that ensures that alternative possibilities for prehending the world are immanent to every situation. The real challenge with such a conception is imagining what those alternatives would be in the context of contemporary forms of social power and the scientific media societies that this book investigates. If Whitehead shores up the idea that alternatives "lurk" in every instance of social infection (in the form of propositional feelings), then what we require is a speculative framework (a "culture of interstices" [Stengers]) that would allow these alternatives to enter into the subject's constitution in the mode of entertainment (qua propositional feeling) instead of being eliminated and negatively prehended outright by an a priori form of relevance (Stengers 2011b, 421). But of course there is no universal valuation that could be made ("secular human values," etc.) to develop this interstitial framework. Intensive contrast in feeling cares nothing for the preservation of this or that item of experience. There is no predetermined ideal (or something to hold in common) that would be capable of prescribing an "escape." On the contrary, the entertainment (as opposed to the rejection) of contrast is unique to each situation and might very well undermine every assumption humans have about resistance. I take Whitehead at his word when he says that the primordial envisagement of depth of satisfaction "is unmoved by love for this particular or that particular . . . is indifferent alike to preservation and to novelty. [It] cares not whether an immediate occasion be old or new" (PR, 105).

Let's use this description to return to molecular animation societies. These are societies organized such that all elements positively incorporate a specific visual logic. However, the positive valuation of this logic is not inevitable. If Whitehead is right, then there is a penumbra of potentialities for feeling the other elements in the society; these possible feelings may have been propositionally entertained, but were ultimately deemed nonconformal to the actual world of social infection and thus relegated to irrelevancy (negatively prehended). The question becomes, What might those nonconformal visual logics be, and under what conditions could they be positively admitted into feeling, and then what would these false propositions change in terms of the intensity of experience? If the regime of visuality valued across the diverse practices of molecular animation is hypercinematic, then an animetic logic of the image would offer us a countertendency (per chapter 1). These are not dialectically opposed logics (since they very often cohere in the same moving image), though they do present qualitatively different ways of organizing the visual. If mastery, decentralization, and flexibility reign supreme in hypercinematic viewing, then hesitation, limitation, and even discord have primacy in the other tendency (LaMarre 2009, chapter 11).[14] Whether this tendency or logic already exists within molecular animation societies is another question altogether. It should make us wonder whether the forms of restraint and minimalism exemplified by the short animation *Molecular Motor Struts Like a Drunken Sailor* already present an opening to nonconformal feelings that would promote greater intensity of contrast in molecular animation societies.

Even if there exist alternative possibilities for feeling the relation between and among elements in molecular animation societies, then there is as of yet no reason to think that it is desirable to admit these propositional feelings into experience. Foucault insists that power and governance are everywhere, and as such, there is no getting outside of infectious social power. Feeling other potentials for organizing the visual would only bring about other forms of infection. All that we have shown is that there is a particular form of power that insists within animation societies, and that it steers or governs the distribution of the visual in a certain way. But we do not know whether and how this infectious form aligns with other regimes of power. In other words, the form of social power under consideration has yet to be theorized in terms of the historical production of visual governance, and so we cannot weigh in on whether other feelings for the visual would produce more depth of satisfaction. Perhaps the most intense satisfactions would come about from maintaining the reproduction of the current visual form (after all, it seems to draw together diverse audiences, histories, and epistemic practices), or maybe it would come about through admitting conceptual feelings that lead to the decay of molecular animation societies.

COSMIC COUNTER-CONDUCT: TOWARD AN AESTHETICS OF CONTRAST

Whitehead's conception of the antisocial lure is an important resource for thinking resistance to social power dynamics. Admittedly, this only comes into focus once we have redirected Whitehead's own thinking. In this context, Foucault's general description of power and government seem to have surprising resonances with Whitehead's characterization of social order. If power functions in such a way that "certain actions modify others" in order to "guid[e] the possibility of conduct and [put] in order the possible outcome," then this general description parallels the way in which Whitehead speaks of the "imposition" of social form as a mode of "infection." It then became possible to conceive of Whitehead's conception of social infection as a theory of power relations, in the broadest possible sense.

If there is something worth retaining about Whitehead's God in the context of social power relations, then it is this: it elaborates a conception of immanent resistance that does not resort to notions of human emancipation. Whitehead's God function anchors the idea that every instance of social infection in the world is accompanied by a penumbra of other potentials for definiteness. If Foucault insists that agonism and counter-conduct are immanent to the power relationship but is unable to articulate what exactly resistance to twentieth-century forms of power means, then I think Whitehead pushes this conversation in new directions. But Foucault's inability does not necessarily have to do with a limitation in his own thinking, but has more to do with the nature of the questions and concerns that animate his thought.

This brings me to a challenge that confronts any attempt to bring Whitehead and Foucault into conversation: they are dealing with different problems and with very different scales of experience. Whitehead develops a speculative experiment that resists the temptation to privilege or exclude any dimension of experience, from the experiences of the dust particle to the experiences of the solar system. His architecture of speculative notions includes abstractions like actual occasions, eternal objects, prehensions, and God. These notions do not index the world as it is but are supposed to lure us to think differently about it. Foucault, by contrast, uncovers relations of power and governance within the purview of human history. Although there is important scholarship on the profound effects that power relations have on nonhumans—animals and plants, for instance (Wolfe 2012; Nealon 2015)—these dynamics of social infection have taken shape in the history of *anthropos.* We should take Foucault at his word when he characterizes his genealogy of power in terms of the history of the "government of men by other men" (1982, 221). However, I also think it would be a mistake to view Foucault's analytics of power as anthropocentric, or at least in the way that it has become popular to critique his work in the intellectual climate of speculative realisms and materialisms.[15] Foucault challenges us to conceive the complex permutations of social infection within the history of *anthropos,* and moreover, how *anthropos* has come to mean different things in the context of these various permutations. But this is precisely Foucault's strength from a Whiteheadian perspective: he unveils historical routes of infection without overextending their domains of relevance.[16] Even Whitehead, who has often been celebrated as one of the preeminent thinkers of nonhuman experience (Debaise 2017; Shaviro 2014), cannot separate his speculative concepts from the very specific problems that they are supposed to address (Stengers 2011b). Whitehead's "nonhuman speculations" are also situated. The problem that Whitehead addresses is more general than the one Foucault attends to, and this precipitates a different mode of response. Whitehead proposes to think about the infectious dynamics between actual occasions from a human perspective, and these dynamics are supposed to be exemplified in all experience, no matter how insignificant. Foucault, on the other hand, makes no such general claims, because the problems he addresses do not oblige him to generalize in the way that Whitehead's do.

The challenge is not to see Foucault's specificity as opposed to Whitehead's generality. Whitehead is adamant in *Process and Reality* that the success of a general scheme of notions is dependent upon whether it can be empirically verified (PR, 5). This means that the system must be exemplified in every practice, in every and all domains of experience. Short of this, the general scheme has to be revised since it excludes, disqualifies, or reduces some element of experience and has managed to bifurcate nature once again. In this spirit, we might think of Foucault's work on power and government as a test for Whiteheadian description, one that asks whether general description is actually exemplified in human social and political affairs. On this reading, what underlies the differences (of problem and method) in Whitehead and Foucault is a shared commitment to understanding the operations of social infection. If Whitehead proposes a generic description of power, then Foucault takes up the task of tracing specific routes of power dynamics within human societies.

But in claiming that a meaningful confluence exists amid the many divergences, I am also proposing that there is an aesthetic relation between Foucault and Whitehead. Divergence does not have to mean incompatibility. This is the great lesson of Whitehead's work on aesthetics. An aesthetic feeling, he reminds us in *Adventure of Ideas,* consists of the interweaving of "patterned contrasts" (AI, 252), so that "all aesthetic experience," he discusses in *Process and Reality,* "is feeling arising out of the realization of contrast under identity" (PR, 280). Thus, "an intense experience," Whitehead explains just above, "is an aesthetic fact" (279). Steven Shaviro (2014) has written at length about aesthetics in Whitehead and, in particular, on the importance of beauty in his metaphysics. He explains that aesthetic contrast operates below the threshold of cognitive recognition and emerges in a feeling of resonance that cannot be subsumed under pregiven conceptual frameworks. This very Kantian understanding of the aesthetic (in the mode of beauty rather than the sublime) guarantees that there is the potential to move and be moved by differences that are irreducible to conscious awareness or formal systems of knowledge and recognition. The aesthetic shores up for Whitehead, on Shaviro's reading, a way to push beyond the cognition of irresolvable differences (presentational immediacy) toward a theory of contrasts that are intensifying and mutually reinforcing when they lie below the threshold of cognition (causal efficacy).

In this spirit, we might think of Whitehead and Foucault as offering two very different philosophical propositions, one oriented toward generality and the other toward specificity. The challenge is not to read these differences as oppositional and only overcome through some Hegelian *Aufhebung.* Whitehead urges us to perform a "shift in meaning which converts the opposition into a contrast" (PR, 348). To read Whitehead and Foucault aesthetically, as I am proposing here, means that we have forged enough "narrowness" and "width" that there is neither "irrelevance" nor "vagueness" in their relation (112). The differences become productive differences: through mutual enhancement, we are able to say and feel something that we were not able to before.[17]

By way of conclusion, let me try to expound how the contrasts between Foucault and Whitehead are woven together into a mutually enhancing aesthetic achievement. By drawing on Foucault's late work, it becomes possible to frame Whitehead's description of the reproduction of social environments through a defining characteristic in terms of a (nascent) theory of power. This becomes even more tenable when we see Whitehead's conception of "imposition" through the lens of "infectious social dynamics." The challenge here is not to see Whitehead as needing supplementation or even correction. Rather, and more aesthetically satisfying, Foucault helps to intensify elements in Whitehead's work so that it is possible to see them through the frame of power and governance. The enhancement of theoretical frameworks works in the other direction as well. Of course, Whitehead, much like Foucault, does not offer a blueprint for what immanent resistance looks like. But what Whitehead provides is a metaphysical explanation for why there can be no blueprint of such content. As we have seen, we cannot predetermine this content because the intensive contrasts that would offset infection are indifferent to the preservation of "particulars" (PR, 105). The endurance of human social orders, along with all other social orders, is incidental to the aim of maximizing the intensity of contrast. If counter-conduct were to be conceived in these terms, then it would mean that all enduring social orders have pragmatic value: one mode of social togetherness may enhance contrast on certain occasions, while it may detract from it on others.

If Foucault is able to articulate the historical routes of social infection that have led to the determination of human societies in the present (which is the task of genealogy and will be discussed in the next chapter), then conceiving alternatives to this determination cannot be secured by upholding the inherent value of the "commons," the "human," or any other form of social infection. Counter-conduct will have to be conceived in terms that are indifferent to such modalities of togetherness.

Concretely, this means cultivating the conditions under which alternatives can be imagined without having them dismissed, reduced, or explained away because they do not uphold what is already taken to be the most valuable form of social belonging. In her work on politics and activism, Stengers has shown how such alternative propositions are easily dismissed in the climate of contemporary capitalisms because they risk undermining the legitimacy of those forms of value that proclaim to be capable of answering on behalf of all others (Stengers 2015; Pignarre and Stengers 2011).[18] For Whitehead, as for Stengers, antisocial counter-conduct undermines the assumption that there exist social orders inherently more valuable than others and that would allow us to know in advance how to answer on behalf of what a situation requires (Stengers 2015, 55). There is no particular "cosmos" capable of serving as an outside that would guarantee emancipation from the suffering wrought by pervasive forms of social infection. The "cosmos" capable of securing contrast has to be forged in each instance.[19] It is unclear at this point whether we would want to call this a "posthuman" resistance to political power. What is clearer is that Whitehead and Stengers bring out a cosmic dimension to Foucauldian power dynamics that has yet to be imagined. Perhaps the task that lies before us is to enhance Foucault's conception of resistance through a cosmopolitical reframing. However, this book cannot speculate about what such cosmic counter-conduct would look like until the Postscript. At this stage, it is merely intended as a lure for feeling.

Using this conceptual reframing, what we need to get clear on in Part III are the historical conditions that produced modalities of infection circulating within molecular animation societies. The wager is that with this (aesthetic) reading of Whitehead and Foucault in hand, we are in a much better position to see how the social infection operative in this media ensemble is inextricably related to the emergence of neoliberalized modes of power and governance. In particular, we examine how the convergence of entertainment and scientific visual cultures in the positive valuation of a hypercinematic logic in molecular animation is the result of government strategically designing this alliance since the nineteenth century. It will remain for the final section of this book to imagine those propositional feelings capable of introducing needed contrast into the forms of power gripping molecular animation practices today.

PART III

> The true method of discovery . . . lands for renewed observation rendered acute by rational interpretation.
>
> —Alfred North Whitehead, *Process and Reality*

6

THE ANIMATION APPARATUS

In chapter 4 we opened our highly focused genealogies of molecular animation to a speculative philosophy capable of accommodating the wider systems of material, semiotic, political, and epistemological force that converge in computer animation practices. After an extended consideration of Whitehead's work, and with a particular eye to his social theory, we began chapter 5 in a much better position to understand how radically divergent processes are drawn together and cohere over time in a media ecosystem. With the outlines of a speculative media philosophy in hand, we could explain the convergence of visual knowledge practices in the life sciences and the visual logics of popular screen media in molecular animation. It became metaphysically conceivable that a "common element of form," manifesting as a specific pattern or logic of visuality, could bind these cultures together for a finite period of time. Whitehead provides an eco-social framework for understanding the emergence of such an element of form, which the focused histories in Part I could not. This was the first adjustment that had to be made to our conception of molecular animation.

The problem is that Whitehead does not give us a way to think concretely about how and why this element of form would take hold. Under what conditions is it generated, and why would it continue to reproduce itself? And what are its prospects? Does molecular animation promise to be the first of many instantiations of this form? Whitehead's work needs to be reframed in terms of a theory that can weigh in on the complexities of human social systems, and in particular how they take shape across myriad institutional, discursive, political, and economic contexts throughout history. In other words, there is no obvious way to think about social and political power as well as more concrete instances of "capture" in Whitehead. Foucault's work stood out at the outset of chapter 5 as up to this task, or in any case, equipped to address the problems of power and governance, but it did not, at least initially, seem to lend itself to the Whiteheadian interpretive framework we set up for ourselves. With the exception of superficial similarities, Whitehead and Foucault appear to be on separate intellectual continents. Despite these vast distances, I exposed ways in which their work could communicate in valuable ways for our project.

A second theoretical adjustment therefore had to be made: Whitehead and Foucault needed to be brought into closer proximity. I argue that this convergence is not designed to reduce one framework to the other, but it is forged to enhance the one through its encounter with the other. In the course of chapter 5, I unearthed mutually reinforcing contrasts between Whitehead and Foucault. These are differences that are enriching and ultimately aesthetic instead of contradictory and irreconcilable. What is at stake in Part III is determining whether and to what extent our enriched conceptual frame can transform our understanding of what's going on in molecular animation societies. For Whitehead, the efficacy of a conceptual framework is determined by what it makes matter, by what it makes legible and meaningful. In this spirit, we need to examine whether the "patterned contrasts" that Whitehead and Foucault generate catalyze a fresh encounter with the complex histories and practices of molecular animation.

To determine this, Part III demonstrates how both Foucault and Whitehead sharpen our perception of the convergence between the social environments of popular screen culture and visual epistemology in the nineteenth, twentieth, and twenty-first centuries. Where Whitehead helps us to grasp the coincidence of visual cultures in eco-social terms, Foucault enhances our conception by showing how such a convergence cannot be thought outside of the unfolding dynamics of power and governance in the history of the West. What this means is that the intersection of popular and scientific visual societies in molecular animation has to be framed in terms of shifting strategies of governmentality. My hypothesis in Part III is that there are similar modes of governance working across both visual societies in the nineteenth and twentieth centuries, and this is what makes their specific convergence in molecular animation possible in the twenty-first century. In what follows, I demonstrate not only how certain techniques of government seized control of visual perception in and through popular screen media (chapter 6) but also how visual perception in the life sciences has come under that same form of governmental power (chapter 7). The suggestion is that these visual cultures are designed by the same regime of governmental reason, and it is this regime, which spans bioscientific and popular media cultures, that makes molecular animation possible. As will become clear, Part III functions as a companion to Part I, but in such a way that Part III deepens our conception of popular screen media and visual epistemology (Part I) by means of the enriched theoretical scaffolding we erected in Part II.

This argument is made apparent in two separate but interrelated conceptual moves that will occupy us for the remainder of the book. In this chapter, I trace a conceptual genealogy through nineteenth, twentieth, and twenty-first century media history and theory that exposes how popular image technologies, from the stereoscope to mobile screens, perpetuate specific forms of political and economic power. In other words, this chapter expresses the ways in which questions of power have been fundamental to the study of modern visual media from the beginning. Thus, if I start this chapter by recounting Jonathan Crary's analysis of abstraction and disembodied perception in pre-photographic media, then it is in an effort to chase an intellectual through-line that exposes the power and economic reason at work in the history of visual and screen media—from early nineteenth-century parlor tricks to online streaming in the early twenty-first century. And in the history of screen studies, it is the Marxists who have largely paved the way for thinking about the political and economic dimensions of popular screen culture. It is for this reason that I turn to Jonathan Beller's important work, *The Cinematic Mode of Production,* since it makes the economics of the image explicit in a way that few other works do (the cinema *is* capital in Beller's view). It also sharpens our perception of how moving images design the human sensorium that they need in order to increase their circulation (attention economics).

But where Beller and other Marxists give us useful tools for critiquing the interface between images and capital, there is an important sense in which their analyses are not well equipped to handle the subtleties involved in the design and control of perceptual economies in contemporary screen culture. While I retain aspects of this Marxian frame, I ultimately show how Foucault's genealogy of governmentality offers a more subtle and less reductive conception of the ways in which image media and perceptual subjectivities are folded into this world-historical moment of advanced capitalism. In other words, the history of governmentality illustrates how power uses image media to design the perceptual subjectivities it needs to propagate. It is in this context that it becomes possible to conceive of the "common element of form" (in Whitehead), which organizes both molecular animation and popular screen culture, in terms of a dispositif that is a part of a general design strategy of neoliberal governance. At this point, Whitehead's eco-sociology will prove to be especially important for us, in conjunction with a theory of apparatus design, as we attempt to unpack how screen-media dispositifs share complex interdependencies with other social assemblages to expand and intensify neoliberal rationality.

In sum: this chapter offers a preliminary sketch of how the techniques of government deployed in popular screen culture are also what make molecular animation practices possible. In this context, I argue that molecular animation becomes legible as a part of an arsenal of techniques and technologies leveraged in popular screen culture to generate new forms of economic productivity. I leave it to chapter 7 to draft how those very same techniques have captured various domains of visual epistemology in the life sciences and are what make it possible to conceptualize the specific coincidence of science and entertainment societies that occurs in molecular animation apparatuses.

SPECTATOR ECONOMIES

Much has been written on the intersections among moving images, the human sensorium, and twentieth- and twenty-first-century political economies. Walter Benjamin (2008) famously voices his concern over mechanical reproduction—photography and cinema—in the age of industrialized capitalism and shows how these dynamics shift what it means to be a perceiving subject in the twentieth century. Jonathan Crary goes back even further: he charts the emergence of an industrialized observer from out of optical technologies in the early nineteenth century. Although Crary's ambitious genealogy in *Techniques of the Observer* takes us back still further (well into the Renaissance), it is really the mutation in nineteenth-century optical technologies that are at stake for him. But what's especially striking about this work is that Crary's goal is to situate the abstract mobile observers produced through rapidly developing digital imaging technologies in the late twentieth century in terms of much earlier historical developments (Crary 1990, 1–2). To do so, he demonstrates, and against prevailing trends in art historical research, that in the first few decades of the nineteenth century a new kind of observer began to take shape that broke with the dominant models of classical vision in Europe. In these early decades of the nineteenth century, there was a dramatic change in the way the observer figured in an array of discourses and practices. By focusing on optical devices and technologies such as the camera obscura and the stereoscope, he brings these transformations into sharp relief. Where the camera obscura was "paradigmatic of the dominant status of the observer in the seventeenth and eighteenth centuries" and instantiated a "stable and fixed observer," the stereoscope and phenakistiscope, by contrast, catalyzed a different perceptual logic. "What occurs," Crary continues, "is a new valuation of visual experience: it is given an unprecedented mobility and exchangeability, abstracted from any founding site or referent" (14).

What's fundamental to this permutation of the observer in the early nineteenth century is that it happens in concert with a whole series of discursive and epistemic shifts in science and culture due to rapid industrialization. Much more than a simple modification in visual art practices, the abstract and mobile observer instantiated in the stereoscope and other optical technologies in the early 1800s signaled a massive overhaul in the organization of knowledge and perception. For this reason, Foucault's conception of the disciplinary subject is of paramount importance to Crary insofar as his concern is the disciplinary formation and mobilization of an *observer-subject* well equipped to meet the demands of industrial capitalism: "New modes of circulation, communication, production, consumption, and rationalization all demanded and shaped a new kind of observer-consumer" (14).[1]

Crucially, this observer-subject predates and prefigures the photographic industrialization of the eye that has been well rehearsed in media and cultural theory (Benjamin 2008; Bazin 2005; Flusser 2000; Doane 2002). With the introduction of the photographic apparatus, visual perception is mechanized and embedded in circuits of production and consumption abstracted from the spatial and temporal limitations of the human body (Doane 2002). As a consequence, a rough equivalence is established between photography and money:

> They [photography and money] are equally totalizing systems for binding and unifying all subjects within a single global network of valuation and desire. As Marx said of money, photography is also a great leveler, a democratizer, a "mere symbol," a fiction "sanctioned by the so-called universal consent of Mankind." Both are magical forms that establish a new set of abstract relations between individuals and things and impose those relations as the real. (Crary 1990, 14)

The connection between the image and money will be pursued in more detail below. What matters here is that optical technologies, such as the stereoscope and phenakistiscope, have been industrializing the eye well before the advent of photography. These earlier technologies were overhauling the practices of observation and aligning them with the demands of an industrializing economy early into the nineteenth century; in so doing, they were preparing the ground for the modes of abstraction that would be introduced by photography, the cinema, and digital imaging.[2]

In many ways, Crary's important work on the normalization of subject-observers in the nineteenth century sits nicely alongside Benjamin's assertion about the development of the cinema: the primary function of the cinema, he explains, is "to train human beings in the apperceptions and reactions needed to deal with a vast apparatus, whose role in daily lives is expanding almost daily" (Benjamin 2008, 26). The cinema, like certain other optical technologies that came before it, attempts to strike a balance between the accelerated pace of industrial capitalism, which steadily encroaches on all aspects of modern life, and the human sensorium. Similarly, although with a slightly different emphasis, Siegfried Kracauer analyzes the abstract industrial bodies displayed in footage of the Tiller Girls—showgirls who became popular because of their "precision dance" (1995, 75–86). In "The Mass Ornament" (1995), Kracauer discusses how the bodies of the Tiller Girls have become abstract and recombinable parts, whose precision and coordination are the visual analogs of elements on a Fordist assembly line.[3]

These early intuitions about the correspondence between photographic technologies and capitalist modes of production have been deepened, modified, and pushed in new directions in the last several decades. In the 1970s and early 1980s, apparatus theory emerged in the wake of psychoanalytical film theory and maintained that the cinema is an ideological apparatus, reproducing the dominant ideology for the spectator and disguising how the film is put together frame by frame (Baudry 1974–75; Comolli 1985; Mulvey 1975; Rosen 1986). More recently, theorists have returned to the economies of cinematic production—even giving rise to a so-called "new apparatus theory" (Kessler 2006)—using a variety of theoretical lenses. Crary's *Suspensions of Perception,* for example, charts the rise of the entertainment industry through the fabrication of images, sounds, and subjects as measurable commodities that can be exchanged; and Mary Ann Doane has shown how cinematic technologies structure time and contingency in capitalist modernity (2002, 4). Hito Steyerl (2012), for her part, has brought the relation among today's accelerated capitalism, screens, and labor into sharp focus, and so have numerous other media theorists, including Patricia Pisters (2012), Pasi Väliaho (2014), and others. But perhaps the most noteworthy example of a work that unites screens, sensations, and capital is Beller's *The Cinematic Mode of Production: Attention Economy and the Society of the Spectacle* (2006). There, Beller contends, and in no uncertain terms, that the cinema, as well as its many offspring (television, videogames, the internet, etc.) do not simply correspond to, reflect, or represent factory work, they *are* factory work. The cinema is an extended factory for the senses to do work for industrial and now late-stage capitalism: "To look is to labor" (Beller 2006, 2). And while the history of cinema and media studies has exploited the shared histories of the cinema and capitalism, Beller pushes this history further than perhaps any other media theorist to date. In doing so, he demonstrates how the cinema was a necessary development for capital: it turns human sensual and affective life into a commodity.

Of course, Max Horkheimer and Theodor Adorno had already critiqued the culture industry's commodification of everyday experience; and decades later, Fredric Jameson would show how postmodern capitalism brings all domains of culture—"high" and "low"—under its control (Horkheimer and Adorno 2002; Jameson 1991). In the Italian context, Mario Tronti, Antonio Negri, Franco "Bifo" Berardi, Paolo Virno, and other postwar autonomists critique the transition from the "formal" to the "real subsumption" of labor (the commodification of cognitive and affective spheres) and the more general production of a "social factory" in late-stage capitalism (Tronti 1973; Negri 2004; Berardi 2009; Virno 2001). Working within this tradition, Tiziana Terranova and a handful of other "digital Marxists" have exposed the ways in which the digital economy is folded into this social factory (T. Terranova 2003; 2004; Dyer-Witheford 1999). Beller's intervention in this genealogy is substantial: he cashes out what many others could not—namely, (1) the mechanisms by which capitalism *becomes cinematic* and requires human sensuous labor for its development, and (2) how this is the foundation for later developments in digital labor.

THE CINEMATICITY OF CAPITAL

For Beller, the cinema does not represent or symbolize the endless circulation of commodities; nor does it simply become another commodity form entering into circuits of exchange. Rather, and far more provocatively, the cinema becomes the primary scene of twentieth-century capitalist production—this is what he refers to as the "cinematicity of capital." "The cinematicity of capital," he writes, "dialectically re-ordains the categories of political economy, meaning that it leaves its older forms extant (wage-labor, circulation, capital, use-value, exchange-value, etc.), while bringing them to a higher level of articulation and abstraction" (Beller 2006, 28). "The logistics of cinematic production," Beller explains elsewhere, "embod[y] the logistics of capitalist production. . . . Cinema is, in effect, the very form of late capital, the highest stage of twentieth century capitalism"; it is the *mise-en-scène* of late-capitalist production (57).

This is not mere provocation. Beller shows all of this in a number of interconnected arguments. His first move is to demonstrate how the circulation of commodities is accomplished cinematically. Using Dziga Vertov's 1929 film *Man with a Movie Camera* as his primary, if slightly surprising, touchstone, he argues that it is in Vertov that we first glimpse a self-conscious and critical (if not yet revolutionary) expression of urban society stitched together through the circulation of commodities (40). Again, this is not because the film represents or symbolizes commodity exchange. The film *is* the self-conscious circulation of commodities (42). This is due to the fact that moving images, for Vertov, are not at a distance or at a remove from the material organization of society. On the contrary, they are "on a technological continuum with the rest of industrial society and are the means in which a society at a certain level of development achieves a higher level of self-articulation" (42).

To fully appreciate the connection between image and commodity and in particular how the commodity *becomes image* in order to achieve a "higher level of self-articulation," we need to return to the basic distinction between use value and exchange value in Marx. Recall that for Marx, there is an inherent duality in the commodity form: it is at once a use value and an exchange value. Where a use value is what a commodity possesses when it satisfies some human need, an exchange value, by contrast, is not directly perceptible in the commodity. The exchange value, Marx will say, is based on the nonsensuous aspect of a commodity that renders it equivalent to other commodities (1977, 143, 126). And in capitalist societies, exchange values are believed to "inhere" in objects, as nonsensuous essences of them, and are even thought to be what "animate" them (126, 50).[4] Marx's classic example of the linen and coat relation is instructive: the sensuous differences between these two commodities (qua use values) are superseded by an equation that renders them exchangeable: "20 yards of linen = 1 coat or = 20 coats or = *x* coats . . . linen = coat on the basis of the equation" (141). In short, the equation is what permits us to abstract from sensuous bodies and attend to their nonsensuous equivalences, which can be extended to any and all commodities (1 coat can be rendered equivalent to 10 lb. of steel, and so on [Bracken 2007, 156–58]).

The important point is that commodities have a "double life" (Bracken 2007). Each commodity has both a sensuous appearance that renders it distinct from every other commodity, and a nonsensuous, inner life, which makes it similar to every other one. It is this inner life, made possible by abstracting from a commodity's sensuous features, that Beller exploits in his work on the cinema. We do not consume objects because they satisfy a basic human need. Rather, we consume abstractions or *images* of them as a part of a system of equivalences. In just this way, commodities have a tendency to *become-image.* We consume a commodity's image, what it seems to be, an abstraction of it, instead of what it actually is in its sensuous reality (of course, Guy Debord [2002] had already intimated this). It is for this reason that Beller describes capitalism as "proto-cinematic," which is to say, there is already a becoming-image of the commodity form in capitalism, and the cinema is what exploits this tendency: it makes the becoming-image of the commodity manifest like no other technology (Beller 2006, 76).

And yet for a commodity to become image—which is to say, for it to become other than itself, and be put into communication with and rendered equivalent to what is sensuously different from it (exchange value)— there needs to be some form of mediation capable of bringing about this communication. Beller recalls Marx's words in the *Grundrisse:*

> At first sight, circulation appears to be a never-ending process. The commodity is exchanged for money; money is exchanged for the commodity, and this is repeated ad infinitum. The constant renewal of an identical process does indeed constitute an essential feature of circulation. . . . The commodity is exchanged for commodity; money is exchanged for the commodity. So, commodity is exchanged for commodity, except that this exchange is a mediated one. (Marx 1986, 132)

Money mediates commodities; it allows them to be exchanged for one another and to circulate. However, Beller shows that in Vertov, and in the cinema more generally, image(-commodity) is exchanged for image(-commodity) through the film frame: "It is the film frame (the screen) that allows images to circulate." What this means for Beller is that "film is the money of cinema (and the frame is its unit)" (2006, 58). Thus "the cinema is the concrete realization of what is already implicit in Marx: All things pass through the frame of capital" (105). In this context, Beller notes, although in passing, that Deleuze had realized something similar about the frame in the 1980s: "The frame ensures a deterritorialization of the image," Deleuze writes, because it "gives a common standard of measurement to things which do not have one—long shots of countryside and close-ups of the face, an astronomical system and a single drop of water" (Deleuze, quoted in Beller 2006, 105). What Deleuze realized, but was unable to give full expression to at the time, is that the frame is money, the "general equivalent."

Beller's argument does not stop there. He also contends that the cinema is the new and essential site of production for twentieth-century capitalism. This is an essential element in his overall argument, because if he is actually going to show how the cinema brings capitalism to a "higher level of articulation and abstraction," then he needs to explain how the cinema is productive, how work is done there, and how surplus value can be extracted from it. But instead of Vertov, Beller now looks to Sergei Eisenstein's *Strike* (1924–25). This is because *Strike* exemplifies the way in which a spectator-laborer is produced to "[sustain] cinematic circulation" (Beller 2006, 69). As Beller writes, "If capital is really becoming cinematic then the film itself must achieve a relation with spectators that can sustain its form" (69). *Strike,* like *Man with a Movie Camera,* is a watershed moment in early-modern cinema for capital: *Strike* expresses the "new regime of the sensorium" *that ensures the circulation of image-commodities,* and in doing so, exemplifies how industrial capitalism overcomes old distinctions between manual and mental labor and paves the way for the "sensuous laborer," and thus the "full commodification of the sensorium" (69, 95). "Beginning in Eisenstein's period," Beller writes,

> cinema dematerializes industrial processes, making them over into fodder for consciousness and sensuality, and projecting them into bodies through visual means. Film internalizes industrial movement and manifests it as forms of consciousness and sensuality. These forms are projected into bodies through the eye. Bodies then internalize industrial movement, making it inform their own consciousness. Film turns movement into thoughts and feelings, or, more generally, affect. (95)

Although I won't go into more detail concerning how Eisenstein's work accomplishes the control of sensual life in industrial capitalism through the "organization of attraction" ("incorporating the spectator into its very movement, its moving" [96]), or how it perfectly coincides, and not coincidently, with the emergence of organization theory (in both Pavlov and Taylor, see Beller 2006, 115–28), it's important to mark it as a turning point in industrial capitalism, and more specifically, as the point where capitalism actualizes its cinematicity by deterritorializing not only circulation (Vertov) but also production (Eisenstein). This sets the stage for understanding the emergence of new sensuous pathways for production in twentieth-century capitalism, as well as their expansion and intensification under late-stage capitalism.

In Beller's view, the new regime of production—where "to look is to labor"—has two different and mutually informing aspects. On the one hand, by viewing an image, the spectator-worker is actually increasing the value of the image. This is because every act of attention—whether we are watching a film, or looking at a website, a YouTube video, or an advertisement on Facebook—"accrue[s] on the image" and increases the image-commodity's circulation, and thus, its value. In short, our sensorium is doing the work of adding value to images by increasing their circulation (Beller 2006, 208–13). But on the other hand, the cinema and its many offspring are training the human sensorium to be productive. Screen media are designing the sensuous faculties they need for work in the extended factory. I will return to this below, but suffice it to say that in consuming images, the sensorium is also in training. It is being designed to do the productive work of paying attention (a scarce resource when information is abundant): "Willing to see almost anything, willing to believe almost anything . . . and needing to be a little different every day, we modify ourselves as we watch" (138). And in return for attention, spectators receive payment in the form of pleasure and affects. However, the scales are always tipped in favor of capital: spectatorial labor ensures that Amazon, Alphabet, Facebook, and other multinational media conglomerates extend their domination while the sensorium is more thoroughly in their service (138). As a consequence, "the theory of the productive value of human attention," continues Beller, "is the theoretical expression of the robbery of the spectator's vision in particular, and of his or her sensuality, perceptivity, activity, and consciousness in general." This leads to the conclusion that, "historically speaking, capital no longer posits merely 'labor-time' as alienated and productive for capital, but all *human* time" (177).

FROM ECONOMISM TO GOVERNMENTALITY

There are a number of reasons to both praise and criticize Beller's provocative book. Steven Shaviro notes that while it stays incredibly close to Marx and Marxist theory, the book also makes an important revision to it: circulation is now directly productive. Beller revises Marx's (outdated?) thesis that circulation is, in most cases, inefficient and costly (*Capital,* vol. 2, chapter 6) and claims that in neoliberal capitalism, where to look is to labor, increased circulation is productive for capital.[5] Despite Beller's important revision to Marx, Shaviro is also critical of the foundational role he assigns to the cinema and questions whether we are in a different media age altogether. Where the cinema corresponds to Fordist modes of production, we are currently in a post-Fordist or neoliberal era where a very different logic seems to have captured the popular mediasphere (Shaviro 2010). But it is on neither of these points that I wish to dwell. Instead, I want push back on Beller's Marxism. Although one of the chief virtues of *The Cinematic Mode of Production* is that it exposes the mechanisms by which the cinema produces value for capitalism, the wager is that the cinema is the primary scene of capitalist production and that a revised Marxism is necessary and sufficient for its critique. The development of the cinema, its various and diverse offspring, as well as the design and control of the human's entire sensual and cognitive life, is reduced to a single and totalizing economic logic. "From a systematic point of view," Beller writes,

> the cinema arises out of a need for the intensification of the extraction of value from human bodies beyond normal physical and spatial limits and beyond normal working hours—it is an innovation that will combat the generalized falling rate of profit. It realizes capitalist tendencies toward the extension of the workday, . . . the deterritorialization of the factory, . . . the marketing of attention, . . . the building of media pathways, . . . and the retooling of subjects. (2006, 13)[6]

While Beller's theory of economic causation may be compelling, I worry that it cannot capture the multiple and complex dimensions of image control in the twentieth and twenty-first centuries. Criticisms of Marxist readings of culture are not new, of course. "Post-Marxists" often cite the widespread social upheaval in the 1960s—spanning many domains of social and cultural struggle—as evidence that class struggle cannot capture the variegated textures of social and cultural unrest in the mid-twentieth century (Touraine 1971; Melucci 1989; Laclau and Mouffe 1985). The problem, for many, is that Marxists reduce questions of patriarchy, ethnicity, and racism to a totalizing economic theory, when they simply cannot be. Well known for their critique of crude economic essentialisms, Ernesto Laclau and Chantal Mouffe argue that the economic sphere is not privileged or exceptional, as Marxists tend to think, but it is one among many sites of control and suffering that demand diverse modes of analysis and critique (Laclau and Mouffe 1985).

These criticisms of Marxist cultural theory have not gone unanswered. Nick Dyer-Witheford acknowledges the limitations in Marx and Marxist theory—that there are "major blind spots to issues of gender, ethnicity, and the destruction of nature" (1999, 9)—but insists that these blind spots should not cause us to lose sight of the real problem: the reductionism of global capitalism. "This reductionism . . . has today a totalizing grip on the planet," Dyer-Witheford insists. "Other dominations, too, are reductive—sexism reduces women to objects for men, racism negates the humanity of people of color. But neither patriarchy nor racism has succeeded in knitting the planet together into an integrated, coordinated system of interdependencies" (10). This enlarged frame for economics justifies the economic critique of all other domains. For even though "patriarchal and racist logics are older than capital," capital's total grip on all planetary forces is what compels these other domains to "manifest themselves within and mediated through capital's larger, overarching structure of domination: as market-racism, commodity-sexism" (10). Translation: all social, cultural, and even natural forms are mediated through the economic sphere; they are reduced to economics, even if they weren't originally commodities (which is precisely what Beller forcefully argues). It is for this reason that Marxism (in its revised forms) is immune to post-Marxist critique and remains the privileged theoretical framework for social criticism.

The issue with this argument is not that neoliberal, post-Fordist, or information capitalisms fail to reduce nearly all aspects of mental and physical life, as well as the global struggles over racial, ethnic, and gender inequalities, to a market-driven calculus. This is not the point of tempering the Marxist critique. Rather, what is at stake is thinking about market rationality as more expansive and all-encompassing than economics alone (Brown 2019; Dardot and Laval 2017). This is precisely what Foucault argues in his late work on neoliberal governmentality: neoliberalism is irreducible to economics (economism) and is therefore not synonymous with the triumph of economics over the state. For Foucault, neoliberalism is a modality of government (and not economics) that cannot operate without the state; it simply uses new and often unrecognizable techniques to ensure the expansion of market rationality to all spheres of human (and nonhuman) life, even those spheres that seem to have nothing to do with economic reason (2008, 243). A theory of governmentality is better equipped to deal with capacious forms of control that use manifold techniques (including the state, the church, derivatives, computational technologies, etc.) to elicit a range of effects whose critique and reimagination cannot be prescribed in advance.

My argument will be that governmentality is an especially useful frame through which to see how contemporary media societies are designed by deploying techniques that include but are irreducible to economic ones. Despite the importance of Foucault's work for my argument, my claim is not that his genealogy of neoliberal governmentality is sufficient to account for the shifting field of relations among sovereignty, political right, financialization, and digital labor that we encounter today. Gregg Lambert, for one, is skeptical that Foucault's work on neoliberalism, which does not extend beyond the Carter administration, can grapple with developments in governmentality leading up to 2008 and beyond; and so for this reason, it might very well be considered a dated theory (2015, 3). And I, for one, am not going to hazard anything much different, and I certainly would not want to suggest that there have not been significant shifts in governmental reason since Foucault's lectures of 1978–79. Instead, my claim is more modest—namely, that Foucault's conception of governance offers a framework that is better equipped than Marxian forms of critique for theorizing the ways in which economies, states, institutions, bodies, and technologies conspire to shape the perceptual economies of viewing subjects in the twenty-first century. And in what follows, I examine how Foucault's lectures on neoliberalism, complemented by Deleuze's essay on societies of control as well as select critical media theories, help to identify certain trajectories of perceptual governance that take shape throughout the twentieth and twenty-first centuries. It is these tendencies of power and governance, I argue, that make specific forms of image and perception control in popular screen culture visible, but in such a way that the logic of the visual that characterizes molecular animation practices comes into view as a mode of perceptual governmentality. Along the way, we will also encounter a theory of entrepreneurial epistemology (via Friedrich Hayek) that will be particularly relevant to our theorization of governance in later sections.

GOVERNANCE AND THE ECONOMIC SELF

In his 1978–79 lectures, Foucault traces, among other things, a genealogy of German ordoliberal and American neoliberal governance and weighs in on their different conceptions of the state-form. This is a task of considerable importance and difficulty, especially given that older notions of the sovereign state no longer seem to hold sway in neoliberal societies (at least from Foucault's vantage point in 1979). It is in order to make certain aspects of Foucault's conception of the state in relation to then-emerging forms of neoliberal power explicit that I now discuss these two theories of governance. This will also serve to further differentiate Foucauldian from neo-Marxian notions of neoliberal capitalism and reinforce the former's importance for theorizing contemporary media societies.

According to Foucault, ordoliberals (whose primary thinkers were Walter Eucken, Franz Böhm, Alexander Rüstow, Wilhelm Röpke, and Alfred Müller-Armack, and is a movement that emerged in the context of postwar West Germany) believe that there is nothing "natural" about market competition. Instead, competitive markets are only cultivated and kept alive through government intervention. In contrast to the negative theory of the state one encounters in liberal theory, for ordoliberals, market competition is furnished through political action (Foucault 2008, Lecture Seven; see also Lemke 2001). Lambert shows how Rüstow describes the capitalism of the nineteenth and twentieth centuries as a "pathological degeneration" characterized by a "subsidy-ridden, monopolist, protectionist, pluralist economy" that must be sharply distinguished from the "free-market economy of perfect competition which constitutes the normal object of liberal economic theory" (Rüstow 1980, 468; Lambert 2015). According to Thomas Lemke, one of the consequences of this conception is that

> there is no capitalism because there is no logic to capital. What is called capitalism is not the product of a pure economic process and historical capitalism cannot be derived from a "logic of capital." We do not have to do here with a firmly circumscribed and defined structure . . . but instead with something that is historically singular (*one* form of capitalism among possible other forms). (2001, 194)

The free market, for the ordoliberals, is not one thing, with one contradictory logic (leading to totalitarianism), but it is a historically specific phenomenon open to political change by altering social conditions (i.e., social policies). "Ordoliberalism," Foucault explains, "envisions a competitive market economy accompanied by a social interventionism that entails an institutional reform around the revaluation of the 'enterprise' unit as the basic economic agent" (2008, 176). And social intervention is precisely what avoids the tendency that "[Joseph] Schumpeter identifies in capitalism toward organization, centralization, and absorption of the economic process in the state, which he saw was not a tendency of the economic process but of its social consequences." Moreover, intervention "will enable us [ordoliberals] to maintain the logic of capital in its purity and get the strictly competitive market to work without the risk of it ending up in the phenomena of monopoly" (178–79).

Still, the ordoliberal notion of the "social market economy" produces a highly ambiguous relation to the state-form. So, for instance, in his lecture from March 21, 1979, Foucault speaks about the difficulty surrounding the generalization of the enterprise from "within the social body or social fabric" according to the ordoliberal notion of *Gesellschaftspolitik* (241). The generalization of the market via social interventions presupposed the ordoliberal ideal of actively creating the conditions for "'warm' moral and cultural values which are presented precisely as antithetical to the 'cold' mechanism of competition" (242). Now, this "warm" instead of "cold" notion of the market spread throughout the entire social sphere is meant to counter the harsh calculating reason of competition and integrate enterprise into the life world of subjects—Rüstow calls this *Vitalpolitik* (242). The difficulty of creating an enterprise society, according to Röpke, is that while it may work in the abstract, competition tends to fragment social relations. As such, it requires a state-form that can "maintain itself above the different competing groups and enterprises" as a political and moral sphere charged with guaranteeing community bonds and social integration (243).[7] But for Foucault, there is an inherent ambiguity in this proposal: the state is both for and against the market (242); it promotes competition but also compensates for it, and in this sense the state seems to have a compensatory function.

American neoliberals, on the other hand, are far less hesitant about instituting a "cold" economic calculus into the social sphere, which is why their theory "appears much more radical or much more complete and exhaustive" (243).[8] On Foucault's reading, Chicago neoliberals do not have the same soft spot for a state-form that ensures a "well-cooked capitalism" (Lambert 2015), since what's important for them, generally speaking, is that market rationality is not one domain among others: it is believed to cover the entire social field. All social and individual behaviors are now analyzable in terms of market competition: "competition, and only competition, can ensure economic rationality" for neoliberals (2008, 19). "This means," Foucault continues, "analysis in terms of the market or, in other words, of supply and demand, can function as a schema which is applicable to non-economic domains" (243). This schema is a "grid of intelligibility" that makes it possible for all relations, behaviors, and interactions to show up as economic relations—"a sort of economic analysis of the *non-economic*" (243; emphasis added).

Now, insofar as all relationships "are analyzed in terms of investment, capital costs, and profit . . . on the capital invested," then this analysis extends into subjective and affective spheres (244). It is within this framework that Foucault's theory of "human capital" becomes legible. Central to the neoliberal conception, according to Foucault, is rethinking labor in classical economic theory (Smith, Marx, Ricardo). Where classical economic theory regards human labor as relevant, for instance, in terms of the "price of labor, or what it produces technically, or what is the value added by labor," it does not actually bring labor "into the field of economic analysis" (223). Neoliberals revise this conception and assume the position of the worker: the wage a laborer receives for his/her work is "by no means the price for selling his/her labor power, but instead represents an income from a special type of capital" (Lemke 2001, 199). And "to that extent," Foucault insists, "it is not like other capitals. Ability to work, skill, the ability to do something cannot be separated from the person who is skilled and who can do this particular thing. In other words, the worker's skill, really is a machine, but a machine which cannot be separated from the worker himself" (2008, 224). For this reason, the neoliberal perspective on labor represents an "epistemological transition," where emphasis is placed on the subject who is engaged in the market: "The worker is not present in the economic analysis as an object . . . but as an active economic subject" (223). What this means is that "the worker himself appears as a sort of enterprise for himself" (225).

What emerges from this is a new image of *homo œconomicus*: from a "partner of exchange" (liberalism) to an "entrepreneur of himself" (neoliberalism). Neoliberal *homo œconomicus* is "an entrepreneur, an entrepreneur of himself . . . being for himself his own capital, being for himself his own producer, being for himself the source of [his] earnings" (226). Citing Gary Becker's (1965) theory of consumption, Foucault insists that "the man of consumption, insofar as he consumes, is a producer. . . . We should think of consumption as an enterprise activity by which the individual, precisely on the basis of the capital he has at his disposal, will produce something that will be his own satisfaction" (2008, 226). All forms of human activity and consumption—from developing specialized skills in various domains of physical or cognitive work to taking a yoga class and watching TV—are possible modes of production and investment in the self.

A crucial part of the neoliberal thesis is that enterprising activities are not confined to specific social groups. Everyone is an entrepreneur. This is why Foucault says that the subject no longer has the history it once had—there are no more criminals or perverse subjects in need of the same old forms of rehabilitation and institutionalization. Here's Foucault: "All the distinctions that have been made between born criminals, occasional criminals, the perverse and the not perverse, and recidivists," or in other words, all those distinctions that mattered to a disciplinary society, "are not important. We must be prepared to accept that, in any case, however pathological a subject may be, . . . he is nevertheless 'responsive' to some extent to possible gains and losses. . . . We must act on the market milieu in which the individual makes his supply of crime and encounters a positive or negative demand" (259). Thus, in neoliberal societies, as opposed to disciplinary and industrial ones, all subjects, no matter their background, are enterprises, bundles of attributes and capacities that respond to economic opportunity. If everyone is an entrepreneur of the self, and this is normalized throughout the social body, then behaviors, even criminal behaviors, are best shaped through market activities. According to Israel Kirzner, a student of Ludwig von Mises, "The entrepreneurial element in the behavior of market participants consists . . . in their alertness to previously unnoticed changes in circumstances which may make it possible to get far more in exchange for whatever they have to offer than was hitherto possible" (1978, 15–16).

What emerges from this notion of the "entrepreneurial element in behavior," and which tends to go unnoticed in the critical literature on neoliberalism, is the epistemology of the entrepreneur. Friedrich A. Hayek, and to some extant Ludwig von Mises (1996), have theorized the unique features of this epistemology. In his 1935 essay "Economics and Knowledge," Hayek writes that the "division of knowledge" is "the really central problem of economics as a social science," and it is at least as important as, and intimately related to, the "division of labor," even if the former has received little theoretical attention (1948a, 50).[9] And then ten years later, Hayek challenges the widely held epistemological assumption of economic theory in his essay "The Use of Knowledge in Society"— namely, that the construction of a rational economic order is a problem to be solved by logic and mathematics alone. Such abstract theories *would be true,* Hayek contends, if the following conditions were met: "*If* we possess all the relevant information, *if* we can start out from a given system of preferences, and *if* we command complete knowledge of available means, the problem which remains is purely one of logic" (1948b, 77). But "this is emphatically *not* the economic problem which society faces" (77). Quite the contrary: "The peculiar character of the problem of a rational economic order," Hayek insists,

> is determined precisely by the fact that the knowledge of the circumstances of which we must make use never exists in concentrated or integrated form but solely as the dispersed bits of incomplete and frequently contradictory knowledge which all the separate individuals possess. (77)

For Hayek, what's distinctive about knowledge acquisition in an enterprising society is that there is no centralized bank of information that would permit a global or totalizing perspective. If anything, there is an emphasis on ignorance and incomplete knowledge, as well as the "decentralization" of knowledge practices carried out by "many separate persons" (79; see also Dardot and Laval 2017, 110). It is crucial to note just how much Hayek's theory of knowledge departs from the prevailing assumptions about modern scientific epistemology. Hayek himself muses on this and writes that "today it is almost heresy to suggest that scientific knowledge is not the sum of all knowledge. But a little reflection will show that there is beyond question a body of very important but unorganized knowledge which cannot possibly be called scientific in the sense of knowledge of general rules: the knowledge of the particular circumstances of time and place" (1948b, 80).

What's striking here is that Hayek dethrones the transcendental authority of science and challenges its capacity to determine what is valuable or relevant knowledge. This challenge is directed not only at modern science (although its status in the public imagination makes it one of the most important targets) but at all monopolies on knowledge. For Hayek, any specific, partial, and incomplete bit of information could be leveraged into an opportunity for economic gain. Knowledge is distributed, partial, and always evolving. This means that no individual, no circumstance, no set of beliefs, attitudes, or feelings can be left outside of the sphere of economic knowledge production: "Practically every individual," Hayek writes, "has some advantage over all others because he possesses unique information of which beneficial use might be made" (80). Information is not already given, but it is rather given through evolving circumstances that cannot be determined in advance. For Hayek, "the economic problem of society is mainly one of rapid adaptation to changes in the particular circumstances of time and place," which is why decision-making must be decentralized (83); only then can "the knowledge of the particular circumstances of time and place . . . be promptly used" (84).

But if economic knowledge production in neoliberal societies cannot be contained and moreover is no longer mediated through a central authority, then what role is there for the state? It is at this point that critical theories of neoliberalism often retreat into some form of economic reductionism (Dyer-Witheford 1999). It appears that the sovereign state no longer exists, or if it does, it does so in a considerably weakened form, which leaves the economic sphere to drive all others. But if Foucault maintains a role for the state-form, and at the same time makes certain ambiguities about the ordoliberal relation to it explicit, then it is still not clear what role the state would have in the neoliberal society that he maps for us. And there is more: Foucault provides no framework for conceptualizing how the emergence of digital and information technologies, which go hand-in-hand with the rise of neoliberalism, and were beginning to take shape at the end of the twentieth century, have seamlessly merged with financial markets and have fundamentally transformed the state's function.

My suspicion, which I share with many others, is that this is largely due to the fact that Foucault's own project on neoliberal governmentality was incomplete at the time of his early death in 1984. Lambert weighs in on just this question and argues that the state-form does not simply disappear from Foucault or from twenty-first-century capitalism, for that matter. But insofar as the state exists, and insofar as it is still effective in shaping the behavior and actions of individuals, then it looks very different than it did in its previous eras. In order to cash out what this means, especially in light of the fact that the twenty-first century has seen the rise of modes of computational governance that Foucault simply could not have foreseen, many theorists, including Lambert, have looked to Deleuze's late reflections on societies of control for answers (Hardt and Negri 2000; Galloway 2004). This is not because Deleuze spells out the state's function within postdisciplinary societies any more than Foucault does. Rather, Deleuze sketches how the diagram of power operative in neoliberal societies works together with computational networks in order to regulate the life of its subjects. And it is in this context that we can begin to map the kinds of techniques the state would have to acquire in order to shape the behavior and actions of "in/dividuals."

IN/DIVIDUALITY

In his short essay from 1990 titled "Postscript on Societies of Control," Deleuze marks a key transition in power dynamics: "Societies of control," he writes, "are in the process of replacing the disciplinary societies" (1995, 178). If disciplinary power, in Foucault, operates by molding individual bodies through confinement and hierarchy, then "control," a term coined by William S. Burroughs, is an emerging form of power that acts by "modulation, like a self-transmuting molding continually changing from one moment to the next" (178–79). And if "it's easy to set up a correspondence between any society and some kind of machine," then the machines that correspond to the control society are digital machines, "information technology and computers" (180). Where disciplinary societies worked with "cybernetic machines" to organize the thermodynamic flows of physical bodies, control societies use information technologies to abstract bodies into endlessly recombinable bits of data. (This is not to say that the old, disciplinary machines no longer exist—it is to say that they are organized differently with information technologies.)

What Deleuze saw at the close of the twentieth century is how human bodies are beginning to be regulated through new techniques of data management. In other words, power dynamics adjusted in order to address the emergence of a new figure: the *dividual.* In place of subjects and bodies sculpted through disciplinary technologies in enclosed spaces (striated space), we now have *dividuals,* or bodies and movements abstracted into endlessly divisible bits of data that can be recombined and distributed over an incalculable number of planetary networks. As Gerald Raunig explains over two and a half decades after Deleuze wrote the Postscript, "The reality of today's dividual data sets, enormous accumulations of data that can be divided, recomposed and valorized in endless ways, is one of worldwide streams, of deterritorialization and machinic expansion, most succinctly expressed as Big Data" (2016, 123). Of course, Deleuze does not predict the rise of Big Data or AI in his short essay. And so if we are still looking to the Postscript today for answers, then it is not because it is particularly good at predicting future technologies. In fact, Deleuze's reference to Guattari's future town accessed using a "(dividual) electronic card" seems naïve in hindsight (Deleuze 1995a, 181–82). With advancements in machine learning and IoT, there are few spheres of human activity and engagement that cannot be mined, divided, and recombined for the purposes of deploying increasingly more supple and effective modes of governance (Greenfield 2017; Bratton 2015). Instead, the reason Deleuze remains important for us today, and the same could be said of Foucault, is that he identifies a particular logic of power, a logic that was beginning to take shape at the close of the twentieth century and has only intensified its hold.

It is from this perspective that we can appreciate the way in which the dividual supplements Foucault's conception of the entrepreneurial self. If Foucault sees this self as a bundle of attributes and qualities, whose knowledge practices, I have been arguing, are partial, distributed, and ongoing (Hayek), then Deleuze pushes this even further. He grasps (if not completely, then at least partially) how entrepreneurial subjectivization is not only territorializing but also deterritorializing. In the age of digital machines, subjects are not simply individuated selves, but they are also, and continuously, being divided, recombined, and distributed across planetary information networks that do not stop transforming themselves. Maurizio Lazzarato offers important insight here when he describes how subjectivity works in two directions in contemporary capitalism. Drawing on Guattari, Deleuze, and Foucault, he explains that on the one hand, there is "social subjection," which is Foucault's individuated subject, the subject defined as a bundle of attributes, in short, the entrepreneur. "Social subjection," writes Lazzarato, "mobilizes signifying semiotics, in particular language, aimed at consciousness and mobilizes representations with a view to constituting an individuated subject ('human capital')" (2014, 39). But on the other hand, there is "machinic enslavement," which is a completely different tendency of subjectivization, or rather, *desubjectivization.* "Enslavement," writes Lazzarato, referencing Guattari, "does not work with 'subjects' and 'objects,' it works on their deterritorialization (or their decodification), that is, with their molecular components, the non-individuated intensive, subhuman potentialities of subjectivity" (27). If social subjection is signifying and representational, then the desubjectivizing force of enslavement involves an "asignifying semiotics (stock market indices, currency, mathematical equations, diagrams, computer languages, national and corporate accounting, etc.) which do not involve consciousness and representations and do not have the subject as referent" (39; Guattari 1984).

With machinic enslavement the individuated subject is divided into parts, and each part is "a gear, a cog, a component part in the 'business' and 'financial system' assemblages, in the media assemblage, in the 'welfare state assemblage' and its collective institutions (schools, hospitals, museums, theaters, television, Internet, etc.)" (Lazzarato 2014, 25).

Machinic enslavement brings out what I believe Deleuze was getting at with his (underdeveloped) conception of the dividual in the Postscript. The notion underscores a different tendency of subjectivity than the one Foucault identifies with entrepreneurial subjectivity. Enslavement expresses the divisibility of the subject, its division into component parts for digital and financial machines, within accelerated networks of planetary computation. The architectures of subjectivity are therefore designed according to the twin tendencies of territorialization and deterritorialization. However, the dividual and the entrepreneur are not competing designs of the self. As Guattari writes, "In reality, there are only mixed semiotics, which partake of both to varying degrees. Signifying semiology is always haunted by a sign machine and, conversely, a sign machine is always on the verge of being reclaimed by signifying semiology" (1984, 140). Thus the two tendencies of subjectivity, "social subjection" (the entrepreneur of the self) and "machinic enslavement" (the dividual), need to be thought together. As Lazzarato writes, "The strength of capitalism lies in its articulation of processes of social subjection and machinic enslavement. . . . Both apparatuses play a fundamental role in controlling processes of capitalist deterritorialization and reterritorialization" (2014, 66). In other words, these are complementary tendencies of the subject; the entrepreneurial subject is dispersed across planetary networks of computational media. But this deterritorialized self (as bits of data fed into Amazon, Google, Facebook, etc.) is reterritorialized into new forms of economic knowledge and human capital: it is recoded and leveraged into a form of economic value for the self. The dividual does not do away with human capital but enriches our conceptualization of it. Today's *homo œconomicus* must be thought together with the emergence of *homo digitalis.*

Returning to the larger issue of the state-form within this new diagram of subjective governance, it's clear that the state can no longer function according to its previous commitments to centralization and strength. If the state does not disappear, then it is because it has been redesigned to deploy subtle and flexible techniques of control well attuned to the subjectivizing and desubjectivizing forces of human capital. The state apparatus does not resemble the old sovereign forms of domination. According to Lambert, what we see is the emergence of a state that is "no longer premised on a universal form of right that belongs to 'all men,' nor even to members of the same species being, much less to other species beings." Instead, under this new state-form we witness the proliferation of new "subjects of right," or rather, "subjects of economic right," who have political rights, not insofar as they participate in the "polis," but insofar as they participate "in the market" (2015, 10). The state that emerges within this new regime aims to "secure" and "protect" the twin tendencies of human capital, which is to say, the vicissitudes of economic life rather than political life (9).

But as Lemke notes, while the state has an arsenal of new techniques well equipped to design and shape the economic life of in/dividuals, it also shirks all responsibility for it. In fact, the state actively creates the conditions under which the subject assumes the responsibility for all of the "risks" of economic activity, and in so doing turns this responsibility into a moral question of "self-care." "The key feature of the neo-liberal rationality," Lemke continues, "is the congruence it endeavors to achieve between a responsible and moral individual and an economic-rational actor" (2001, 201). This idea also sits well alongside Lazzarato's contention in *The Making of the Indebted Man* that one of the consequences of entrepreneurial subjectivization is that the subject "assum[es] the costs as well as the risks of a flexible and financialised economy, costs and risks which are . . . especially those of precariousness, poverty, unemployment, a failing health system, housing shortages, etcetera. . . . In the current crisis, the most that capitalism demands and compels, in every area, is less knowledge than that one *take upon oneself* the costs and risks externalized by the State and corporations" (2012b, 51). In this way, the formation of debt, for Lazzarato, is an ethico-political project, just as much as it is an economic one to generate subjectivities that are "accountable and guilty" (49). One of the essential functions of the state, then, is to ensure that the reterritorilization of subjectivity into human capital is experienced as debt. "The injunction to become human capital," he writes elsewhere, "has been transformed into the negative and regressive figure of the indebted man.

The individual is still responsible for his fate except that today he is fated to debt" (2014, 24).

While this proposal has a wide range of consequences that bear directly on larger questions of political right, debt, and shifting notions of the biopolitical, I won't pursue this line of inquiry any further here. One of the attendant challenges that this discussion raises is whether and to what extent this notion of the state realigns with conceptions of sovereignty (in this context, see Benjamin Bratton's *The Stack: On Software and Sovereignty*). Lambert ventures the hypothesis that what we are witnessing today is the return of a form of sovereignty, which neither Foucault nor Deleuze were necessarily equipped to theorize, even if they may have feared it, and which may very well reference "earlier and even more archaic forms of social control and sovereignty" (2015, 13). I will leave the particulars of this argument for another occasion, since what matters is that the state does not disappear or withdraw in the face of a computationally driven economic rationality that subsumes all aspects of social and subjective existence. Rather, it takes on new forms and must acquire new techniques in order to govern effectively.

BRAINS, SCREENS, AND MOLECULES

Although it may seem that we have strayed from our initial focus on popular screen culture and molecular animation, we have actually stayed quite close to them all along. My hypothesis is that we now have a conception of political and economic power that is subtle and agile enough to conceptualize the formation of these visual media throughout the twentieth and twenty-first centuries. In the perspective we have been developing, neoliberal capitalism is not reducible to economics (economism), but it is rather an inherently ethico-political project, inextricably tied to the state and its institutions, that aims to transform the social and subjective spheres of human activity into something that is eminently governable. It is in this context that moving images, and the popular mediasphere more generally, are essential to the governance of in/dividuals. This means that the development of the cinematic apparatus cannot simply be traced back to a crisis of accumulation (Beller). This is far too blunt of a theoretical instrument. Instead, the cinema and its many offspring are among the techniques and technologies deployed as a part of a transformation in governmental reason. This is not to diminish the importance of the cinema and other screen media for capitalism (since it is no longer *the* solution to the falling rate of profit), but it is to fold these media into a wider set of interwoven techniques of government used to redesign social, cultural, and subjective life.

With these theoretical adjustments we can return to many of Beller's insights with fresh eyes. For example, the suggestion that moving images train our sensuous and cognitive faculties to do productive work for capitalism needs to be nuanced. Recall that the subject's consumption of images is also productive because it gives attention to them, which increases their value (by increasing their circulation). What the subject receives in exchange for her attention are "pleasure and affects," while companies such as Amazon and Google increase their profits. Although "pleasure and affects" are certainly a part of the experiential horizon of entrepreneurial subjectivity—indeed, Foucault notes that the "man of consumption" produces his own "satisfaction" (2008, 226)—we also require a more fine-grained analysis of how human capital is folded into the landscape of visual media consumption and production. We need to unpack how the deterritorialization of attention through the ubiquity of screens and cameras today is reterritorialized into forms of economic knowledge and subjectivation.

In the remainder of the chapter I provide such an account not only to ground a more comprehensive theory of in/dividuation through media practices but much more specifically to uncover how molecular animation societies find themselves situated squarely within this media landscape. I discuss a number of conclusions that can be drawn from our expanded discussion of neoliberal governance thus far: (1) that popular screen culture is an important site of in/dividuation and can be seen through the lens of entrepreneurial knowledge practices, (2) that the regulation of biological bodies through brain–media interfaces ensures the reproduction of these practices, and (3) that molecular animation is folded into this embodied neoliberal knowledge economy. What will emerge from this extended discussion is how the conception of "infectious social power" that we developed in relation to Whitehead and Foucault in Part II is what anchors our understanding of molecular animation sociality within a neoliberal perceptual-knowledge economy.

Entrepreneurial Perception

For Hayek, entrepreneurial epistemology is specific and partial, but is also distributed and incomplete. What is especially striking about this epistemic framework is that no element is deemed irrelevant data. In short, there is no limit to what may become knowledge. What may seem like a trivial, fleeting, or irrelevant bit of experience today may suddenly be of use and leveraged into some form of economic gain. This is why Hayek insists, and against the prevailing economic wisdom, that the kind of economic knowledge which interests him is the one that by "its nature cannot enter into statistics and therefore cannot be conveyed to any central authority in statistical form" (1948b, 83). This is because the statistics on which a central authority would have to rely would necessarily "[abstract] from minor differences between the things, by lumping together, as resources of one kind, items which differ as regards location, quality, and other particulars, in a way which may be very significant for the specific decision" (83). Thus what Hayek saw in 1945, and may very well be commonplace today, is that economic calculation is not restricted to abstract quantities that can be measured in advance, but it rather descends into the fine-grained details of everyday life, into the microactivities and contingencies of human experience. The affective spheres of human life are sites of economic knowledge. What this means is that our various and diverse habits of attention are also important sites of economic knowledge. And this necessarily includes the attention humans devote to screens and online entertainment. As Beller and many other media theorists have noted, screen time is not downtime, far from it: the countless hours spent in front of screens are valuable working hours (Crary 2013; T. Terranova 2012; Goldhaber 2006). What is essential to underscore in the already well-established discourse on attention economics is that there is another kind of knowledge work being done here: the consumption of visual content translates into valuable knowledge for the stylization of *homo œconomicus.* In other words, visual work is also knowledge work on the in/dividual self.

Let's return to our discussion of popular digital image culture from chapter 1. There, we examined how rapidly evolving digital imaging technologies are such that there is little, if anything, that cannot be visually captured. Not only do the numerical points comprising a digital image allow visual content to be combined and recombined with any other visual content, but sound and image files can also be converted back and forth, which permits everything from the vibrations of the molecular world to the intimacies of human touch to become image (Min 2015). Over a decade ago, Matthew Fuller argued that media culture is a dynamical system where "any one part is always multiply connected, acting by the virtue of those connections, and always variable, such that it can be regarded as a pattern rather than simply as an object" (2005, 4). Manovich of course explores these and related questions through his taxonomies of software culture and shows how digital images are endlessly modular and variable numerical objects that are not constrained by the spatial and temporal coordinates of a profilmic world. He examines the subjective dimensions of this computational landscape in the context of what he calls "soft[ware] cinema," a project that studies the "data-subjectivities" that belong to the culture of ubiquitous screens, databases, and algorithms by automatically generating a potentially endless number of films from out of a database of media elements (Manovich and Kratky 2005). Taking this notion of "data-subjectivities" further, Byung-Chul Han meditates on the implications of technologies such as Google Glass for subject formation. He contends that by connecting the internet directly to the eye, Google Glass and other smart lenses are inaugurating "the age of total information" where "human perception achieves total efficiency" since "seeing the world and grasping the world coincide" (2017a, 42–43).[10]

The larger point is that perceptual subjectivities are forged at the intersection of proliferating screens and cameras operating in conjunction with expanding databases, deep learning methods for computer vision, as well as other forms of machine learning. For almost a century now, media and cultural theorists such as Benjamin, Kracauer, Crary, Doane, and others have analyzed the ways in which optical technologies deterritorialize perception by constructing abstract observers. What I am proposing is that today's perceiving subjects are likewise abstract observers, although in a different way than they were in previous eras. If nineteenth- and twentieth-century observers were more or less disconnected from their felt and lived worlds—since they were disciplined and mechanized perceivers, and so on—then what's different about today's economy of computerized perception is that the situated body, the felt, phenomenological body, is precisely what is incorporated into this perceptual economy. The intimacies of touch, the peculiarities of body position: these are all folded into the field of vision through a variety of AR, VR, RR (Remote Reality), and other responsive and interactive media technologies (Begole 2017).[11] Today's visual media do not simply discipline the body but they also redesign themselves according to the peculiarities of it. In other words, the affective body is no longer at a distance or alienated from the technics of visual perception, but it is precisely what is at stake for it. Through rapidly evolving responsive media technologies, the lived body is no longer external to the technologies of visual knowledge production, but it has become so internal to these technologies that they redesign themselves according the peculiarities of it (Sha 2013).[12]

Byung-Chul Han coined the phrase "transparency society" to refer to a society in which everything must be made visible: "Everything must lie open, ready and available for everyone" (2017a, 31; see also Han 2015). The truth of this condition now becomes shockingly clear: all bodies are put on display. Human and nonhuman bodies, cellular and molecular bodies, even quantum bodies—these and every other body are made available for visual consumption in a transparency society. The interactive tools used by Ribosome Studio (chapter 1) give the viewer a fully "immersive" experience of the molecular world on mobile screens. These technologies of mobile perception are folded into a much wider tendency in today's media culture that mandates even the processual dynamics of the molecular world be felt and lived by a human perceiver. Visual perspectives do not stop multiplying for today's perceiver; they are continuously being divided and recombined without end so that there is no limit to what can be made visible.

And yet, in the conception we have been developing, deterritorialized perceptual practices need to be reterritorialized into economic knowledge.[13] In other words, the deterritorializing tendencies of the visual actualized by computational technologies cannot be dissociated from the production of new forms of visual knowledge that augment human capital. The very same tendencies of the image that we examined in Part I and identified as hypercinematic cannot be dissociated from wider systems of power and governance that demand the production of new forms of economic value. As Deleuze writes, "The machines don't explain anything, you have to analyze the collective apparatuses of which the machines are just one component" (1995a, 175). In the context of twenty-first-century image production, this means that perceptual subjectivities distributed across globally networked screen media create new opportunities for reterritorialization in the form of entrepreneurial knowledge. The bombardment of the senses, the all-out assault waged by screen media on sensuous life, has to be optimized: the various and scattered acts of visual consumption need to be reterritorialized into forms of economic knowledge (Hayek). Diversification must be transformed into an economic opportunity.

Deterritorialized acts of viewing are never in isolation; they are also being viewed, surveilled. The so-called participatory culture of internet use has radically altered how we think about media consumption. We are "prosumers." When we watch a video on YouTube, like a Facebook post, scroll through someone's Twitter feed, or visit a website, we are actively broadcasting units of attention that have fluctuating value for reterritorialized selves. Membership to certain sites—for example, the live-streaming video platform Twitch—can turn out to be incredibly valuable for some users (T. Clark 2017); while other media consumption activities may have less obvious value today, but may turn out to be valuable in the near future. Algorithms may also deem viewing patterns to be suspicious, in which case they are programmed to update government databases. In these instances, patterns of online activity are reterritorialized to have long-term repercussions ranging from being searched, interrogated, and detained in airports and other checkpoints to having one's bank accounts frozen, property seized, and so on.

Han suggests that in this hypervisible era, the subject has lost all sense of interiority: it is "its own advertising object . . . stripped, exposed, and undressed, and put on show," or in other words, it is a "pornographic subject" in an "exhibition society" (2015, 11). This means that all our viewing practices (in a narrow sense) are in some sense also being "viewed" (in a wider sense). This is one of the consequences of living in such an enterprising exhibition society. If everything is on display, if all aspects of subjective life are viewed as bearers of potential economic value, then this includes the diverse and scattered practices of viewing itself (in the narrow sense). It is precisely the display of perceptual activities that the entrepreneur must design and curate to her advantage. The entrepreneur is responsible for reterritorializing her perceptual practices, for finding economic opportunity in the flood of deterritorialized viewing activities, so that she may count them as items of economic knowledge (even if this is to pay off an unpayable debt [Lazzarato 2012b]).

Neurodesign

But to turn the consumption of visual content into an opportunity for the subject, to see economic possibility where others do not, requires a specific kind of embodied attunement. Beller notes that the consumption of images is also a kind of training for the body. The senses are shaped and designed to act and react in a way that is productive for capital. In other words, the malleability of the body's sensuous surfaces enables it to be a conduit for economic value. What we have been exploring thus far is a perceptual subject situated at the intersection of twin tendencies of subjectivization. But these twin forces do not operate in isolation of the body, and more specifically, the biological body (hence, we are far beyond so-called disembodied information [Hayles 1999; Hansen 2004; Mitchell and Thurtle 2003]). What the wealth of scholarship on distributed cognition and enactivism has shown is that perceptual subjectivity emerges out of the dynamic coordination of brains, bodies, and worlds (Noë 2004; E. Thompson 2007; Di Paolo 2005). The implication is that the deterritorializing and reterritorializing activities of perceptual subjectivity in the computational landscape of emergent capitalisms are also embodied practices. But we cannot stop at the phenomenological body, as Francisco Varela, Evan Thompson, Alva Noë, and others have argued (see especially, Thompson 2007; Noë 2004; Varela, Thompson, and Rosch 1991). The phenomenological body cannot be isolated from biology, especially the biochemistry of the brain (as the history of phenomenology has often done). In other words, the perceptual practices of subjects are also and fundamentally cerebral practices.

It is the brain-screen interface in the context of networked media culture that media and cultural theorists have examined in recent years. As Patricia Pisters (2012) has shown, and in conversation with Deleuze, Guattari, neuroscientists, and other neuroculture theorists, our distributed and complex digital image culture actively shapes cerebral matter in ways that reflect the control society in which it is situated. Using carefully selected case studies, combined with examples from film and philosophy, Pisters demonstrates how brains and screens are yoked together in contemporary media culture (drawing heavily on Deleuze's powerful assertion that the "brain is the screen"), and even sketches how the brain-screen might function as an important site of resistance to control (99). Pisters's important work echoes a more general shift toward thinking about "neuroplasticity," "brainhood," and the "synaptic self" in the contemporary discourse on neuroscience and culture (LeDoux 2003; Damasio 2003; Vidal 2009), as well as more pressing arguments about the fabrication of a "cerebral subjectivity" in contemporary screen culture captured by neoliberal modes of governance. Similarly, Pasi Väliaho argues in *Biopolitical Screens: Image, Power, and the Neoliberal Brain* that "the brain, and not the eye, becomes the primary locus of mediation and 'capture' by which individual bodies and persons are woven into the political reality of contemporary life" (2014, 12). Studying videogames, such as first-person shooter games like *Modern Warfare 3,* Väliaho argues that "action-oriented images can be considered epigenetic and developmental, functioning as environmental conditions in response to which the brain adjusts and modifies itself" (43).

Such conceptions of neural adaptation and development are in no small part indebted (either explicitly or implicitly) to Francisco Varela and Humberto Maturana's autopoietic systems theory and the profound influence it has had on cognitive science and neuroscience in the last few decades (Maturana and Varela 1980; Varela 1992). As Varela explains in a lecture on intentionality and autopoiesis, "The neuronal dynamics underlying a perceptuo-motor task is, then, a network affair, a highly cooperative, two-way system, and not a sequential stage-to-stage information abstraction" (1992, 9). And it is this network of neuronal interactions that facilitates the operational, and not interactional, closure of the nervous system to "[bring] forth a specific mode of coherence, which is embedded in the organism. This coherence is a cognitive self. . . . The cognitive self is the manner in which the organism, through its own self-produced activity, becomes a distinct entity in space, but always coupled to its corresponding environment from which it remains nevertheless distinct" (10). In other words, the coherence of the nervous system, which conditions the emergence of the cognitive and perceptual self, does not arise in isolation, but only through ongoing and evolving couplings with its environment. Andy Clark and others have explored these "environmental couplings" in terms of "brain-machine interfaces" (BMIs) and have examined in depth the ways in which our brains respond and adapt to technologically mediated environments (Clark 2004, 2007).

Warren Neidich examines these brain–environment adaptations in terms of what he calls "neuropower," a variation of biopower that "acts upon the neural plastic potential of the brain in the living present" and, more specifically, "works to produce changes in the material logics of the brain by affecting the brain's neurons and synapses—its firmware—as well as its dynamic properties such as binding and reentry" (2012, 219, 228; Väliaho 2014). For Neidich, as well as a host of other theorists of "cognitive capitalism," the structures of power in contemporary capitalism are such that they exert control over brain development in order to ensure the design and cultivation of productive forms of cognitive work.[14] In *What Should We Do with Our Brain?* Catherine Malabou extends the idea that neoliberal forms of power require a particular "image" of the brain, although she argues that it is a "flexible" image of the brain, instead of a genuinely "plastic" one, and it is the basis for a new "neuronal ideology." In the end, this ideology, which has completely overtaken neuroscientific research, provides scientific justification for the ethico-political project of creating flexible economic subjects (2008, 11).

While this barely scratches the surface of the "neuro-turn" in contemporary critical and cultural theory, it helps to orient our discussion toward the ways in which screen environments, operating in conjunction with contemporary forms of power, are designing the neuroperceptual subjectivities that they require. What is also especially important about the so-called neuro-turn in discourses of power is that it fills out the picture of entrepreneurial knowledge practices in potentially interesting ways. The proliferation of screens, cameras, and networked media systems not only de/re-territorialize perceptual practices but they also de/re-territorialize brain activities: they diversify neural pathways and expand neural networks in order to optimize the plasticity of the brain. In this context, studies in the growing field of "neuroentrepreneurship" suggest that successful entrepreneurs have more diverse neural pathways than the so-called general public. A recent handbook devoted to the topic, titled *Handbook of Research Methodologies and Design in Neuroentrepreneurship* (2017), testifies to the growing interest in the field. Essays in the handbook cover a full range of topics, from neuroscience fundamentals and the physicochemical basis of entrepreneurship to the design and implementation of techniques to cultivate the neural dispositions of entrepreneurs. The takeaway is that neural diversity allows entrepreneurs to see connections (i.e., opportunities) where others do not, and thus take risks where others would not. These pathways also have an impact on the number of dopamine receptors in the brain—the hormone released as a reward for successful risk-taking—which increases an individual's willingness to take risks (Zald et al. 2008).[15]

Whether this research stands up to careful scientific scrutiny is not the point. Following Malabou, we might say that it is part of today's neuronal ideology: neural deterritorializations allow entrepreneurs to foster new combinations, to connect seemingly unrelated material, and to find useful information and economic knowledge where others do not. Neuroentrepreneurship and associated fields are leveraging the authority of science to legitimize the idea that the transformation of cerebral matter that neoliberalism catalyzes (neuroflexibility) is grounded in good science (instead of creating psychopathologies [Berardi 2009]). This permits the normalization of neoliberal science's flexible brain. In the end, this is the conception of the de/re-territorialized brain on which Hayek's entrepreneurial knowledge practices depend.

Molecular Animate Media

It is in the context of neuroperceptual subjectivities and networked media systems that many of the unique features of molecular animation become legible. Not unlike other forms of screen media today, animate images of molecular structures and processes are constantly altering the sensory input that they deliver to cerebral matter. Returning to *The Inner Life of the Cell, Protein Packing,* and *X Inactivation and Epigenetics,* these films showcase no single perspective on molecular processes. They express how there are countless perspectives on and multiple scales folded into the molecular interactions involved in a routine process such as leukocyte migration. What *The Inner Life of the Cell* makes apparent through camera movement and other cinematographic techniques is that there are an unlimited number of shots of this molecular interaction and that there is no aspect of it that cannot be captured by the mobile frame. The film broadcasts that molecular imaging is inexhaustible—that is, there is no global or final perspective on this molecular process—and that the mobile camera is capable of discovering each new perspective.

This sense of discovery is brought into sharp focus in *Protein Packing,* where the camera explicitly breaks down visual barriers. By establishing the cell wall as a visual barrier in the opening shots, and then piercing through it in order to access a molecular world that has a very different spatial and temporal logic, the film implicitly references the history of biology (from cell theory and classical genetics to molecular biology) and its various struggles with visual epistemology (as we saw in chapter 2). There, we examined how the history of visual epistemology in the experimental life sciences can be read as an attempt to secure visual evidence for the molecular unfolding of living processes. *Protein Packing* stages this history by directing our desire off-screen, to the inner workings of the cell, as if to suggest that off-screen space is what drives bioscientific knowledge practices. By piercing through the cell wall, the camera satisfies scientific desire. And yet desire is only temporarily relieved, since beyond the cell wall there is a vast and teeming world of molecular movement and processes that present ever-new challenges (or opportunities) for the molecular eye.

We also examined the penetration of the cell wall in terms of Deleuze's theory of virtuality and then related it to his work in *Cinema 2* on the loss of the cinematic "outside." This reading of the animation can now be supplemented by a more comprehensive understanding of the broader tendencies of contemporary screen culture and embodied entrepreneurial epistemology. If, for the entrepreneurial viewer, there is no limit to what can be made into an opportunity, if opportunism has become the new norm, then molecular animation presents a wealth of opportunities for the contemporary viewer. At the very heart of molecular animation is a radical innovation in the world of moving images: molecular processes unfold in *time.* Computer animation brings the complex temporal processes of the molecular and cellular world into contact with the neuroperceptual subject, animating the molecular world for her. But with the addition of time, the molecular world also incites new and complex feelings: the perceiving subject begins to feel what it's like to be a molecule, to see from the perspective of a white blood cell, and to be overwhelmed by molecular crowding. Molecular animation folds new affective dimensions into the teeming molecular world; this world is infused with an emotional tonality, as Whitehead would say. What's more, the technology continues to deliver innovations: new dimensions of the temporally unfolding molecular world are constantly discovered, and seemingly without end. The endless clips of molecular movement produced today (and which may one day soon be fully automated) do not stop offering opportunities for the perceiving subject.

Notice how this discussion takes us back to Jason Sharpe and his colleagues' contention in chapter 3—namely, that molecular animation mediates molecular dynamics and neuroperception. There, we examined how visual perception for these animator-scientists is a high-speed pattern recognizer made possible by a supercomputer behind our eyes (Sharpe, Woolridge, and Lumsden 2008, 4). And because molecular animation translates the computing power of the chemical and physical world into mathematically based patterns that can be processed by visual perception (a more complex computer), animation is thought to yoke these different computing systems together (molecules and brains). Of course, there is a way in which we are in agreement with Sharpe and his colleagues, although with a twist: molecular animation no longer stiches brains and molecules together through the natural language of computation. Rather, animation is a medium that governs the communication between molecules and neuroperceptual subjectivities to ensure that it reproduces the new "natural" language of neoliberalism.

This perceptual mediation is not only occurring in the sterile environments of the laboratory. The animate suturing of molecules and cerebral matter is occurring in the general public. Molecular animations not only showcase many of the same formal characteristics as popular screen media but also circulate through the same media channels. They are consumed in bits and pieces, in snapshots, stills, and clips on YouTube, Facebook, TED, and other video-sharing platforms. They are liked and shared like any other media content today. The TED talk that molecular animator Drew Berry gave in 2011, for instance, has received (as of June 2020) a remarkable 2,757,464 views, while David Bolinsky's talk has garnered an astonishing 2,232,901, and the YouTube video for *The Inner Life of the Cell*: 2,258,459 views.[16] This latter animation inspired the Hoogenraad research laboratory's rather humorous *A Day*

Figure 29. *A Day in the Life of a Motor Protein*. Directed by Casper Hoogenraad's lab, Utrecht University, 2013.

in the Life of a Motor Protein (2013), which has picked up a respectable 997,120 views.

Animating molecular worlds has also proven to be fruitful in areas of business entrepreneurship. Gaël McGill's animation company, Digizyme, has an impressive client list, which includes many prominent pharmaceutical companies. Similarly, Jason Sharpe's company, AXS Studio, also has a notable list of Big Pharma clients, as well as the eye-health product company Bausch + Lomb.[17] Molecular animations have multiple economic affordances, catalyzing diverse forms of economic production. Viewing a stunning clip of a motor protein online can be leveraged into diverse opportunities. Content producers, on the other hand, have also used the mass consumption of science (in terms of snapshots of innovation, sound bites of technical breakthrough, etc.) to generate other forms of human capital—e.g., celebrity (via TED). This is in addition to the many business entrepreneurship opportunities of which many molecular animators are beginning to take advantage.

What this discussion makes clear is that molecular animation is more than the digital technologies on which it runs: it is a medium for the communication of heterogeneous elements that enables new forms of perceptual subjectivization to emerge. This animation technology mediates the relation among entities in a way that is significantly different from how we typically think about media—as technical devices that store and transmit information. The medium in question is closer to "medium" in the (older) sense in which life scientists use the term: as substances, such as nutrient broths, that allow microorganisms to grow. As John Durham Peters argues, media are "enabling environments that provide habitats for diverse forms of life, including other media" (2015, 3). Similarly, Robert Mitchell examines how media for biologists working in the eighteenth and nineteenth centuries "accounted . . . for the emergence of new kinds of living beings" (2011, 96). Drawing on this biological conception of media, Mitchell looks to Gilbert Simondon's philosophy of technology to advance a theory in which media are what "enable 'communication' between previously unconnected potentials . . . such that new fields or environments can come into being" (102). In our context, this means that computer animation facilitates the communication among diverse spheres of molecular and cellular data production, networked screen media, and neurosubjectivity to enable new forms of *homo œconomicus* to take shape. Molecular animations are medial elements: they catalyze a relation among components to create other medial situations (Geiger and Littau 2013).

The mediality of molecular animation provides a critical supplement to how we have conceptualized animation thus far. If we initially theorized molecular animation as an eco-social assemblage that integrates diversity through the positive valuation of a common element of form (chapter 4), then we came to see how this integration takes shape through infectious social power (chapter 5). What is imposed on each member of the social ensemble through infection is a logic that allows the members of the ensemble to be meaningful to one another. But this logic does not spread throughout the social group in isolation of wider eco-social dynamics; it spreads for the purposes of making other social regimes possible—namely, those connected to entrepreneurial subjectivity. The common element of form creates a mediating situation that connects heterogeneous elements in order to produce a subjectivizing logic that satisfies the production of perceptual subjectivity in visual media culture generally. This mediating situation therefore captures heterogeneous forces for particular ends. With this characterization, certain parallels between Whiteheadian and Foucauldian conceptions of assemblage become apparent. In particular, common elements of form are media that appear to function as apparatuses of capture, dispositifs. What remains to be seen is whether these two conceptions of assembly (elements of form and dispositifs) can be reconciled to provide an enriched conception of the medial function of molecular animation within neoliberal perceptual economies.

THE DISPOSITIF AND MOLECULAR ANIMATE MEDIA

Foucault introduces the notion of the "dispositif" in the mid-1970s, and it first appears in *The History of Sexuality, Vol. 1.* Since then, the term has generated a wide range of debate over its meaning, use, and continued relevance for understanding contemporary systems of power and governance. As numerous commentators have explained, "dispositif" is exceptionally difficult to translate into other languages, which has led many simply to retain the French, *dispositif,* although it is not uncommon to see it translated into English as "apparatus" (Kessler 2006). Contributing to this general uncertainty is the fact that Foucault himself never provides an exhaustive overview of the concept and leaves us to sort out what its conceptual genealogy might be, as well as the precise mechanisms through which it is deployed in the late twentieth and early twenty-first centuries.

Although both Deleuze (1992) and Giorgio Agamben (2009) provide commentaries on the dispositif, neither of them gives us anything more than general descriptions of what arguably deserves a much finer-grained analysis. As Lambert notes, Deleuze and Agamben "preferred to remain in the haven of abstraction rather than going through the pains of a concrete analysis of the workings of such mechanisms" (2016a).[18] Despite these shortcomings, Deleuze and Agamben agree that the dispositif is best characterized as a specific coordination and control of heterogeneous materials—agents, bodies, discourses, institutions, subjectivities, and so on. Agamben, for his part, characterizes the concept broadly and explains that it is "literally anything that has in some way the capacity to capture, orient, determine, intercept, model, control, or secure the gestures, behaviors, opinions, or discourses of living beings" (2009, 14). He then proclaims that "today there is not even a single instant in which the life of individuals is not modeled, contaminated, or controlled by some apparatus" (15). What is more, the dispositif does not operate in isolation; indeed, Deleuze characterizes dispositif in the plural, as social apparatuses (dispositifs) that make heterogeneities "appear" in a certain way (1992, 160; Lambert 2016a). This sentiment is echoed by Agamben, who notes that what we have today is a "massive accumulation and proliferation of apparatuses" (2009, 15).

This conception of the dispositif is taken straight out of Foucault's arguably most succinct characterization of the notion in an interview from 1977: "What I am trying to single out with this term *[dispositif]* is," Foucault explains,

> first and foremost, a thoroughly heterogeneous set consisting of discourses, institutions, architectural forms, regulatory decisions, laws, administrative measures, scientific statements, philosophical, moral, and philanthropic propositions—in short, the said as much as the unsaid. The apparatus itself is the network that can be established between these elements. (1980, 194)

The dispositif is a kind of device that connects disparate elements: it is the connective tissue that draws components together, and in so doing causes or "makes [them] ready" to appear. Thus the apparatus makes it possible for a heterogeneity of things to become visible *as something* (Lambert 2016a). But this "making ready to appear" cannot be understood without putting it in the context of what Foucault calls a "general strategy." As he explains later in the interview, "I said that the dispositive is by nature essentially strategic, which indicates that it deals with a certain manipulation of forces, of a rational and concerted intervention in the relations of force, to orient them in a certain direction, to block them, or to fix and utilize them" (Foucault 1980, 196). The dispositif therefore organizes and assembles heterogeneities to achieve a "certain end" (Lambert 2016a). The heterogeneity becomes "visible" as a part of the advancement of a strategic plan.

In choosing to characterize the organization and control of heterogeneous forces in terms of the apparatus, Foucault does not mean to suggest that it is a part of the same conceptual lineage as mechanism (even if the disciplinary "mechanisms" are still operative). Lambert pays special attention to the way in which Foucault's use of the term derives from Georges Canguilhem (who derives it from the neurologist Kurt

Goldstein [Pasquinelli 2015]), who makes a distinction between how the apparatus functions in living and mechanical systems (Lambert 2016b; Pasquinelli 2015; and Canguilhem 2009). Foucault borrows Canguilhem's (and by extension, Goldstein's) understanding of "biological normalization" as a way to describe the operations of power outside of the frame of "mechanical causality" (the centralized mechanisms of the state, and so on).[19] Just as environmental perturbations to an organism require that it search for a way to establish balance, to find homeostasis, and to normalize its relation to the environment, so too the dispositif has a normalizing effect in society:[20] it does not operate through a central mechanism, but through multiple and evolving channels that bring heterogeneous elements together to find equilibrium, a new normal state, that fits into an overall strategy (Lambert 2016b).[21]

Although Whitehead never uses the concepts of "apparatus" or "device" to describe the organization of heterogeneity, he would agree with Canguilhem and Foucault that this organization cannot be subsumed under a conception of mechanism. Thus if a new conception of the dispositif for Canguilhem partly emerges out of an attempt to overcome "Descartes and his transcendent mechanistic program for the human body and for the living organism in general" (Pasquinelli 2015, 84; Canguilhem 2009, "Machine and Organism"), then so too Whitehead manages to conceive of the organization of heterogeneity outside of the confines of mechanistic philosophy. But for Whitehead, this occurs in the other direction: not through the extension of the organismic to the social but through a rethinking of the social as organismic (Halewood 2011); or rather, everything that endures is social and is organized through the self-imposition of a common element of form. This form is what allows a diversity to experience itself as more than mere diversity; it is what enables diverse elements to be visible to one another and to the wider environment as *this* instead of *that.* Hence, the common element of form, much like the apparatus, serves to draw a diversity of forces together and allows them to be visible *as* something. And for Whitehead, the common form is what expresses the quality or manner of relation; it expresses *how* elements relate. It is not a noun but an adverb (Kraus 1998, 30).

More than grouping occasions under a form that comes from the outside (hylomorphism), the element of form is both external and internal to the occasion. It is that aspect of the other that enters into the constitution of the subject. This is why Whitehead says that it is genetically inherited among the members (PR, 34). And as we saw in chapter 5, this means that the form is imposed on an occasion in the mode of infection, where one occasion infects another with its form and structures what it is possible to feel and experience. If social infection lends itself to a Foucauldian understanding of power relations, then it is because the form specifies how subjects govern their relations to one another. It is in this sense that the common form functions as a communicative medium, an apparatus, that brings heterogeneities together and allows for interoccasional communication.

Although there are similarities between Whitehead's notion of social infection and Foucault's conception of the apparatus, it's critical to note that the dispositif is always connected to a larger strategy. As Foucault explains, "The apparatus is essentially of a *strategic* nature, which means assuming that it is a matter of a certain manipulation of relations of forces" (Foucault 1980, 196). Heterogeneities are captured in an apparatus to achieve certain ends. This characterization is not altogether foreign to Whitehead's conception of sociality. Social orders never arise simply for the sake of their own enjoyment; they always have a concern for the larger social ecology. As Whitehead never tires of explaining, each social order interpenetrates all others. There is no form of togetherness that is not also and in some way in the service of larger ends (molecule societies are organized in such a way that they help achieve the end of maintaining the cell's integrity, and so on).[22] Indeed, molecular animation is an instance where the common element of form draws heterogeneities together (the visual logic that coordinates the animation assemblage) to serve other forms of social togetherness (a new form of neoliberal in/dividualization). It is this interpenetration of social spheres that resonates with Foucault's conception of the apparatus: the medium of communicative assembly in molecular animation is an apparatus that coordinates elements for the purposes of generating new forms of political–economic value.

Before we conclude, there are a several critical points to make about setting up an equivalency between the element of form and the apparatus. First, it means that apparatuses of capture are also communicative media. The coordination of disparate materialities in a Foucauldian apparatus makes a specific form of communication possible. For Whitehead, when one entity enters into another (causal efficacy), then a communication occurs with a certain tonality. Molecular dataset societies are brought together with societies of cerebral matter, networked media technologies, and other social orders to form one larger social order united by its manner of communicating. If we take any set of elements within the molecular animation society (e.g., different chunks of code), then their relation within the molecular animation society is governed by their manner of communicating—namely, by means of a particular visual logic (or in any case, it is the dominant form that makes all other forms subsidiary); this is what they "infect" each other with. Of course, code is incredibly modular today and can be combined into chunks of functionality for which it was not originally designed (cut and paste), but this does not change the fact that its execution of a function is occurring in *this* society and not *that* one. What this means is that a bundle of code comes to be relevant to the other elements in the molecular animation ensemble because of its particular contribution to the (re)production of this logic of the visual and not another one.[23] All the elements prehend one another through this form. (This does not preclude these elements from communicating in other ways and from having simultaneous sub-communications, etc.) In the end, there is no simple transference of molecular data points to brains; it is the mediated relation between them that is valuable to wider perceptual economies. Molecular animation's communicative relation is what makes it an apparatus of capture.

The second point is that this conception of the social medial apparatus intervenes in larger questions about the social and political impact of Big Data and visualization in biology. On the one hand, our theory of computer animation accounts for the fact that data analytics and machine learning are essential to the animation of cellular and molecular processes in the postgenomics era. (Sharpe and his colleagues [2008] contend that the reverse is also true). As we have seen, molecular animations rely on structural data about proteins and other biomolecules acquired through the data analytical methods used in bioinformatics; these data are then uploaded and animated in 3D visualization software (tasks that are also increasingly automated). There is a good reason that Sharpe and his colleagues place molecular and cellular animation in the same lineage as biocomputing. On the other hand, our theory of the social media apparatus illustrates how Big Data do not operate in isolation of wider social, political, economic, and subjective spheres. There is no innocent collection, analysis, and visualization of biomolecular data; there is an entire field of eco-social relationality woven into apparatuses that design and ensure the production of certain outcomes. Put in other terms: between Big Data and their visualization there is neoliberal control.

This is not unrelated to what Alexander Galloway argues in *The Interface Effect* (2012). There he contends that in the age of data visualization and information design we have lost the meaning of "data" and "information," or in any case, we've conflated the two and use them interchangeably when they are separate. Where data denote "the things that have been given" (which is to say, numbers and pure mathematical values), information, on the other hand, designates the act of "giving form to," or the in-formation of the given. In short, data gains information (such as visual information) that does not belong to it by right (Galloway 2012, 82). Although I disagree that pure or raw data exist (this is impossible in a Whiteheadian frame, since there are no "bare facts" that are not infused with "value"), I nevertheless take his larger point to be that there is no necessary connection between data and visual organization. And yet he observes that phenomena such as networks (which are what interest him) have only one visual form—namely, the node and edge aesthetic. A dialectical tension emerges for Galloway: there is no representation of data and there is only one representation of data. What is it, then, that prohibits other representations of data? Galloway's thesis is that control societies mandate only one aesthetic form.

"One of the key consequences of a control society," Galloway continues, "is that we have moved from a condition in which singular machines produce proliferations of images, into a condition in which multitudes of machines produce singular images" (91). What Galloway forcefully illustrates is that representation is not irrelevant in the age of Big Data: control governs the subtle workings of visual representation. In other words, data are *informed* through control. Galloway's point about network visualization is germane to the one I am making about the computer animation of biomolecules: a plurality of elements (machines and materialities) that have no necessary relation to one another seem to nevertheless have one, or in any case, there is only one relation that seems possible for them. And it is one that is governed by a visual logic (an element of form) that ties heterogeneous forces together, prescribing how they are informed or mediated. In the end, the visualization of Big Data cannot be isolated from broader dynamics of political and economic power.

This is where Whitehead is especially helpful. And this is my third point: by reading Foucault's apparatus alongside Whitehead's conception of the common element of form, we gain an especially powerful frame for conceiving how every apparatus (and not just molecular animation) is a part of a dense web of ecological dependency that does not respect modern divisions between "nature" and "culture." So-called modes of cultural production are intimately tied to biochemical spheres through their prehensions of one another. Likewise, there are no data societies that are not already infused with cultural, political, and historical values. These social orders are thought to actually enter into the constitution of one another. Whitehead is absolutely clear on this point: for any social order, "the environment, together with the society in question, must form a larger society in respect to some more general characters than those defining the society from which we started. Thus we arrive at the principle that every society requires a social background, of which it is itself a part" (PR, 90).

One of the consequences of this is that none of the social orders that molecular animation assembles belong to it exclusively. The software packages, machine learning algorithms, human labor practices, and so on are a part of multiple societies at once. They are simultaneously prehended by other *nexūs* of occasions. Likewise, the animation software (Maya), programming languages (MEL, Python), and programming norms (object-oriented programming) are a part of countless other social arrangements (entertainment societies, computational architecture societies, etc.); and so is human cerebral matter, which has been pushed to find new neural connections and reorganize itself in response to new environments. Put another way, these components are "autonomous" in the sense that Deleuze and Guattari mean in "Of the Refrain" (Deleuze and Guattari 1987), if only because they are constantly being deterritorialized and reterritorialized into new social formations. For this reason, Whitehead thinks about the coordination of heterogeneities as a creative achievement. Their communication is not given in advance, but it is forged through the fabrication of mutual feelings; the molecular animation apparatus is therefore an achievement of social reterritorialization that is ecologically fragile and dependent upon its wider environment for stability (PR, 100). Neoliberal control societies do not just offer the right environmental setting for the emergence and growth of molecular animation (a medium for the production of another medium), they also serve to benefit from computer animation's successes: molecular animation strengthens and expands the hold of neoliberal techniques of governance. This is why Foucault's conception of the dispositif is crucial: it articulates how social assemblages are in the service of larger ends; they are a part of general strategies. Molecular animation is an apparatus designed to expand and intensify governmentality.

Ultimately, Whitehead makes it possible to think about the fate of molecular animation in slightly different terms. Insofar as each social order finds itself integrated into other modes of togetherness, then it is conceivable that the same element of form that structures the molecular animation society could be felt differently. That is, it could be incorporated into another nexus of actualities, one that is at a safe distance from, or possibly adjacent to, the modalities of governance that have gripped popular media culture. Whitehead leaves the door open to the possibility that molecular animation might not be completely determined by this all-encompassing frame. In the previous chapter, we examined how Whitehead's thought makes it possible to conceive of other possibilities for feeling a form of definiteness. Propositional feelings are those hypothetical spaces where a subject can feel other potentials for relevance, other ways that a form of relation can enter into the actual world and create sites of resistance to dominant forms of social infection.

This question cuts to the heart of what we were dealing with in Part I. We saw that molecular animation is not only a popular entertainment medium but also a visual epistemic tool in the biosciences. More than this, it serves to bridge the very different epistemological cultures of theoretical and experimental biology. But at that time, we did not have the theoretical framework in place to understand how the same ensemble could be relevant to two different cultures at the same time. At stake is whether and to what extent visual epistemology in the life sciences functions as a social nexus that lies outside of neoliberal forms of perceptual governance. Might the visual culture of the biosciences provide an actual world for this logic of the visual that is outside of neoliberalizing forces that have gripped the popular mediasphere? And if it does not, then we need to understand how the visual culture of the biosciences came to be under the same form of governance as popular media culture. We need to inquire into how these visual cultures function as different aspects of the same phenomena, and how molecular animation gives us a lens through which to see their convergence. It is in the final chapter of this book that I will explore these questions.

7

EPISTEMIC CAPTURE

We ended the previous chapter suggesting that Whitehead's sociology leaves open the possibility that molecular animation is not completely determined by the operations of neoliberal governmentality. Where chapter 6 demonstrated how popular screen media are folded into the techniques of today's dominant systems of governance and how molecular animation becomes intelligible within this matrix, it also became clear that there may yet be other possibilities for this medium. By bringing Whitehead's work to bear on our framing of computer animation as a dispositif, we came to appreciate how each society comes to have different relevancies. There is not one, all-encompassing perspective from which to judge or determine the efficacy of a mode of togetherness.

One could argue that Whitehead's entire philosophical career was devoted to saving us from the crippling habit of thought that safeguards the dominance of one sphere of meaning over another. When Whitehead wrote about the problem of modern scientific professionalization in *Science and the Modern World,* and how it creates "minds in a groove" that are incapable of entertaining the wider environments into which a matter of fact enters, he was insisting that every fact of togetherness is taken up into other adventures of relevance. His daring attempt to overcome the bifurcation of nature is a testament to this: he wanted to ensure that whatever phenomenon enters into our experience is irreducible to a single frame of reference. The scientist's description of the sunset does not disqualify or explain away the poet's (CN). "Philosophy destroys its usefulness," Whitehead insists, "when it indulges in brilliant feats of explaining away" (PR, 17).

Whitehead's concerns are not lost on us. Our discussion of molecular animation would not be complete if it failed to account for its other possible adventures. Although the frame of neoliberal governmentality has a great deal of critical purchase, and it is even capable of confirming the suspicions of some scientists that animation is not suitable for scientific research, we also have to resist the temptation to assume that it determines every possibility of meaning and relevance. It is in this spirit that one might return to our discussion in chapters 2 and 3 as evidence that molecular animation can be taken up into other adventures of thought. There it became clear to us that molecular animation practices, and indeed, the very same ones that circulate in the popular mediasphere, have an essential role to play in the visual culture of the life sciences. And at that time, what seemed to be an especially important, if undertheorized, aspect of the medium is that it is situated in two different visual cultures at once. The pressing concerns of theoretical and experimental biologists seemed to us far enough removed from the concerns of popular media consumers. Molecular animation appeared to be part of (at least) "two cultures," giving concrete expression to Whitehead's conviction that the elements of experience are "analyzable in an indefinite number of ways" (PR, 19).

The question we are left with then is the following one: What do we mean when we say that the visual culture of the life sciences is *separate* from the culture of popular screen media? Are they two cultures in the reductive sense that C. P. Snow (2012) meant by "two cultures," or are we talking about something else? Of course, the visual culture of the life sciences has its own complex history, with its own materials, discourses, and epistemic practices, and we spent a significant amount of time in Part I trying to understand these, especially as they played out across experimental and theoretical biology. By all accounts, this would seem to testify to the fact that the technologies of visual perception in the life sciences have their own lineage, with their own materials and concerns, which do not wholly coincide with the history and practices of popular media consumers in the twentieth and twenty-first centuries. Even though there is a convergence of value, Whitehead does not preclude that value from being taken up in different adventures of relevance. Eternal objects can be actualized in any number of different ways, and propositional feelings are what allow us to feel these alternative worlds.

But if Foucault's archeological period taught us anything about the study of history, then it is that historiography cannot be limited to the linear history of conscious subjects. The conceptual possibilities of a given era are defined by unconscious rules that do not respect disciplinary divides or linear histories. What appears to be a separation of domains of knowledge might actually conceal deeper discursive alignments. It is in this spirit that I inquire into whether there are meaningful overlaps between these two visual cultures, using the interpretive frame of governmentality. If chapter 6 demonstrated that developments in popular visual culture are legible in terms of the perpetuation of neoliberal techniques of governance, then what is at stake in this chapter is determining whether the visual economies of the life sciences are governed by similar techniques.

My argument is that the visual economies of life science research have also been captured by wider systems of governmental regulation. This is not meant to be a totalizing thesis, however. I do not argue that all visual practices in the life sciences are dispositifs. My claim is more circumspect, which is that the domains of visual epistemology in which molecular animation practices are embedded are a part of a longer history of techniques designed to govern scientific perceptual practices. Whether this system of capture could be extended to other visual practices in the biosciences is for another discussion. What will become clear is that the "separation of cultures" is meaningful only as a kind of "abstraction" in Whitehead's sense of the term. For Whitehead, there is no getting rid of our abstractions to discover something more real or concrete behind them (SMW; Stengers 2011b). Instead, and against certain genealogies of critique, we are called upon to be attentive to our modes of abstraction, and in our context this means that overvaluing the separation of cultures would be to "mistake the abstract for the concrete" (the fallacy of misplaced concreteness) and to conceal other modes of attunement or abstraction. It is these latter modes that interest me most and are what showcase the ways that the visual cultures of science and entertainment find themselves caught up in the same system of governance. And it is molecular animation that has emerged as a result of this shared, rather than separate, adventure of abstraction.

But before I go on to discuss how this chapter will unfold, I want to situate this proposal within the larger discourse on neoliberalism and science. To contend that a technoscientific practice like molecular animation is subjected to regimes of political and economic governance is not especially new or transformative. Although I do think that we are investigating this intersection in a new way. To appreciate this, note that within the last fifteen years there has been a wealth of scholarship on the neoliberalization of the biosciences: from neoliberalism's development and transformation of fields such as biotechnology, bioinformatics, synthetic biology, bioengineering, and genomics to the effect it has had on conceptions of labor, exchange, life, and epistemology within technoscience more broadly. For instance, Kaushik Sunder Rajan's *Biocapital: The Constitution of Postgenomic Life* (2006) uses ethnographic methods to trace the rise of genomics and drug development in the United States and India and draws on a combined Marxian and Foucauldian discourse to develop a notion of "biocapital" to examine the economic forces at work in the genomic sciences. Similarly, Melinda Cooper's *Life as Surplus: Biotechnology and Capital in the Neoliberal Era* (2008) shows how the proliferation of biotechnologies in the 1970s and 1980s in the United States is inextricably linked to the rise of neoliberal capitalism and then demonstrates how life itself comes to function as surplus value within this emergent "bioeconomy." And in *Tissue Economies: Blood, Organs, and Cell Lines in Late Capitalism* (2006), Catherine Waldby and Robert Mitchell examine how the global exchange of human tissues has shaped a new "tissue economy" in late capitalism that troubles classical distinctions between the circulation of tissues as gifts and as commodities.

Echoing many of these concerns, Paul Rabinow and Nikolas Rose refine notions of biopower and the biopolitical in the context of the rapidly commercializing biosciences. Rabinow, for his part, introduces the notion of "biosociality" to capture the transformation of the biopolitical through the Human Genome Project and other biotechnological innovations. Biosociality addresses how "nature" and "culture" can no longer be kept apart, since in this new biopolitical landscape, social relations are understood through the lens of biological and medical knowledge production (Rabinow 2005). Likewise, Rose reworks biopolitical thought to fit the growth of human genetic and medical knowledge and introduces the notion of "ethopolitics" to describe how today's biopolitical rationalities, working in concert with neoliberal policies, no longer regulate the body of the population but control the genetic risk of individual members of society (Rose 2001). This makes possible a wide range of new techniques (operating at the level of genetic intervention) for self-governance and the production of human capital (Rose 2001, 18). At roughly the same time, Eugene Thacker's early work grapples with biopolitics within the shifting media and information landscape of the biological sciences. In *The Global Genome* (2005), he addresses the global distribution and exchange of biological data within a wider biopolitical field and shows how our conception of "life itself" is now inextricably linked to the global biotech industry.

The point is that theorists are reflecting on how broader notions of the social, the political, and even life itself are undergoing profound transformation due to the tight coupling of technoscience and neoliberal policies within the last several decades. And to bring these transformations into sharper focus, new concepts are introduced—"bioeconomy," "tissue economy," "biovalue," "biocapital," and "biosociality" are among them. And since the publication of these works, scholars have sought to deepen, extend, and intervene in many of their important propositions. And in this context, there can be little doubt that much of the work I am doing here falls squarely into this genealogy, especially insofar as it aims to develop a critical framework to examine how certain shifts in technoscientific production are tied to the spread and intensification of neoliberal modes of governance.

However, my work departs from this lineage in at least a few important ways. One striking difference is that it is not especially interested in biotechnology (classically defined) or the genomic and medical sciences. Another difference is that it is not directly concerned with shifting conceptions of "life" (only in chapter 2), and it is even less interested in the business of technoscience. Instead, what I am concerned with are mutations in the knowledge practices connected to the technologies of visual perception in the life sciences. For this reason, this chapter is a departure from much of the previous scholarship on neoliberalism and technoscience: it investigates domains of experience and practice that are not typically subjected to critical analysis in the history of science (as will become evident below). It is in order to gain a fuller picture of these transformations that I focus on the diagrams of power operating within them.

But to examine these diagrams I will need to return to chapters 2 and 3 with fresh eyes. If my initial examination of visual epistemology in the life sciences was sheltered from the vicissitudes of political, economic, and ecological thought, then in this chapter I aim to bring these dimensions to bear on our understanding of this history of scientific perception and knowledge. Just as chapter 6 rehearsed many of the themes in chapter 1, and in this sense functioned as a companion chapter to the previous one, likewise, chapter 7 aims to complement the work done in chapters 2 and 3. This is possible because the theoretical framework we have been developing allows us to discern aspects in the history of knowledge and perception that did not seem especially relevant to the linear history that I was telling in chapters 2 and 3. But now that we are equipped with a better understanding of how governmentality functions within today's networked media culture, it is possible to identify key incidents that make the intrusion of neoliberal techniques of governance within the visual culture of the life sciences especially clear. In other words, my contention in this chapter is that there are moments within the history of visual media that only now seem relevant to the historical formation of scientific perception and epistemology. It is for this reason that this chapter is genealogical in the full sense that Foucault meant by this term: we are writing a history from the perspective the present.[1]

In what follows, I will look at two such moments. I consider early twentieth-century experiments with microcinematography and mid-twentieth-century visual and graphic design. These are episodes in the history of twentieth-century visual technology that have an ambiguous relation to knowledge practices in the life sciences. They are also research practices that are rarely, if ever, thought together as a part of the same history. But from the perspective of our present, these practices become visible as essential components in this genealogy. They reveal much deeper continuities cutting across domains of scientific knowledge production and popular media culture, exposing how governmentality has been hard at work merging these spheres of media culture. Molecular animation signifies an achievement of governmental reason: popular media consumption and scientific knowledge production are fused in the same dispositif. I will end this chapter by deepening the growing suspicion that this achievement is part of a larger design strategy: governmentality is a designing agent that shapes the apparatuses it needs to extend and intensify itself. And molecular animation is one of these apparatuses. This will prime us to propose in the Postscript that the entire book can be read as a speculative design theory of perceptual governance.

CELLULAR TIME-LAPSE

In chapter 2 I told a selective history of visual knowledge production in the experimental life sciences. In the course of this history, we covered work by cell theorists, classical geneticists, X-ray crystallographers, and others. This story gave shape to the ways in which visual technologies were essential to knowledge production, and this is because of how "knowing" and "seeing" collide in the history of the experimental life sciences. This is why Evelyn Fox Keller provocatively asks whether there are not "circumstances when the mere observation of a phenomenon is itself so satisfying and compelling that no further explanation seems to be required? When seeing itself is a kind of understanding" (2002, 206). Her answer is an unequivocal yes, and, with some modification to her bold thesis, we saw this play out in the history of biology: the technologies of visual perception are often what make explanation and knowledge possible. Visual practices are also epistemic practices.

In the context of chapter 2, I was primarily concerned with demonstrating how knowing in biology depends upon seeing. Seeing not only provides the necessary verification for a theory but also helps generate the theory itself. But what I did not consider at that time, since it fell outside of the scope of our discussion, are those instances when seeing is not enough. In other words, these are cases when certain aspects of living systems are visible through technical mediation but it is unclear whether anything can be said about those systems with confidence. This is an important supplement to Keller's bold assertion that "seeing is a kind of understanding." What was not considered in our previous discussion, and what I will pay special attention to here, are those cases that fall into the vague or perhaps shadowy region of visual knowledge practices in the life sciences. These are cases that tend to be left out of history textbooks and thus are not usually cited among the most important moments of scientific knowledge, but they are essential to its production.

Microcinematography (or the combination of microscopy and cinematography) emerged in the beginning of the twentieth century at right about the time when cell theory was going out of fashion and classical genetics was getting into full swing. In recent years, scholars from the history of science and film and media studies have come to appreciate the importance of microcinematography. But as previously mentioned, there is a reason why microcinematography was not included in our history of visual knowledge practices in the early twentieth century. In contrast to prevailing trends in scholarship on science media, I argue that it is only after the architectures of power and governance are adequately posed that we can return to visual practices that were on the margins of scientific acceptability and understand their deeper significance for the design and formation of our present. Thus if I rely on the careful scholarship of Hannah Landecker, Oliver Gaycken, and Scott Curtis, then I also argue that it needs to be supplemented in important ways. In particular, what I call the "cinematization of biological knowledge" needs to be connected to wider strategies of governance in the twentieth century.

In general, microcinematography is a practice that uses time-lapse photography in conjunction with microscopy in order to witness living cells develop in time. Although the French medical doctor Jean Comandon and businessman Charles Pathé are often credited with inventing microcinematography in 1909, the technology had already been in use well before their popularization of it. In fact, the first wave of microcinematographic experiments was produced in 1903 by the joint efforts of the naturalist Francis Martin Duncan and the British film producer Charles Urban, founder of the Charles Urban Trading Company (Gaycken 2015). (Although Étienne-Jules Marey and Alfred Nachet experimented with microcinematography as early 1891.) As Gaycken discusses in *Devices of Curiosity: Early Cinema and Popular Science,* Duncan and Urban produced the first public program of microcinematography that showed at the prestigious Alhambra Theater in London on August 17, 1903. The program was called "'The Unseen World': A Series of Microscopic Studies, Photographed by Means of the Urban-Duncan Micro-Bioscope" and included a series of short films that had titles such as *Cheese Mites, The Circulation of Protoplasm in the Waterweed, The Fresh-Water Hydra,* and so on.[2] According Gaycken's research on the program's reception history, "The Unseen World" attracted a considerable amount of public attention and provoked a wide range of responses from the media. Suffice it to say that the majority of them expressed the sense of "marvel," "horrified fascination," and even a return to cinematic "roots" (à la Lumière brothers) that the audience felt upon seeing the mysterious movements of the invisible world of microbes projected before them (Gaycken 2015, 16–53). Gaycken notes how *Cheese Mites* in particular seemed to provoke the audience. With an opening shot of a man (played by Duncan) reading a paper and then inspecting his piece of cheese with a magnifying glass only to find a teeming world of microscopic cheese mites, the film seemed to tap into the fears and fascination of the audience members concerning the "microbic monsters" in their food (30). Despite the tendency to read the film as a kind of "horror," Gaycken argues that there are also important comedic elements in the film (Duncan's own "self-mockery," etc. [33]). There are many other details to note about *Cheese Mites,* as well as the overall program of the "Unseen World," including its mode of presentation in the theater (24) and even its proximity to the genre of "trick films" (37). Taken together, these elements suggest that Duncan and Urban's cinematic experiments fit neatly into the history of early cinema.

As a consequence, these films seem far removed from the serious business of science. If anything, they seem much closer to the "cinema of attractions" (Elsaesser 1990) than to the meticulous work of microscopists and geneticists in these early years of the twentieth century.[3] Thus it is not clear why the history of science would have reason to include them among their important moments of knowledge production. As one reviewer for *Nature* put it: "Those interested in science need not spend the evening there [Alhambra Theater]" (cited in Gaycken 2015, 22). Such skepticism was not limited to "The Unseen World," however: Comandon's 1910 film capturing the movement of *Syphilis spirochetes* was not readily embraced by the wider scientific community either. After screening one of his films, outraged scientists exclaimed that the cinema was not to be taken seriously—"entertainment for the ignorant," one scientist declared (Painlevé 2000, 17). Likewise, prominent scientists voiced their concerns about using the medium for serious science. As Peter Medawar writes in *Memoir of a Thinking Radish,* scientists who used cinematography were "delighted, distracted, and beguiled by the sheer beauty of the cultivated cells," but unfortunately this led them to miss the opportunity to "solve biological problems" (1986, 63; Kelty and Landecker 2006).

Although this skepticism may have been shared by many viewers of early microcinematography, this is not the full story: microcinematography was used for research in cell biology throughout France, Germany, the United States, and the UK in the first few decades of the twentieth century. It is in this context that Curtis's *The Shape of Spectatorship* is interesting. The book develops a critical method for evaluating how the cinema could be appropriated by various scientific disciplines in pre–World War I Germany. In particular, Curtis argues that in order for cinematography to be "useful" outside of the entertainment industry—for example, in early twentieth-century cell biology—the medium had to "fit" into the larger disciplinary system. In other words, what film and media scholars think film is does not necessarily correspond to what scientists think film is. And so, in order to imagine that an entertainment technology could be used by a scientific discipline, certain conditions need to be met. However, the criteria for fitting are never apparent or straightforward: for "'what cinema is' for one group," writes Curtis, "was not necessarily the same as for another. Indeed, the nature of the appropriation often depended on what the agents thought film was" (2015, 2). And for Curtis, "The formal features of the technology [the cinema] matter. . . . The disciplinary legitimacy of a technology or the successful appropriation of a technology by an expert group depends on a correspondence between the logic of the discipline and the formal features of the technology" (4). The cinema's formal features concern "the projected image and the frames on the celluloid filmstrip," while the disciplinary logic "refers to its problem-solving protocols, its investigatory methods, and its ideological (in terms of its discipline) assumptions" (4–5).

Now there are several aspects of Curtis's analysis that interest us. One of them has to do with his discussion of Henri Bergson's argument that scientific knowledge is, by its very nature, cinematic. For Bergson, the cinema and science are internally related because they both cut up continuous processes into instants (abstractions) that can be divided to infinity, and for this reason fail to register the living reality of the changing whole. Neither science nor the cinema understand movement as continuous change and so "to the extent that science followed this tendency to divide movement into stable units and to treat the interval as an enabling ellipsis and not as precisely the problem, modern science for Bergson was essentially cinematic" (34). For Curtis, a cursory reading of Bergson's critique illustrates the natural affinity between science and cinema: the formal features of film fit into the disciplinary logics of science. This would seem to account for why microcinematography was an important research tool for Jean Comandon, Alexis Carrel, Hermann Braus, and many others (35).

However, there is another sense in which this identification misses the point entirely: the biologist's use of cinema had much more in common with Bergson's method of intuition than it did with scientific abstractions (83).[4] This didn't mean that microcinematography was "anti-science" (pace Bergson), rather it meant that scientific knowledge changed through motion pictures: it became time-based (86). To see how the cinema produced new knowledge in cell biology, Landecker's analysis of Comandon and Carrel is especially useful. In particular, she argues that their work opened up new conceptual territory for biology. Their cinematic experiments, along with those of a handful of other working scientists, were not capturing what was already known about the biological world; nor were they spatially dividing up the world into analytical frames, but they were forging new possibilities for thinking about continuous movement (duration) through the cinematic medium.

Let's start with Comandon. His research was centered on diagnosing *Syphilis spirochetes,* and in the early 1900s, this was a considerable challenge since syphilis can lie dormant in the infected patient for years and it often imitates other diseases. The other challenge was technical. With light microscopy, the greater the magnification, the less light illuminates the specimen. This was a serious problem, since identifying this hidden disease required making very precise distinctions among specimens at or below the wavelength of light. To solve this problem, Comandon experimented with the ultramicroscope, which was developed in 1902 by Richard Adolf Zsigmondy and Henry Siedentopf, under the Carl Zeiss AG (Gaycken 2015, 92–93). The ultramicroscope, unlike the traditional light microscope, illuminates from the side through scattering light rays, allowing the specimen to be seen against a dark background. This method enabled Comandon to examine hundreds of specimens and eventually notice that *Syphilis spirochetes* was identifiable by its unique movements. And because the ultramicroscope provided just the right amount of contrast, Comandon could capture these movements using time-lapse photography. As a number of scholars have noted, the technical equipment at the hospital where he was working was insufficient, so he teamed up with Pathé, the French film industry pioneer, who saw a unique investment opportunity (Landecker 2005; Gaycken 2015; Lefebvre 1993).[5]

According to Landecker, the cinema changed how biologists thought about the living world. Where histological methods required living specimens to be fixed and stained—in other words, they had to be dead and motionless to be seen—film allowed researchers to study living systems as the animate systems they really are. No longer was it necessary to infer movement over time from "histological stills"; film made it possible to see movement itself—the dead brought back to life, so to speak (2005, 911–12). What's more, Comandon's films on syphilis and other bacteria acknowledge the various scales of time folded into the experimental setup: "The time-lapse films that Comandon began to make after 1909," Landecker writes, "relied on the difference between the time of the experiment (hours) and the time of projection (minutes) and the resulting acceleration of very slow, otherwise utterly imperceptible movements" (912). Comandon ensured that the viewer registered the multiple timescales folded into the experiment. In the *Movement of Leukocytes,* Comandon shifts between different magnification scales: "The viewer gets the distinct impression of what it is like to flip between objective lenses on the microscope, from lesser to greater magnification. Depending on the choice of lens, one could focus on the undulating membrane—the physiognomy of a single cell's surface—or one could pull back and focus on a teeming mass of white blood cells attacking a clump of bacteria" (914). Comandon's experimental apparatus—ultramicroscopy and the cinematograph—was not only able to capture and study the time of living systems but also able to shift between and among the many timescales of cellular life.

Another key figure in the history of scientific microcinematography is the French surgeon, biologist, and eugenicist Alexis Carrel. In *Culturing Life: How Cells Became Technologies* (2007), Landecker situates Carrel's work not only in terms of its relevance for the experimental life sciences but also in terms of the history of tissue culture and biotechnology generally. The framing question for her book is: How is it that living matter became technical matter, which is to say, matter that persists and is experimented on outside of the host organism in a laboratory setting? In other words, how did life become technology? And in the course of her analysis, Carrel's microcinematographic experiments figure as key ingredients in the emergence of "technical life." This is significant for a number of reasons, chief among them being that the cinema is placed in a genealogy that conditions the biotechnical and bioinformatic life that scientists presume today (2). In this context, Landecker examines how Carrel's research on cell cultures makes it possible for cells to be "induced into a state of immortality using the right techniques" (68). She notes that while the American embryologist Ross Harrison demonstrated the "autonomy" of cells from their host bodies—with the right medium, cells could live in vitro for extended periods of time—Carrel modified Harrison's techniques to radicalize what could be done with cell cultures: "Rather than taking an internal process and making it visible externally [Harrison], in Carrel's system the internal process was rendered permanently external to the animal body" (72). Thus we get a new way of thinking about externalization in Carrel: tissues are no longer related or connected to the original body of the organism; they have their own, artificially induced life.

However, keeping tissues alive for an indefinite period of time, as well as ensuring that cells multiplied at the same size, initially proved difficult for Carrel. Key to tackling the first challenge was the technique of "rejuvenation." Essentially, rejuvenating a tissue required removing all of the elements in the culture that prevent the tissue's growth (waste products) and then placing it in a fresh medium (74). The second problem, multiplication at the same size, was solved once Carrel realized that ground-up embryonic extracts, or what he called "embryo juice," provided the right "nutritive medium" for the continued growth cells and maintenance of size (77). And so in 1913, once Carrel realized that he had to continually refresh tissues in the right nutritive medium, he was able to keep an embryonic chicken heart alive longer than a chicken (79). This experimental result led him to the conclusion (prematurely, perhaps) that "senility and death of tissues are not a necessary fate but an accidental one" (74) and that "immortality" could be attained through the appropriate technological intervention (79–80).[6] Although the First World War interrupted Carrel's research on tissue cultures, he resumed work after the war. According to Landecker, Carrel found his old hanging-drop preparation to be inefficient both for observing tissue growth and for refreshing the medium. To address these challenges, he designed a new flask, which later came to be known as the "Carrel flask," specifically suited to the needs of his working method (82–84). But what is essential to Carrel's new experimental setup is the introduction of time-lapse microcinematography. The camera was "[a] cornerstone of the new tissue culture apparatus" because it permitted the control and manipulation of time (84). Cinematography not only made the organismic time visible but also allowed it to be an object of knowledge, intervention, and control. And so together with the other elements in the apparatus, the camera worked to "materialize biological time as a thing that could be physically intervened in as a part of experimentation with tissue culture" (85). What's crucial is that the cinema enabled the study and control of the very thing that Carrel aimed to examine in cell cultures all along: biological time.

For this reason, Carrel (1931a) called for a "new cytology" that would make the study of the cell's unfolding in time its central concern. Where old cytology focused on inert and abstract forms due in large part to the equipment available, new cytology would focus on functions, and more specifically, on the cells as temporally complex agents. And where old cytology was wedded to microscopy (see chapter 2), new cytology would fold cinematography into microscopy to form one microcinematographic apparatus. What's more, Landecker claims that "Carrel used cell cinematography as the foundation for a theorization of biological time as completely distinct from clock or solar time" (2007, 87). And in this spirit, Carrel made explicit connections between his work and Bergson's theory of duration, even suggesting that his cinematic experiments allowed him to "see duration," and thus to study it, measure it, manipulate it, and even bring about immortality (Carrel 1931b). Carrel therefore enjoins us to see the cinematic apparatus outside of Bergson's early critique of it and view it as a partner in the (bio)philosophical method of intuition.

This brings us right back to Curtis: the cinema and science are indeed deeply intertwined (as Bergson thought), but not in the way

Bergson originally conceived: the cinema is an essential companion in the time-based investigation of matter. This doesn't mean the cinema is alien to science, but rather that the cinema is the right technical medium for the transformation and refinement of biology's problem. If there is a "fit" between biology and the cinema, then it is one that has to evolve. What the cinema and science mean to one another needs to be rethought (Curtis 2015, 86). Citing the German biologist Hermann Braus's use of microcinematography to demonstrate Ross Harrison's research on the development of nerve fibers in tissue culture, Curtis writes, "Motion picture technology presented itself as a potential research instrument to biologists involved in tissue culture, but film became a powerful tool for researchers because the research program in biology had changed to accommodate it. That is, Braus came to film not only because it solved a particular problem but because the problem itself was shaped by changes in conceptions of growth and time that motion picture technology could articulate (indeed, that the moving image might have prompted)" (86).

What's crucial to realize is that immortality is constructed: it is not a given, but rather brought into being through the cinematic apparatus (Landecker 2007, 89–90). In this context, Landecker argues that microcinematography "is a technological portal to another world of time," and because of this, the medium shapes scientific perception and knowledge in very specific ways (2011, 385). Drawing on Don Ihde's assertion that "technologies, particularly instruments . . . should be regarded as means by which our perceptions and our wider experience are modified and transformed" (1998, 1), she acknowledges that perceptual subjects are constructed by means of their tools—and in this case, the cinema. But what also occurs here, and seems more pressing for her, is that cellular worlds are produced by the microcinematographic machine: "The introduction of time-lapse mechanisms and practices," she continues, "significantly transforms the quality of that world from one that is spatial in dimension to one that is another world in time" (Landecker 2011, 386). In other words, the cinema constructs (or is a portal to) a particular kind of cellular world for the scientist-perceiver/subject: one that is dynamically unfolding in time and permits new forms of knowledge about cellular life and death, including notions of immortality, cellular locomotion, endocytosis, and apoptosis (387).[7]

To situate this discussion in terms of the theoretical frame developed in the previous chapter: microcinematography is a medium that brings heterogeneities into a communicative relation. It connects the perceptual subjectivities of scientists and microscopic cellular worlds. Between the perceiving subject (scientist) and object (temporally developing cell) there is no pregiven relation; it has to be constructed though the microcinematic medium. Although Landecker is masterful at examining the cellular worlds that the cinema constructs for its subjects, she seems less interested in cashing out how the perceptual subjectivities of scientists are shaped by technical media, especially in light of the wider developments of media subjectivity in the twentieth century.[8] So if, as Ihde and others claim, scientific perception is "modified and transformed" by its instruments and technologies, then we also need to inquire into the quality of that modification and transformation, instead of imagining that these changes are insulated from how perceptual subjectivities are shaped by the wider media landscape in the twentieth century (see chapter 6).

But before I examine these transformations in greater depth, I want to reflect on the wider implications of our analyses thus far. We noted at the outset that microcinematography does not hold a prominent place in the annals of science. Even though Landecker and Curtis demonstrate that microcinematographers were contributing to the scientific discourse, and in many ways proposing conceptual frameworks that would not have been possible without the cinema, there is an important sense in which their findings did not gain the kind of traction within the wider scientific community that one might have imagined. This claim may seem slightly overblown, since it would be nearly impossible to measure the currency of a set of technologies, practices, and discourses within the scientific community at any given time. Furthermore, Curtis mentions that "while film histories often repeat legends of academic hostility to researchers using motion pictures in their experiments . . . evidence of such protests is actually rare in comparison with the general enthusiasm displayed for the new technology" (2015, 29). What's more, Curtis shows how the cinema was used and celebrated in the context of early twentieth-century biomedical sciences in Germany: indeed, motion pictures fit squarely into what "German doctors considered to be the *proper* mode of observation" in contrast to the lazy spectatorship of the masses (95).

Nevertheless, what do we make of the fact that microcinematography has been largely forgotten by historians of science and media until recently? Christopher Kelty and Landecker write that "microcinematography is not generally recognized as significant to the making of biological science, despite its constant role of reminding biologists that their subjects are alive" (2006, 43). Given this, it should be no surprise that a prominent historian of science like Evelyn Fox Keller, who has devoted considerable attention to the visual cultures of the life sciences, would only make one mention of microcinematography in her entire work (Keller 2002, chapter 2). She notes, and somewhat dismissively, that it missed the real prize in life science research in the early years of the twentieth century: "Even with cinematography," writes Keller, "the interior processes by which a fertilized egg gives rise to a complex organism remained invisible" (216). In other words, microcinematography was largely beside the point. Further evidence of this can be gleaned from the fact that no "new cytological paradigm" was initiated by a cinematic revolution in the experimental life sciences. While film continued to be used in the biomedical sciences throughout the twentieth century and even thrived in some arenas and continues to be combined with light microscopy practices today (see chapter 2), it did not have the kind of impact on the practices of life science research that Carrel and others imagined (Hentschel 2014; Gaycken 2015).

But why was microcinematography often overlooked in these early years, especially when, according to Landecker, Curtis, and others, it was making such important contributions to the biological sciences? Keller's words, which echo our discussion in chapter 2, are instructive: "Interior processes by which a fertilized egg gives rise to a complex organism remained invisible." For experimental biologists, the molecular world concealed in the chromosome was the true prize. And by the 1910s, when microcinematography was maturing, Mendel's laws had already been rediscovered and genetics research was in full swing. Johannsen, Correns, and Morgan were among those leading the charge to hunt down the material basis of the gene. This magical entity promised to answer the mysteries of both heredity and development. Landecker notes microcinematography's unfortunate timing in the introduction to *Culturing Life*: "If we look at the historical literature, it would seem that the cell was of little importance after the rise of genetics and molecular biology in the 1920s and 1930s" (2007, 6). And while chapter 2 complicated overly simplistic readings of classical genetics—as if this period of experimental research could be summed up as narrowly focused, wildly reductionist, and therefore worthy of our dismissal—it is no less true that cytology was no longer popular among experimentalists and that research on the cell had been largely supplanted by the experimental study of the gene and then eventually the DNA molecule. Carrel's "new cytology" would likely have seemed like a futile call to revive a dead paradigm. And so while there may be truth to Curtis's claim that the cinema was relevant to the medical sciences in pre–World War I Germany, the same cannot be said for the biological sciences: the criteria for relevance had shifted—cell biology was old biology. Time-based knowledge about cellular development (produced via microcinematography) was peripheral to gene-centered investigations of heredity and development.

But the story of the cell is far from over. The molecular paradigm in biology is fading and cellular function has returned to the biosciences and does not seem to be going anywhere soon (Keller 2000). After years of isolating DNA molecules, and then finally sequencing the human genome, scientists have begun to realize how little they actually understand about gene function. Genes are not isolated from the rest of the organism and can only do their work as a part of an overall system. In chapter 3 we discussed how this shift in orientation helps to account for the rise of systems biology, as well as the excitement over epigenetics in recent years (Deichmann 2016; Goldberg, Allis, and Bernstein 2007). Now we can add more texture to this history: one way to understand the rise of systems biology is in terms of the return of the cell to bioscientific research. Landecker celebrates this so-called "return," noting that "after decades of attention to other biological entities, particularly The Gene, such a return to the livelier, complex, and indeed entirely more personable entity of the cell seems refreshing" (2007, 6). The cell's "reappearance as a central actor in today's biomedical, biological, and biotechnical settings" is enthusiastically greeted by Landecker, since it signals a move away from the mechanistic operations of molecular biology and a return to the dynamic complexities of the living.

If microcinematography was often overlooked by scientists in the early twentieth century, then it was not because the technology was anachronistic. Quite the contrary: it may very well be that microcinematography was before its time. Researchers working with microcinematography were pursuing questions about the temporal development of the cell (and even going so far as to make comparisons to Bergson's duration) that would only come into focus for scientists many decades later. Scientists would have to wait until certain dead ends in molecular reductionism were discovered before they could be reacquainted with the temporal complexities of the cell. There is a reason that historians of science are returning to the temporal "portals" opened up by microcinematography over a century ago: similar portals are being opened up for biologists today.

SENSORIAL DESIGN: THE CINEMATIZATION OF BIOLOGICAL KNOWLEDGE

Such prolepsis in biological thought should not distract us from asking other pertinent questions—namely, how is the perceiving subject modified, transformed, and designed by its perceptual apparatuses? If Landecker acknowledges, following Ihde and others, that the cinema transforms and modifies the scientific perceiver, then she leaves this question for others to pursue and almost dismisses it as intuitive and a given (2011, 385). But I do not think it is as intuitive as she claims, and it is certainly not obvious how to deal with the fact that the modes of perception that the cinema makes possible within the domains of biology cannot be wholly disassociated from the modes produced by popular media practices.

Curtis, for his part, addresses scientific and popular viewing practices but focuses on how medical perceivers differentiated themselves from spectators:

> Most obviously, researchers, unlike the average movie-going audience, were in control of the image and its projection—they could literally manipulate their films however they pleased. But this ability to control the rate of projection went hand in hand with a mode of viewing that was very much part of a nineteenth-century researcher's training and identity, which distinguished itself from another mode of viewing attributed to the lay public, especially the cinema audience. (2015, 95)

The cinema was thought to have a "hypnotic effect" on the spectator, and "the masculine, medical professional was distinguished," Curtis continues, "not only by his will, but by his training in scientific observation. Indeed, this training was his primary defense against the onslaught and seductive power of (moving) images. The scientific appropriation of motion pictures depended on the ability of the researcher to halt this onslaught, to forestall its assault and to read slowly" (140). The scientist's ability to enact the "enlightened" practices of cinematic perception depended, in large part, on his capacity to differentiate them from "unenlightened" perceptual practices (141). Enlightened practices of scientific observation display strict command over sensuous intake (the rush of modern life), thanks to the well-honed training of attention through scientific or medical education (132–33, 110–25). This Kantian subject is defined in sharp opposition to the mere spectator who is hypnotized by the speeds of modern life. But is it actually possible for this imagined Kantian subject able to immunize itself from the "hypnotic agent" of microcinematography?[9] If it is true that scientific perceivers are shaped by their technologies, then what are the ways in which perceiving subjects are shaped by microcinematographic experiments in particular? Do they parallel how subjects are controlled in the wider cultural sphere, or are there other modes of perceptual design operative?

The crossover between science and popular entertainment culture is explicitly addressed in Gaycken's work. This is one of the reasons he opts for the label "popular-science film," a phrase that straddles the thorny epistemic and cultural terrain of films that are made by scientists, and had scientific aspirations (in most cases), but which are consumed by the wider public (2015, 8–9).[10] Despite the clean differentiation medical practitioners made between scientific and popular perception in the context of pre–World War I Germany, in reality the relation between these two domains is not so neat and tidy. Comandon's films are good examples. Although the general public may not have been concerned with the quantification of organismic time (Comandon's interest), they were interested in consuming cinematic expressions of organismic time. With Pathé's backing, Comandon's films were shown widely and were met with rave reviews in the press and even made whole new markets for the French film industry possible. "Comandon's films," insists Gaycken, "fulfilled certain crucial roles for an expanding cinema industry, allowing

Pathé to integrate popular-science films into the studio's increasingly rationalized output" (97). Gaycken speculates that the success of these films may also have had something to do with their resemblance to trick films (because of their black background), their original subject matter, and even their visualization of something "taboo"—namely, syphilis (96–97).

The larger question is: Does the influence between science and entertainment flow only in one direction? Or does the influence go both ways? We know that motion pictures made with scientific audiences in mind were also readily consumed by the moviegoing public; we also know that this public was subject to a subtle but no less significant rewiring of its sensual apparatus in order to better equip it for the modes of perceptual subjectivity required by then-emerging paradigms of governmentality. But was the sensorium of the scientist not also being redesigned, sculpted, and shaped in ways that would prove advantageous for emergent regimes of power? Or was it the case that the scientist was a good Kantian subject capable of protecting himself from the moving images' capacity to redesign his sensorium?

My claim is that this is clearly not the case, and that one only need to look to the knowledge produced by microcinematography to understand how it fits into wider programs of sensorial governance. In other words, scientific perceivers were not at a safe distance from wider circuits of power. Biologists working with microcinematography had no illusions about peering into an unmediated biological world. They were not epistemological realists in a naive sense. Carrel, Comandon, Duncan, Ries, and others were all too aware that the realities brought before them were conditioned by the apparatuses that made them possible. Landecker makes this fact startlingly clear when she remarks that "the notion of a microscopic world exists well before microcinematography, but the introduction of time-lapse mechanisms and practices significantly transforms the quality of that world from one that is spatial in dimension to one that is another world in time" (2007, 386). "What changed with microcinematography at the beginning of the twentieth century," Landecker continues, "was the constitution of the microworld as a distinctive *temporal* domain" (390). And this is echoed in Carrel's own writing when he describes how scientific concepts, such as immortality, are "operational concepts; in other words, concepts equivalent to the set of operations by which they were acquired" (Carrel, quoted in Landecker 2007, 89; Carrel 1938).

The scientific perceiver of the microscopic world exists only in relation to the mechanisms that give rise to her practices of perception. And in the experimental culture of early twentieth-century biology, the microscope was largely in charge of engineering the possibilities for biological perception. The light microscope constructed a biological world framed through space: the structure of the chromosome and the location of the gene on it, these were the animating questions for a biological perception designed to relate to the living through the logics of space (see chapter 2; also see Hacking 1983, chapter 10). By incorporating the cinematograph into the experimental apparatus, the biologist's perceptual practices changed dramatically: they were oriented by time. Time instead of space was the organizing principle for a biological perception shaped by the cinema. Thus many of the invisible processes and functions that plagued experimental biologists were now visible. What seemed entirely heterogeneous and alien to human perception—the temporal unfolding of living systems—is now coterminous with it. The elusive temporalities of the living are sped up and made human (Pauly 1994). Microcinematography therefore marks a profound moment in the history of scientific perception: it is when the temporality of living systems is available for sensuous intake. Not even "immortality" is out of the reach of perception. The scientific perceiver is no longer spatially bound: she has been pried open and is capable of consuming multiple scales of time at once (Comandon). She can even touch what is altogether outside of time: the immortal (Carrel). As Sergei Eisenstein understood all too well, this happens because the sensuous faculties are reengineered; the nervous system, the cinematic apparatus, and the time of the living are all "speaking the same language" and are a part of the same communicative mechanism. Thanks to the cinema, cellular time is no longer exogenous to perception; it has entered into circuits of exchange with it.

The cinematic redesign of scientific perception thus shares important features with Eisenstein's conception of cinematic management and control. According to Jonathan Beller, we should situate Eisenstein's *Strike,* as well his thinking about the function of cinema more generally, in terms of Ivan Petrovich Pavlov's work on the mechanized control of the psyche, as well as Fredrick Winslow Taylor's theories of the industrial management of work (2006, 117). Eisenstein's revolutionary cinema perfectly expresses their theories of management and control: the cinema is a mechanical system for redesigning the psyche of the proletariat. As Eisenstein writes, "I am a civil engineer and mathematician by training. I approach the making of a motion picture in much the same way as I would approach the equipment of the poultry farm or the installation of a water system" (Eisenstein, quoted in Beller, 124). For Eisenstein, the sensorium of the audience merges with the cinematic machine; it becomes a component in a cybernetic system engineered to achieve revolutionary ends. The qualitative differences of the audience, explains Beller, are rendered quantitatively equivalent to the cinematic machine in order to achieve the scientific reorganization of a population (126). Radically heterogeneous systems and forces are made compatible; "In computer talk: Industry, the nervous system, and representations are all beginning to speak a compatible language, a compatible machine—or systems—language" (137). And while Beller admits that this sounds like a far cry from the revolutionary cinema Eisenstein envisioned, and especially the "autonomy" of the worker associated with it, "in the historical long view, the revolution of which he [Eisenstein] was a part largely differs from the one thematized in his early films" (127). And so the revolution Eisenstein actually helped instigate is "the production of film as a factory for laboring sensoriums" (127). What Eisenstein realized very early on is that the "spectator was conditioned by the cinematic apparatus," that the cinema trained and molded the sensorium, and for this reason it was of paramount importance to place "the cinematic apparatus in the capable hands of a politically correct director" (127). The spectator's nervous system and the cinematic machine "speak compatible languages," they are a part of the same cybernetic system—the spectator becomes a component in a much larger mechanism.

If we were to use Eisenstein's theory to explain the redesign of scientists' sensuous faculties, then we would end up with a very different conception of scientific perception than we get in the history of science: the cinematic molding of perception that happens in microcinematographic experiments is not exterior to the governance of perceptual practices occurring in the wider public; it is part of a coordinated effort to make heterogeneous systems compatible. Just as Eisenstein drew attention to how the apparatus reengineers the psyche, so too the microcinematographic machine makes a whole new sphere of information inputs available for the scientist. With microcinematography, an elusive data field—namely, the temporal patterns of living systems—is opened up for sensuous consumption. If experimentalists in the twentieth century found themselves largely bound to spatial information, then microcinematography was indeed a portal: it was a portal into an entirely different world of visual information. Suddenly, the scientific brain and cellular time could interface through the medium of cinema. They could speak the same language. If biological perception and cellular time were once exogenous systems in experimental biology, then the cinematic machine renders them compatible.

But insofar as the microcinematographic apparatus reveals a temporally conditioned cellular world, then this is a world in process; it is a world that is flexible, open-ended, and unpredictable. Landecker (2007) argues that cinematic time and plasticity are intimately related in the history of biology. But notice that it is just this flexible, process-based, open-ended learning that is a hallmark of neoliberal epistemology (see chapter 6). Indeed, the ongoing deterritorialization and reterritorialization of perceptual knowledge that scientific cinema makes possible is in many ways a cornerstone of Friedrich Hayek's theory of economic knowledge. Here's Hayek: "Today it is almost heresy to suggest that scientific knowledge is not the sum of all knowledge. But a little reflection will show that there is beyond question a body of very important but unorganized knowledge which cannot possibly be called scientific in the sense of knowledge of general rules: the knowledge of the particular circumstances of time and place" (1948b, 80). These are knowledge practices that go well beyond the spatially bounded practices of disciplined knowers (the space of Kantian subjects), even if they do not leave them totally behind. They are open to the particularity of circumstance, to the vicissitudes of sensuous environments, to the ongoing transformation of situations, and to the reality of constant change. In short, they are open to time-based phenomena. Are these not, then, the same architectures of perceptual knowledge designed through microcinematographic apparatus? Does microcinematography not anticipate neoliberal styles of knowing well before neoliberalism would become a theoretical and practical reality?

Within this frame, microcinematography is as much a deterritorializing force within scientific epistemology as it is a reterritorializing one.

Insofar as microcinematography is a scientific knowledge practice, then it also poses a nascent threat to its power within the cultural imaginary, which is to say, it represents a countertendency that undermines the authority of scientific epistemology. For Hayek, neoliberal epistemology undoes the privilege of scientific knowledge. A true economic epistemology is one that expresses knowledge outside of the "general rules" of science, and it applies to "the knowledge of the particular circumstances of time and place." Microcinematography does not simply undo the dominance of spatial knowledge practices within experimental biology (and thus reterritorialize it on time) but also poses a threat to the hierarchy of scientific knowledge as such (deterritorializing it from the authority of science) and anticipates various mergers of neoliberalism and science in the twentieth and twenty-first centuries (molecular animation).

What this means is that microcinematography marks a crucial moment in the genealogy of knowledge practices in the biosciences: it is when biological knowledge *becomes-cinematic.* The cinema makes a temporal way of knowing the biological world possible, if not sanctioned immediately, then eventually. (It is in retrospect, after the reemergence of the cell dynamics in biology, that we can affirm cinematic ways of knowing living systems to be important moments of twentieth-century biological knowledge production.) The essential work that Landecker, Cartwright, Curtis, Thierry Lefebvre, and others have done in this area is a testament to this fact. But these ways of knowing cannot be divorced from wider systems of cinematic capture. As we saw in chapter 6, the cinematic apparatus is part of an elaborate network of interwoven techniques used to govern perceptual economies. The fact that microcinematography was popular among mass audiences is a symptom of a much deeper transformation: biological knowledge was being subjected to a form of cinematic governance similar to the one controlling the attention of popular audiences. What this means is that there was a simultaneous *cinematization* of biological knowledge in the twentieth century. This cinematization had its own historical rhythms and would take decades for it to work itself out (and is most fully expressed in molecular animation), but in the early years of the twentieth century, biological modes of knowing seemed to merge with popular forms of consuming. Together, they would prime the sensorium for the neoliberalization of perceptual practices that would take hold in the coming decades.

What I am suggesting here challenges much of what has been argued about microcinematography in the history of science and film and media studies. This undervalued practice is read as a watershed moment of scientific knowledge production that contributes to conversations about the time-based development and evolution of living systems. We now know that living systems are irreducible to gene mechanisms and cannot be boxed into neo-Darwinian causal schemes (Henning and Scarfe 2013). The temporality of evolution and development is nonlinear and cannot be pre-stated (Kauffman 2016; Longo, Montévil, Kauffman 2012). Microcinematography was already tackling the temporality of cellular life well before biology was hijacked by molecular reductionism. Some have even alluded to the ways in which microcinematography bears on contemporary conversations about neovital materialism, something that was already intuited by Carrel in his work on Bergson (Carrel 1931b). What I am proposing does not so much contradict these arguments as much as it offers a necessary supplement to them. It implicitly channels decades of STS scholars who argue that scientific knowledge cannot be isolated from wider spheres of social and political knowledge (Fleck 1979; Jasanoff 2004; Latour 1987; Hacking 1999; among others); and it makes this plain in a domain where popular media practices intersect with scientific knowledge practices. I have attempted to shed light on how wider systems of governance are folded into what counts (or came to count) as knowledge in science. Lisa Cartwright's *Screening the Body: Tracing Medicine's Visual Culture* (1995) maps similar intersections of visual media, science, and power. However, she is concerned with broader practices of cinema and science than I am here and does not move outside of Foucault's disciplinary framework. I certainly would not want to claim that scholars working on the history of microcinematography are not also engaging with this tradition of STS and also channeling contemporary developments in media theory. Nonetheless, I do think that there is a tendency to uncritically affirm microcinematographic practices without placing them into wider systems of political and economic power. I worry that this trend is symptomatic of what Cary Wolfe (2012), David M. Berry and Alexander Galloway (2015), and others have diagnosed as our era's unfortunate tendency to affirm networks, vitality, and becoming at the expense of interrogating the political conditions that make these notions possible.[11]

What I am arguing for is a theory of microcinematography that on the one hand affirms the scientific value of the practice: this technology makes time, plasticity, and vitality central to visual knowledge practices in the life sciences. But on the other hand, microcinematography introduces these notions because it is subject to the same modes of governmentality that have captured the visual economies of popular media consumers. The implication is that microcinematography introduces new regimes of governance into the visual knowledge practices of the biosciences. It is for this reason that microcinematography functions as an apparatus in the Foucauldian sense: it is a medium that draws the heterogeneous domains of scientific perception and cellular activity together (makes them communicate thorough a compatible language) to produce knowledge and achieve specific ends. The cinema introduces ways of knowing into the biological sciences that are (proto-)neoliberal.

Microcinematography is a precursor to molecular animation. But where the latter is digital and molecular, the former is analog and cellular. It would take almost a century for governance to mature and devise a way to exert control in the biosciences by means of more flexible media (digital), at even smaller scales (molecules), and to even greater ends (neurosubjectivity and neuroentrepreneurship). What this means is that our earlier question concerning whether the life sciences provide an adjacent actual world (Whitehead) for molecular animation is beginning to have a clearer answer: the visual practices of experimental biology have not been outside of the purview of modern techniques of government for some time now. Well over a century ago, power was experimenting with ways to capture the psychic spaces of scientists, and microcinematography was one of those experiments. The implication is that molecular animation is an apparatus firmly situated in the biosciences because there was already the appropriate infrastructure in place for it to take root. Molecular animation represents a variation on a similar theme instead of an abrupt historical and epistemological transformation. The much more interesting question at this point becomes: In what way does molecular animation function as the evolution of a dispositif within the biosciences rather than emergence of something entirely new? I cannot pretend to answer this question in full here. And in fact this is more of a question that this book leaves us with instead of answering in any satisfactory way.

To help us frame the evolution of this dispositif, I want to address another moment in the history of visualization. This is a visual practice that is even less closely aligned with those of working scientists than microcinematography, but it is no less instructive for how we understand the evolution of those practices. I am referring to the mid-twentieth-century work of the Hungarian artists, designers, and educators László Moholy-Nagy and György Kepes. My interest in their research stems from how it begins to clarify the way in which visual perception comes to be controlled as part of an evolving information system, and how this is explicitly addressed as a problem of design. When we get to molecular animation in the twenty-first century, cerebral architectures and molecular databases are thoroughly entwined through the digital screen. With Moholy-Nagy and Kepes, we catch a glimpse of the early attempts to forge a convergence between visual perception and complex systems through the design of visual interfaces. And while Moholy-Nagy and Kepes do not concern themselves with the molecular world per se, they are explicitly tackling the problem of merging vision with the abstract patterns of biological organization through visual media—which *is* foundational to the practices of molecular animation. In our genealogy, these designers are not only continuing the work of microcinematographers, they are also preparing the way for molecular animators. They help us fine-tune how we frame the question of an evolving dispositif within the visual culture of the biosciences.

DESIGN AND THE EVOLVING SENSORIUM

There is already a wealth of excellent scholarship on Moholy-Nagy and Kepes. As fellow Hungarians, friends, and influential art and design theorists who both settled in the United States in the 1930s, their work has proven to be indispensable for understanding the emergence of a quasi-scientific, design-based discourse centered on questions of information theory, visual design, and social engineering in the mid-twentieth century. Two works in particular stand out as offering crucial insight into their research, especially as they relate to notions of control and governance in Deleuze and Foucault: Reinhold Martin's *The Organizational Complex: Architecture, Media, and Corporate Space* (2003) and Orit Halpern's *Beautiful Data: A History of Vision and Reason Since 1945* (2014).

Both books trace overlapping genealogies of information, control, and design to shed light on the governance of today's visual and epistemic spaces. I follow their extensive archival work closely in order to discern the ways in which Moholy-Nagy and Kepes (in conversation with a host of scientists and designers) are also a part of a genealogy of visual governance in the life sciences that begins to take shape in microcinematography and then matures in molecular animation.

Moholy-Nagy (1895–1946) was born in Bácsborsód, Hungary, and was trained as a painter and photographer. In 1923 he became an instructor at the famed Bauhaus, where he, along with Joseph Albers, was appointed to teach the foundational course. At the Bauhaus, Moholy-Nagy would teach and experiment in many different areas—from industrial design and sculpture to printmaking—but really honed his work in the photographic arts in connection with what he called the New Vision *(Neues Sehen).* The details of this new vision are outlined in a series of publications, but especially in *The New Vision: From Material to Architecture,* which proposes to design and cultivate an improved and better-adapted human sensorium in response to a quickly modernizing world. As he explains, "We are faced today with nothing less than the reconquest of the biological bases of human life. Only when we go back to these can we reach the maximum utilization of technical progress in the fields of physical culture, nutrition, housing and industry—a thoroughgoing rearrangement of our whole scheme of life" (L. Moholy-Nagy 1932, 30). According to Martin, "The ubiquity of rapid movement in all aspects of modern life necessitated," for Moholy-Nagy, "a biological adaptation of the human visual apparatus, aided by experimental photography, in order to process visual information received at even greater speed than ever before" (2003, 22). Experimental photography, working together with film and architecture, was essential for designing a superior sensing apparatus that could respond to and thrive in the modern technological world. Here's Moholy-Nagy:

> Without wishing to solve in a sentence all the imponderables of human life, we may say that the composition of a man is the synthesis of all his functional mechanism; i.e., that the man of a given period is most perfect when the functional mechanism of which he is composed—the cells as much as the complex organs—is being

> used to the limits of its biological capacity. Art brings this about—*and this is one of its most important missions,* for the whole complex of effects depends upon the perfection of the functioning. Art attempts to establish far-reaching **new relationships** between the known and the as yet unknown optical, acoustical, and other functional phenomena so that these are absorbed in increasing abundance by the functional apparatus. (1969, 30)

For Moholy-Nagy, photography was not a representational medium in the traditional sense. As he outlines in *Painting, Photography, Film* (1969), photography concerns the "manipulation of light" and generates new "optical relationships" instead of reproducing what is already given in the world (Martin 2003, 47). "To proclaim," writes Moholy-Nagy, "that our optical senses have been fundamentally enriched, by new creative principles in painting and the film still produces a revolutionary effect" (1969, 34). The function of the photographic medium, which is enhanced in various ways by film, is to expose the human sensorium to new connections instead of rehashing old ones. These visual media are capable of redirecting the *evolution* of the sense apparatus to be better adapted to its rapidly mechanizing environment. As Alain Findeli explains, Moholy-Nagy thought that "one should take nature, the great designer, as an example; however, whereas nature had plenty of time to operate and to make the right selection after many trials and errors, the designer has to operate more quickly with his or her project. This reality is why techno-scientific knowledge and esthetic judgment have to be developed and strengthened" (1990, 10–11). Consequently, it is imperative that humans intervene in the evolution of perception to overcome the "atrophy" and "deterioration" of sensuous capabilities, and photography is an indispensable tool for making this intervention (Martin 2003, 57).

In many ways this is a call to arms that justifies the redesign of the visual apparatus, as well as the creation of an entirely new design pedagogy to support it (Findeli 1990). To achieve these ambitious goals, Moholy-Nagy was in conversation with the evolutionary biologist and eugenicist Julian Huxley. According to Martin, Moholy-Nagy met Huxley in the 1930s and was compelled by many of his theories, especially his so-called secular humanism. In *Evolution in Action,* Huxley explains that "Man's development is potentially open. . . . He has developed a new method of evolution: the transmission of organized experience by way of tradition, which . . . largely overrides the automatic process of natural selection as the agent of change" (1953, 132). Out of this view, Huxley's eugenicist program took shape, which sought to breed a genetically superior human race by eliminating undesirable characteristics through managing birth rates, employment, diet, and so on, among the lowest classes of society (or in other words, eugenics via biopolitical racism [Foucault 2003]).[12] Moholy-Nagy, for his part, was encouraged by Huxley's proposal and sought to implement it in his own work on the evolution of human sense (Martin 2003, 54). In *Vision in Motion* (1947), Huxley is cited twice in the context of Moholy-Nagy's diagnosis of the poverty of sensual, physical, and intellectual life under industrial capitalism. Drawing on Huxley's *Evolution: The Modern Synthesis,* Moholy-Nagy writes that Huxley "sees the future of man in the development of man's capacities. 'There are many obvious ways in which the brain's level of performance could be genetically raised—in acuteness of perception, memory, synthetic grasp and intuition, analytic capacity, mental energy, creative power, balance and judgment'" (1947, 14). Moholy-Nagy then references Huxley's argument that not all selected-for characteristics are necessarily beneficial and that "they may be a dangerous balance of useful and harmful, or they may be definitely deleterious. For the statesman or the eugenicist to copy its [natural selection's] methods is both foolish and wicked" (14). Essentially, Huxley, and by extension Moholy-Nagy, assert that the elimination of certain characteristics (even if they have been selected for) through a combination of political (statesman) and scientific (eugenicist) measures is beneficial for the survival of the species.

What's especially troubling is that the biopolitical (or thanatopolitical) side of Moholy-Nagy's work is almost never mentioned in the scholarship (save a very brief note by Martin in *The Organizational Complex*). And when the connection between Moholy-Nagy and Huxley is referenced—for example, in an online article for the Tate website—the encounter becomes illustrative of the Bauhaus designers' affinity for ecological design and progressive views about the convergence of science and design in the achievement of ecological stability.[13] Others simply underscore the designer's "biofunctionalist" agenda and even play up its anticipation of much later theories of distributed cognition, affect, and emergence (C. Terranova 2016). While I have little doubt that Moholy-Nagy prefigures many of these developments, there is also a darker side to his thought: it is part of a wider conversation in the early twentieth century aimed at controlling the evolution of the human species to achieve an "ideal type."[14] And for Moholy-Nagy, the sensorium could be redirected through the appropriate use of visual art and design. He thought about the photographic apparatus in the way that eugenicists thought about their tools for managing the germ-plasm of "human stocks": it was a biotechnology for designing superior sensing machines.

A crucial feature of Moholy-Nagy's work on the reengineering of vision and media is curriculum design. If visual media could redirect the evolution of the human species, then a whole new breed of designers was needed (Martin 2003, 53). And much of this curriculum design took shape in the context of his work with Kepes in Chicago. According to Martin, in 1937 Walter Gropius (who was originally offered the position) recommended Moholy-Nagy for a project conceived by "industrialists" to create a Bauhaus-inspired school in Chicago that would be less experimental and have more "practical applications" (52). Upon accepting the position, Moholy-Nagy invited his friend Kepes to join him. Kepes, who had been with Moholy-Nagy in London in the mid-1930s, would be instrumental in shaping design education in Chicago. The New Bauhaus was only in operation until 1938; it was then reconfigured into the School of Design in 1939, and then into the Institute of Design in 1944 (Moholy-Nagy 1947).[15] But in its various permutations, this Chicago-based design initiative remained no less faithful to Bauhaus principles. Although as Martin describes it, there was a distinct influx of "protocybernetic" discourse in those years, especially with the addition of faculty such as the physiologist Ralph Gerard, who would later become a cybernetics promulgator (55).

The influence that cybernetics had on Kepes's work is profound. Kepes's own critically acclaimed volume, *The New Landscape in Art and Science* (1956), was in large part shaped by Norbert Wiener's work, especially his seminal *Cybernetics: Or Control and Communication in the Animal and the Machine.* In fact, Wiener contributed an essay to *The New Landscape* titled "Pure Patterns in a Natural World." What caught Kepes's eye was not only that Wiener had proposed a common communicational language for machines and organisms, and that he applied this to the human nervous system in particular, but also that Wiener proposed a conception of the organism that was a purely abstract "organizational pattern," which was not ontologically distinct from any other informational pattern (Martin 2003, 24–25; Wiener 1954, 132–33). What Kepes saw in Wiener's work was the possibility to build on what his longtime friend Moholy-Nagy had already conceived in terms of the redesign of sensuous surfaces. For Kepes also thought that the human sensuous life was in dire need of redirection. "Today," he asserts, "we experience chaos. The waste of human and material resources and the canalization of almost all creative effort into blind alleys bear witness to the fact that our common life has lost its coherency" (1969, 12). But for Kepes in particular, at stake was the alignment of the visual apparatus with very particular informational patterns.

Kepes, like Moholy-Nagy, envisioned a technologically and evolutionarily more advanced species of human perceivers. As he explains in *Language of Vision*:

> The language of vision, optical communication, is one of the strongest potential means both to reunite man and his knowledge and to re-form man into an integrated being. The visual language is capable of disseminating knowledge more effectively than almost any other vehicle of communication. . . . Visual language must be readjusted, however, to meet its historical challenge of educating man to a contemporary standard. (1969, 13)

Kepes thought that it was of the utmost importance to equip our visual apparatuses with the tools needed to survive in modern society. This was a rapidly changing postwar society that was dynamic (rather than static) and operating on many physical and symbolic levels at once; such were the textures of modern life that threatened to tear at the seams of perceptual life unless artists and designers intervened in its sensuous training. "Until today," Kepes writes, "the symbolic organization of psychological and intellectual conflicts has been limited in its power because it was fastened to a static system of objects and concept"; however, given the dynamism of contemporary life, "visual language must absorb the dynamic idioms of the visual imagery to mobilize the creative imagination for positive social action." For this reason, creative practitioners must "learn and apply the laws of plastic organization needed for the reestablishing of the created image on a healthy basis" (14). And "The ultimate aim of plastic organization," Kepes writes later in the text, "is a structure of movement that dictates the direction and the progression toward ever new spatial relationships until experience achieves its fullest spatial saturation" (52).

By the time Kepes writes *The New Landscape,* redesigning the sensorium explicitly entails training it to see patterns—"pattern seeing"—instead of isolated objects—"thing-seeing" (Martin 2003, 45). And the patterns at the base of this new and enhanced mode of vision are the ones obtained through scientific imaging of the natural world. *The New Landscape* conceives a perceptual landscape in which the subject shares a common language with the rest of the natural world. Kepes envisions a homogenized communicational field where "everything would be connected to everything else through a cascading series of feedback loops" (1956, 45). The perceiving subject is no longer at an ontological remove from the natural objects it observes; it has an immanent relation to them because subject and object share the same computational language. Seeing, for Kepes, therefore needs to be brought into line with the relational patterns (of binary code) that really exist in the natural world, instead of the isolated things that humans are habituated to see. And it was only through the design of the right visual interfaces that subjects could be trained to see dynamic patterns instead of static objects (not unlike how Sharpe and his colleagues describe computational function of molecular animation).

The images that Kepes had in mind, and which are on full display in *The New Landscape* and also described in his earlier work, *Language of Vision,* are "organic images." For Kepes, the plastic image is really a living image ("for the image to remain a living organism, relationships within it must be constantly changing" [1969, 52]), whose vitality is secured, for example, by the proper balance of rhythm (e.g., Mondrian and Doesburg [58]), continuity, and variety of organization. Living images are drawn from a wide spectrum of science, art, and architecture and display similar organizational patterns: individual components or parts are always a part of a wider dynamical system. In *The New Landscape,* Kepes selects individual images that have their own unique patterns but that also share organizational patterns with images of other temporal and spatial scales of phenomena. As Kepes writes:

> The obvious world that we know on gross levels of sight, sound, taste and touch, can be connected with the subtle world revealed by our scientific instruments and devices. Seen together, aerial maps of river estuaries and road systems, feathers, fern leaves, branching blood vessels, nerve ganglia, electron micrographs of crystals and the tree-like patterns of electrical discharge-figures are connected, although they are vastly different in place, origin and scale. . . . Their similarity of form is by no means accidental. As patterns of energy-gathering and energy-distribution, they are similar graphs generated by similar processes. (1956, 260)

Not unlike D'Arcy Thompson, who found morphological correspondences between the natural and physical worlds (see chapter 3), Kepes aimed to discover deep organizational continuities between seemingly unlike systems. Beneath the apparent differences in form lie much more profound relational patterns. Kepes's images were not static and isolated: they were dynamic and relational and had all the plasticity and flexibility of an organism. "The experience of a plastic image," writes Kepes in the *Language of Vision,*

> is a form evolved through a process of organization. The plastic image has all the characteristics of a living organism. It exists through forces in interaction which are acting in their respective fields, and are conditioned by these fields. It has an organic, spatial unity; that is, it is a whole the behavior of which is not determined by that of its individual components, but where the parts are themselves determined by the intrinsic nature of the whole. (1944, 16; see also Martin 2003, 59, 61)

According to Martin and Halpern, Kepes's notion of a new visual landscape was not only indebted to Wiener's cybernetics and to the attendant possibility of designing an algorithmic regime of seeing that would flatten the differences between inside and outside. It was also profoundly influenced by the work of theoretical biologists (Halpern 2014, 92; Martin 2003, 52; Roach 2010). While in London in the 1930s, Kepes had the occasion to meet and discuss his work with the Theoretical Biology Club, an Oxford group founded by, among others, C. H. Waddington, Joseph Needham, and Paul Weiss (Peterson 2016). As Donna Haraway and others have observed, this group was animated by questions related to the possibility of an organic philosophy, or "organicism," influenced by Whitehead's philosophy of organism, and interested in notions of process, pattern formation, self-regulation, and gestalt psychology, and how to visualize these dynamic relations (Haraway 1976). For Kepes these concerns resonated deeply; and along with cybernetics, they were the intellectual backdrop for his own organicist theory of visual design.

What is intriguing is that Kepes did not think that the reigning paradigms of scientific imaging were sufficient to redesign the human sensorium. Despite his interest in scientific images, and especially those that concerned the Theoretical Biology Club, they lacked the dynamism and continuity needed to shape the new perceptual landscape Kepes had in mind. Martin emphasizes this point when he draws attention to Kepes's interest in popular entertainment. In *The New Landscape,* for example, he writes that "motion-picture, television and related techniques have given us the flexibility we need to mark the time flow, the growth, succession, rhythm and orderly continuity of vision" (1956, 371; Martin 2003, 45). Similarly, in the *Language of Vision,* Kepes is drawn to advertising's capacity to design a "dynamic spatial experience" (1969, 217) precisely because it is "not handicapped by traditional forms" (221). He explains that spectators require more than "a sheer illustration of a fact," for such an illustration lacks the "vitality" required to effectively communicate (221). "To put an advertising message through effectively," Kepes insists, "the most heterogeneous elements—verbal message, drawing, photography, and abstract shapes—were employed. This variety of meaningful signs and symbols, could only be integrated by a dynamic meaning organization" (221, see also 221–28).

The takeaway here is that the revolution in vision, which Kepes believed would put the human sensorium on the same plane as the rest of nature (organic vision), was modeled on the dynamics of entertainment media. "Significantly," writes Martin, "this chain of patterns seen only by a new, technologically enhanced vision finds its most prescient articulation not in static scientific photographs but in the instruments of the mass media" (2003, 45). Popular media, rather than scientific visualization, held out the promise of informationalizing vision. Scientific imaging tends toward stasis and lifelessness, while film, television, and advertising display evolving time-based patterns capable of reengineering the nervous system. If there is some vague resemblance between

Kepes and D'Arcy Thompson, then it is offset by Kepes's reliance on the logics of popular media to generate a landscape of patterned vision necessary to construct a technically enhanced human perceiver. Like microcinematography before it, and molecular animation after it, Kepes's visual communication design used popular media to scaffold the architectures of perception that it wished to design.

It is important to remember the political urgency that underwrites Kepes's work. Not unlike Moholy-Nagy, who sought to meet the demands of a rapidly modernizing world, Kepes also found it necessary to respond to political and economic pressures through the reorganization of vision. But the world Kepes faced was one that had just been devastated by the Second World War (see the introduction to *Language of Vision*) and was on the brink of nuclear disaster in the Cold War era *(The New Landscape).* As Martin stresses, in *The New Landscape* Kepes "appealed to the organizational pattern as a means of overcoming the formlessness of subjectivity exposed to the conditions of the cold war" (2003, 70). He aimed to establish equilibrium through feedback within a control system. For Martin, the design of perceptual subjectivities continuous with the organizational patterns in the natural world cannot be seen outside of the mechanisms driving the military-industrial(*-entertainment*) complex. He argues that the erosion of subject–environment distinctions through computing aligns with the control mechanisms Deleuze outlines in the Postscript, as well as the production of new modes of subjectivity—namely, dividuality. What this means more generally is that if Moholy-Nagy's work exposes the thanatopolitical dimensions of sensorial design, then the visual landscape that Kepes envisions moves away from designing individual perceivers and embraces the production of endlessly dividing and recombining *dividuals* in a control society.

SHARED HISTORIES OF APPARATUS DESIGN

In the first sections of this chapter I proposed that microcinematography is crucial to the epistemic and discursive formation of a scientific culture that would come to find molecular animation valuable. Not only did this early twentieth-century technology pave the way for using time-based media to visualize biological processes but it also helped introduce concepts such as time, plasticity, and process into the visual epistemology of the life sciences. It is partly on this basis that I suggested that microcinematography is the cellular and analog precursor to molecular animation. But in the course of our analysis we also discovered how the medium functions as an experiment in governmentality: it is a dispositif that brings heterogeneities together for the purposes of governing perceptual and epistemic practices in the biosciences. If molecular animation is a dispositif, then it is because it is an evolution of an earlier microcinematographic apparatus.

It is in this context Moholy-Nagy's and Kepes's work becomes relevant. While their own biographies make it difficult to place them in any one disciplinary history, rarely are they thought to be crucial to the formation of the epistemic practices of the biosciences. However, as we piece together a genealogy that begins in microcinematography and ends in molecular animation, their work emerges as deeply relevant. Microcinematography is an effect of various technologies of power that makes time-based knowledge about living systems possible. But if microcinematography is preparatory for molecular animation, then there is still a gulf that separates the two: where the cinema makes it possible for scientific perception and cellular systems to "speak a common language," by the time molecular animation practices are in full swing, the nervous system and the teeming world of molecules are thoroughly enmeshed through the computational language of computer animation. What Moholy-Nagy and Kepes clarify is how these two moments in the history of perception might be stitched together. In Moholy-Nagy, the human sensorium functions as "raw material" that can be sculpted for the purposes of adapting to the rapidly modernizing world. At stake is designing sensuous surfaces that are "fit" for technical modernity—in the evolutionary sense of fitness. Moholy-Nagy's project therefore has deep affinities with the early twentieth-century eugenics movement, although it aims to intervene in the biology of perception using the media of painting, photography, film, and architecture. Kepes, on the other hand, was also concerned with designing an enhanced mode of technical perception that would respond to the (evolutionary) pressures of modernity. However, what he brought to bear on this question was a distinct blend of cybernetics, organicist biology, and popular entertainment media. In this unique cocktail of science, engineering, and popular culture the sensorium merges with the organizational patterns in the natural world. A new equilibrium (or normalization in Canguilhem and Foucault's sense) could be established. The evolution of sense necessitates the design of an algorithmic visual landscape that would unify the dynamic relations in the world with the perceptual practices of subjects. Kepes envisions a computational continuum that would erode distinctions between inside and outside and that would be the basis for the formation of *dividualities* that could restore equilibrium to the disorder experienced by subjects in a Cold War environment.

This mid-twentieth-century design discourse offers an especially compelling way to see how visual practices, computation, and biology are yoked together in the name of greater organization and control (Halpern 2014). It is just such a (hypothesized) computational merger that has continued to develop under a new heading: molecular animation. From the perspective of visual governance, molecular animation is an evolution in the communicational merger of brains and molecules through the computer interface. And it is the quality of the communication forged that is valuable to neoliberal governance. What is staged early on with microcinematography in terms of a new technicity of scientific perception, and then intensified in various ways through the language of computational visual design, is then made explicit in the life sciences through molecular animation practices. Put another way: if microcinematography was instrumental in yoking scientific vision and plastic organisms together, then visual design gives these practices the informational language they need to give birth to molecular animation.

This is anything but a causal history of molecular animation. As I said at the outset of this chapter, as well as throughout this book, I am interested in excavating genealogies of molecular animation. Genealogies do not follow linear timelines and they certainly do not obey disciplinary boundaries. They exhume episodes that are relevant to the formation of the present in light of evolving histories of power. And given molecular animation's configuration as both an apparatus of popular media culture and a valuable tool for research in the biosciences, we require an account of how *this* particular present came into being. What we find is that the visual culture of science is not unrelated to the history of entertainment media. Or in Whitehead's terms: these are not entirely separate social orders with distinct and therefore unaligned modes of organization and togetherness. And while it may seem obvious on one level that science and popular culture are technically and discursively aligned at various points, on another level it is not at all obvious what these alignments are and why they come into being the way they do in molecular animation. This book's hypothesis is that the governance of visual perception is what explains the historical convergence of science and entertainment that led to the emergence of computerized molecular animation in the present. But these overlaps only come into focus once we have a clear picture of the subtle operations of governmentality.

For this reason, governmentality's strategic control of science and popular media was not apparent to us in Part I. If Part I is taken in isolation of Parts II and III, then molecular animation has divergent meanings, uses, and histories. Parts II and III added theoretical and genealogical robustness to these histories. And in this chapter in particular, we came to appreciate how lesser-known practices have a newfound relevance. For if chapter 2 recounted a history of visual epistemology in the experimental life sciences, then this history is incomplete without also showing how moving-image media were combined with the dominant practices of microscopy to yield new knowledge of cellular systems. And if chapter 3 offered a history of theoretical and systems biology, and especially its alienation from the observational practices of experimentalists, then it is just as important to describe how the adjacent field of visual communication design was carrying out experiments on perception and cybernetic systems that would shape later developments in science visualization. These episodes give shape to a history of visual practices that lead to a present in which molecular animation straddles both popular and scientific modes of viewing.

In response to the question that animated this chapter—namely, does molecular animation get taken up in a separate adventure of thought in the domain of life science research?—it is becoming increasingly clear why molecular animation does not get taken up in an alternative social adventure. The visual culture of the life sciences was experiencing many of the same transformations that were gripping the wider spheres of media culture. But to say that the visual culture of the biosciences is also "neoliberal" is in many ways a crude shortcut. Perhaps "neoliberalizing" is better, but it is still inadequate. It is not even clear whether neoliberalism is the best description for what we are currently living through, although the wager is that the governmentality is still a useful analytical frame. In any case, the biosciences had their own methods and techniques for governing visual practices; microcinematography and visual communication design were two such techniques. They were both subtle, but no less crucial, technologies of government that paved the way for developments decades later. What I am suggesting is that molecular animation became an important medium for scientific research because the necessary discursive, material, and epistemic infrastructure was in place. This infrastructure took time to design and perfect and it did not always coincide with the infrastructures designed in the domains of popular entertainment culture. Nevertheless, what is at stake is explaining how the visual logics reproduced in popular media culture are also at home in the epistemic spaces of twenty-first-century life science. My hypothesis is that this is because similar techniques of governance have been deployed throughout the twentieth century to capture the perceptual and psychic spaces of scientists and popular consumers alike. This slow and uneven process of governing is what normalized parallel visual logics within life science research and popular media culture. Molecular animation is therefore not an aberration in the biosciences, but it is the result of a long process of governing two distinct social orders to converge through shared values.

One of the central suppositions of this book is that distributed technologies of government synthesized heterogeneities to form a molecular animation apparatus that brings about certain ends. Put in other terms, the regime of perceptual subjectivization that molecular animation makes possible is an achievement of governmentality. In this regime, scientific subjectivization can no longer be meaningfully distinguished from popular subjectivization. This is a form of (post-)neoliberal subjectivization that is at once continuous with the past (as I have been suggesting throughout this chapter) and a departure from it, inasmuch as a new paradigm of social organization and togetherness is brought into being: a coincidence of biological knowledge production and popular consumption/production (prosumers of biological knowledge?). In Whitehead's perspective, a new society is constructed that is at once related to and distinct from what surrounds it. While this regime represents something new in the overall landscape of perceptual governance, we cannot say that it is arbitrary: *it is by design.* Governmentality designs the apparatuses it needs.

That the question of design emerges at the end of the book may seem strange. For shouldn't apparatus design be addressed at the outset, especially when design involves a sense of deliberate planning and fabrication? Perhaps. But it is only at the end of our analysis that we appreciate the deliberate planning involved in the fabrication of this media apparatus. Molecular animation is designed to bring new perceptual subjectivities into being to expand and intensify the grip of emergent regimes of governmentality. If design is invoked throughout the second half of the book (e.g., neurodesign, sensorial design) then it only now comes into focus as a part of a more elaborate design theory. The trouble with this conception of design is that it overhauls many of our basic assumptions about what design is, who the designer is, and for whom design is intended. In short, molecular animation calls upon us to rethink the very meaning and relevance of design in twenty-first-century practices of perceptual governance. In the Postscript, I explore how many of the resources of critical design theory can be brought to bear on our understanding of the molecular animation apparatus. It is through this encounter with design that we not only push critical design theory in new directions but also gain new insight into molecular animation and how we might design propositional alternatives to it.

POSTSCRIPT
A PROLEGOMENON TO GOVERNMENTAL DESIGN

Whitehead's speculative project does not offer an exhaustive metaphysical description of the world. The concepts he deploys in his elaborate system are speculative propositions that respond to particular problems in our modern experience of the world. As such they are not final descriptions of experience. Recall that at the end of the preface to *Process and Reality,* Whitehead notes that "there remains the final reflection, how shallow, puny, and imperfect are efforts to sound the depths in the nature of things. In philosophical discussion, the merest hint of dogmatic certainty as to finality of statement is an exhibition of folly" (xiv). What this means is that upon landing from the speculative flight of the imagination, and returning to experience with fresh eyes, philosophic thought should not aim at completion. If "philosophy begins in wonder," then "at the end when philosophic thought has done its best the wonder remains" (MT, 168). For this reason, one might say that the success of a philosophic experiment is determined by whether thought is renewed for new speculative adventures. As I have written about elsewhere, and in reference to George Allen's work on the philosophy of education, Whitehead's "Rhythms of Education" deals with the processual nature of learning as an essential component to his speculative method more generally. Education, like philosophical inquiry, is cyclical: it begins in romance, passes through precision and generalization, and then reignites romance again (Allen 2012, 35–58; Nocek 2018). This is not a vicious circle: each adventure is transformed by the adventures that precede it. Every speculative journey should end renewed for new philosophical exploits.

I say this by way of introducing the Postscript to *Molecular Capture.* If this book has loosely followed Whitehead's speculative method—begin in "the ground of a particular observation" (Part I), take flight into the thin air of speculation (Part II), and then land "for renewed observation rendered acute by rational interpretation" (Part III)—then my hope is that we have been able to return to our experience of molecular animation with a new perspective. And yet the goal is for the "wonder" to remain. This Postscript is an attempt to rekindle the romance stage of learning and to begin to ask questions that are left hanging in the air and that beg for new modes of inquiry. This is why I am calling this final section of the book a "Postscript" instead of a conclusion. As opposed to a conclusion, which wraps things up neatly, a Postscript suggests that there is more to be said. The message has been written, and yet there is something else, there is an additional statement. And so the Postscript to this book's message includes the ingredients for a new romance. It is what is still lingering in the mind, and asks for new connections to be drawn, even after the work is largely done. And like any good romance in Whitehead's sense, the discussion will be fresh with enthusiasm, rich in suggestiveness, and filled with vivid connections to adjacent concepts, fields, and modes of inquiry. But much like Whitehead's own Romantic conception of God in *Process and Reality,* this discussion will be incomplete and far more suggestive than final (Kraus 1998). The hope is that it nevertheless brings new questions about the molecular animation apparatus to the table that require new flights of the imagination. This is why the Postscript is also a prolegomenon.

I ended chapter 7 with the following suggestion: that the molecular animation apparatus is designed by distributed operations of modern governmentality. The implication is that governmentality is somehow a designing agent or force. This is a proposition that emerged after seven chapters of investigation and requires thorough examination. It is one thing to use design as a metaphor or as a synonym for planned fabrication (which modern forms of government surely are); but it is quite another thing to use design as a technical term that describes how the molecular animation apparatus is fabricated. It demands a serious engagement with whether and how design methods, theories, and discourses help to understand the relation between governmentality and media construction, as well as how governmentality sharpens our perception of the way in which design operates in the contemporary. It is to these concerns that I now turn in the hope of setting the stage for a more sustained inquiry into the relations among design, animation, and governance in the future.

COMMUNICATION DESIGN

I have made reference to design throughout this book. Indeed, apparatus and sensorium design are a part of the overall conceptual architecture of this project. However, the importance of design as a professional discourse and practice only came into focus in chapter 7. Moholy-Nagy and Kepes make graphic, communication, and architectural design an explicit point of reference for a genealogy of molecular animation that traverses the media practices of biologists and popular entertainment consumers. We witness a clear concern for how the design and fabrication of visual interfaces simultaneously design perceptual subjectivities. It is in this way that Moholy-Nagy and Kepes also anticipate the concerns of media theorists many decades later: mobile screens design neuroarchitectures that reflect the wider political and economic landscape (chapter 6). It is not simply the designer who has agency, but the design object does as well: it designs modes of technical perception well adapted to contemporary forms of power. This echoes what the design theorist Tony Fry has written about modern design: it is not just humans who design, but design designs humans right back—design *designs* (Fry 2012; Willis 2006). But to understand the meaning of design agency in the context of molecular animation is not straightforward. It is also not clear what the benefit of framing the project in terms of design would be in any case. To address these challenges, we need to clarify what is designed in molecular animation, and then who designs, and for what purpose.

One thing is relatively clear: it is not the computer animation technologies that are stake in the design process. Rather, it is the molecular animation apparatus that is designed. And the apparatus is composed of heterogeneous materials held together by their mutual but different integration of the same element of form. The design object is not therefore reducible to the technical materials that compose computerized molecular animation (the algorithms, the software, the datasets, etc.); on the contrary, it is the relational fabric that draws these materials together for a finite period of time. Or rather, what is designed (following Whitehead) is a medium for divergent materials in the apparatus to communicate.[1] Hence apparatus design is also a form of communication design. And what these communicational apparatuses bring into being are new possibilities for perceptual subjectivization that unite bioscientific knowledge practices and popular modes of media consumption/production. These apparatuses are designed to redesign the human sensorium.

There is a sense in which this is not an entirely new proposition for critical design theory. I say this for several different reasons. Increasingly, designers are not only diversifying what traditionally counts as a design artifact—from biological sensors, protocells, and ecological networks to communities, modes of activism, democracies, global financial networks, and even the future—but they are also investigating the relational fields that constitute the design object (Armstrong 2011; Tharp and Tharp 2019; Lukens and DiSalvo 2013; Manzini 2015; Dunne and Raby 2013). Design artifacts (like the apparatuses) emerge at the intersection of a plurality of forces, including relations to much wider political, economic, social, and biophysical spheres (Boehnert 2018). This has led some researchers to draw on complex dynamical systems, as well as systems thinking more generally, to conceive of objects as systems whose emergence, maintenance, and evolution depend upon, but are irreducible to, local interactions, nonlinear causation, and various forms of feedback (Cairns and Tunas 2017; Boy 2017; Winograd and Flores 1986). These engagements with complex systems range from more engineering-focused attempts to model architectural and urban systems as well as efforts to design for complex human-technological entanglements in human-centered design (Boy 2017; Norman and Draper 1986) to more critical and situational work with second-order autopoietic systems in autonomous and pluriversal design (Escobar 2018).

Relational ontologies of design have made it possible to forge meaningful connections to the work of Whitehead, Isabelle Stengers, Deleuze, and Bruno Latour, as well as to science and technology studies more generally, and in so doing, stress not only the ecology of design but also the co-becoming of subject and object in the design process itself (Farías and Wilkie 2015; Wilkie, Savransky, and Rosengarten 2017; Marenko and Brasset 2015). In this frame, the designer is not a Cartesian or Kantian subject standing outside of the unfolding world, but she is inseparable from its emergence. The designer is no longer championed as the modern rational subject (the Promethean designer) who shapes the world in "his" own image (Latour 2009). Critical design theory is therefore beginning to challenge Western, patriarchal notions of subjectivity and to engage with the design of both the artifact and the designer through relational activities operating at the intersection of brains, bodies, affects, and political–economic worlds (Escobar 2018). Any separation between the "designer" and the "artifact" is symptomatic of modern dualistic thinking that conceals the various and complex modes by which they co-emerge in a shared and evolving world. In this way, the designer and artifact shape or design each other. In other words, designer and object are *designed*. What this means for Fry and Anne-Marie Willis is that design is ontological: not only does the designer relinquish control of the design object (dethroned from their role as Promethean designer) but the designer is also the product of design (Fry 2017; Willis 2006).

On the one hand, my proposal for apparatus design fits neatly into this genealogy of design: the design object (the molecular animation apparatus) is a relational assemblage whose design cannot be separated from much wider systems of political and economic force. On the other hand, this overview also raises important questions about design agency: namely, who or what designs the apparatus, and for what purpose? And to what extent does the animation apparatus design the designer in turn?

DESIGN: BETWEEN THE ART AND REASON OF GOVERNMENT

Reconceived as a design object, this book proposes that molecular animation is designed by modern techniques of government. Governmentality is the design agent. Not only does this proposal shed new light on the subtle operations of governmentality—giving it a sense of intentionality, craftiness, and even cunning (Flusser 2000)[2]—but it also responds to certain ambiguities in Foucault's own work on the concept. In his lecture from January 10, 1979, at the Collège de France, he opens by discussing what he calls the "art of government." But in introducing this phrase, he feels compelled to correct any assumptions that one might have concerning his interest in "the way in which governors really governed"; this is decidedly not what the "art of government" is about. Rather, the art of government concerns "the reasoned way of governing best and, at the same time, reflection on the best possible way of governing" (2008, 2). What's difficult here, which Foucault commentators have noted, is that he makes very little, if any, distinction between "the art of government" and the "rationality of government" (Gordon 1991, 3). This is because the governing that interests Foucault is not only an activity or a practice but also a kind of thought, a *technē,* or rather, a thinking about the "best possible way" to govern within the practice of government (Foucault 2008, 2). Hence, Foucault is concerned with the "rationalization of governmental practice" (2). The art of government consists in rationalizing its governing practices.

What is striking about this description, if never spelled out in Foucault or in the commentary, is how this rational art of governing bears a striking resemblance to what many associate with the history of modern design practice. As scholars and practitioners have noted over the years, one of the hallmarks of "design," as opposed to "art," is its "utter usefulness" (Twemlow 2017, 9). Design often distinguishes itself from art through its problem-solving agenda: it introduces objects and services into the world in order to solve problems. For this reason, design is often characterized as a purposeful, reflective, and iterative activity (Boradkar 2010). As Nigel Cross explains in relation to what he calls "design ability," designers have a distinct approach to solving problems: it is through experimenting with different solutions (iterating them) that they come to understand the true nature of the problem that needs to be solved (Cross 1995, 108). Or as Herbert Simon has suggested, design is a kind of science that distinguishes itself from other fields (including science itself) insofar as it solves "ill-defined" problems in order to shape the world according how it should be (Schön 1983, 37–38; Willis 2006). And according to Ezio Manzini (2015), design has a unique role to play in a world obsessed with social innovation: it is at once a problem-solving and sense-making practice that engages experts and nonexperts alike in the facilitation of the design and redesign of social life. Indeed, the revolution in design thinking within the last several decades is an attempt to exploit "designerly" ways of thinking and, in particular, replicate the designer's methods of ideating, prototyping, and iterating possible solutions in every area of life (Cross 2011). And while it would be unfair to say that art is useless and does not iteratively solve problems (as Terry Eagleton suggests [1983]), or to claim that an absolute distinction between art and design holds up once we move beyond banal generalities, there is nonetheless something important to glean from this caricature of design.

The so-called art of government seems to have something designerly about it (at least according dominant, Western conceptions of design). Governmentality is a purposeful and calculating practice, which is why the art of government is just as much about the rationality of government: it is a *technē* that thinks about the best possible way to govern. However, governmentality does not have an a priori grasp of the best technique or set of techniques for exercising its control. The apparatuses it fabricates are experiments in governance. As Foucault emphasizes throughout his late lectures, states, economies, and technologies are not given in advance as universals that are always and forever adequate to the task of governing (Foucault 2008, 3–4). Governmental reason, much like design thinking, experiments with, prototypes, and iterates new techniques and apparatuses for solving its own problem: how best to steer or guide conduct within a wide range of social, economic, and subjective fields. Government poses its own design problem. And while much more could be said about the general parallelism between governmental reason and design thinking (which I will allude to below), suffice it to say that insofar as design expresses the iterative problem-solving dimension of a creative practice, then Foucault's indiscriminate use of the art and reason of government might also be thought of in terms of the design within governmental practice. Or perhaps better still: design is the bridge between the art and reason of government.

What this proposal offers is a different analytical frame for governmentality's direction of perceptual subjectivities. The convergence of perceptual registers did not happen all at once, and they did not happen by accident or by chance. The molecular animation apparatus is a result of governmental reason working historically and in various fields to prototype different alliances (e.g., microcinematography) and iterate on them (e.g., mid-twentieth-century communication design) in order to fabricate an apparatus that responds to a design problem—namely, how to fuse scientific and popular modes of perception. In other words, each episode of visual experimentation in the history of the biosciences (including episodes that might seem unrelated or peripheral) are prototypes in an expanded design practice that aims to extend and intensify its control of the perceptual economy of subjects.

This conception of design not only transforms how we conceive the molecular animation apparatus (the result of an extended design experiment) as well as our thinking about the construction of media apparatuses generally, it also responds to a difficulty within contemporary architectural and design discourse—namely, what are the spatial and temporal scales at which design operates? If designing happens in excess of modern Promethean subjects (who are themselves the products of design) and occurs through various political, economic, and computational channels, then what are the scales and materialities of design agency today? While Benjamin Bratton (2015) proposes that the stack functions as design brief for the new geopolitics of planetary-scale computation, and Keller Easterling puts (2014) forward an ambitious proposal for thinking about infrastructure space as a designing force, I think Fry comes closest to the proposition of governmental design (of media apparatuses) when he conceives of the colonizing force of televisual media design. Here's Fry:

> This designing comes with the arrival of the sum of all visual electronic media within the modern world as they offer up the world of commodified desire. . . . What televisual designing effectively does is to colonise the imagination with a 'world picture' (which is not an image of the world, but the world *as* image, that is the world of modernity. (2014, 82–83)

Governmental design is in many ways an extension and complication of Fry's proposal (through Foucault and Whitehead instead of Heidegger). However, a crucial feature of this proposal is that design neither emerges from nor is for human subjects. This is not to say that government fails to address these scales of experience. Quite the contrary, as we have seen throughout this book, government works with them and through them, in the neural pathways of the brain, in the spatial organization of the body in front of mobile screens, and in the distribution of perception across expansive data fields—these and countless other modes of human and nonhuman experience are the raw materials or media of governmental design practice. But governmentality, especially in its neoliberal forms, favors subtlety over the exertion of brute force. Not unlike those who champion "good design" as unobtrusive and frictionless, governmental design often flies under the radar and goes unnoticed (De Jong 2017). Or then again, sometimes it parades as one thing—liberatory—and it is really another thing: an instrument of control (Deleuze 1995a). The design thinking that emerges from the rationality of government is not human-centered, or at least not in a way that is recognizable to the fields of Human–Computer Interaction (HCI) and Human-Centered Design (HCD). This is a design intelligence, or rather a multiplicity of distributed intelligences, that shapes and is shaped by humans but is not for humans. For this reason, we might say that governmentality displays a *nonhuman design intelligence.* Government makes use of biological and nonbiological materials; it works with them in conjunction with the remnants of the state-form, institutions, digital infrastructures, machine learning algorithms, and global financial markets to fabricate apparatuses for the purposes of extending and intensifying the grip of governance. Government is the designer and user of the product.

In the case of molecular animation, governmentality had to work at multiple temporal and spatial scales (e.g., through institutions, technologies, infrastructures, epistemologies, biochemical systems, and affective bodies over successive generations) to coordinate the perceptual practices of life scientists and popular media consumers into a dispositif. The design of this apparatus could not have been anticipated or seen in advance. Its design required prototyping and iterating, and in domains that seem to have very little to do with one another. The various and diverse successes could then be leveraged and eventually deployed in the design of the molecular animation apparatus. This apparatus, unlike the earlier prototypes, was able to coordinate brains, datasets, networked media systems, and a host of other material systems into a communicative relation capable of merging the epistemic practices of life scientists with viewing patterns of popular media consumers. Human and nonhuman actions and materialities are media for designing new and more intense forms of government. Thus, and in sharp contrast to any notion of human-centered design, governance designs with itself in mind. In this posthuman conception of designing, animating molecules for scientists and mass audiences becomes a key moment in a much longer history of governmental design that aims to control the psychic and perceptual spaces of subjects.

NEW SPECULATIVE FLIGHTS: PROPOSITIONS FOR REDESIGN

Molecular Capture is a book that is and is not about molecular animation. As I noted in the Introduction, the practice is far too important and interesting for the book only to address molecular animation, narrowly construed. Molecular animation is deceptively simple: it conceals the fact that it is a crucial component in a much larger and complex design strategy. Despite the pessimistic tone of this concluding section, it is nonetheless grounded in a more profound commitment to hope and optimism. If the overall aim of *Molecular Capture* has been to excavate the forces responsible for the emergence of molecular animation, then it is also these forces that hold the keys to resistance. As we saw in chapters 5 and 6, Foucault sought to understand the present by means of the power dynamics that give rise to it. Nevertheless, freedom, he maintains, is essential to power: "At the very heart of the power relationship, and constantly provoking it, are the recalcitrance of the will and the intransigence of freedom" (1982, 221–22). Foucault contends that every deployment of power, every act of governance, also contains within it the seeds of "struggle" or "counter-conduct" (2007, 200–203).

In chapter 5 we reframed counter-conduct in Foucault in terms of Whitehead's theory of the proposition. Whitehead shores up a way to conceive of a penumbra of alternative feelings that accompany every instance of social infection. The problem with neoliberal governance in a Whiteheadian frame is that it eclipses all genuine contrast from being entertained. All forms of difference are converted into opportunities for the intensification of neoliberalism's infectious grip. If we are to think alongside Stengers thinking alongside Whitehead, then the real challenge is to cultivate spaces for entertaining propositional feelings that do not conform to the dominant forms of social infection but are also not relegated to irrelevancy (and negatively prehended). This is an especially difficult task today, when neoliberal forms of power operate using so many "infernal alternatives": "if you refuse this bad-sounding thing, then the consequences will be far, far worse"; this has the effect, writes Stengers, of "sapping or capturing the capacities for thinking and resisting of those who were apt to do so" (2015, 55; Nocek 2018).

The question we are left with at the end of this book is how to reclaim propositional modes of entertainment in the context of molecular animation. If it is true that molecular animation is an apparatus designed to fuse the perceptual spaces of scientists and popular consumers, then what we require are propositions that are exogenous to what can be converted into more opportunities for the dominant regime of perceptual governance. What I would like to propose is a slightly different way to frame this problem in the hopes of altering how we focus our research in the future: If molecular animation is a product of governmental design, then what would it mean to cultivate spaces for propositional entertainment that speculatively redirect governmental design strategies?

In this context, I think a lot can be learned from the successes and failures of critical and speculative design (CSD).[3] CSD promises alternative ways to design within a technoscientific present determined by late-stage capitalism. As Anthony Dunne and Fiona Raby write in their 2013 work, *Speculative Everything,* "Rather than giving up altogether, there are other possibilities for design: one is to use design as a means of speculating how things could be—[this is] speculative design. This form of design thrives on imagination and aims to open up new perspectives" (2013, 2). This proposal has captured the imagination of many designers, as well as artists, fiction writers, filmmakers, and others. It is both critical and hopeful: instead of conforming to the dominant regimes of power (commercial design), it leverages design techniques for other purposes—namely, to critique the current technoscientific imaginary by proposing alternatives to it. Design no longer solves problems in the present but works to problematize and reimagine the present (2).

As I have argued elsewhere (Nocek 2019), despite the elegance of this proposal, the question that plagues so much of the work that falls under the heading of CSD today is: Where do these alternatives come from? In response to this question, designers tend to place a high premium on the "imagination." The "imagination," write Dunne and Raby, "gives us entry to abstraction. . . . We gain the ability to conceive alternatives. . . . We gain the ability to think of futures and outcomes, skills of planning. The ability to think ethically also becomes a possibility" (2013, 70). Julian Bleecker, one of the main proponents of design fiction, explains that "design fiction is a way of exploring different approaches to making things, probing the material conclusions of your imagination, removing the usual constraints when designing for massive market commercialization—the ones that people . . . call 'realistic'" (2009, 8). And in a somewhat different key, Bruce Sterling, the science fiction writer turned design fiction advocate, echoes Dunne and Raby's enthusiasm for imaginative redescription but notes, and a bit more pessimistically, that "science fiction is not about the freedom of imagination. It's about a free imagination pinched and howling in a vise that other people call real life" (2005, 34). What's more, the forms of the imagination valued most by critical and speculative designers tend to be those that that emerge out of "technical fluency." As Jonathan Lukens and Carl DiSalvo write, fluency is not the same as literacy. Where literacy entails nothing more than "rote application," fluency, by contrast, "is the ability to be creative with technology, . . . the capability to understand, use, access technology" (2013, 24); it is to be able to "write poetry" with it, instead of just read, write, and speak it. CSD, they even suggest, is a kind of "technical poetics," which manages to explore "the space of possibilities created by technology—a space that can be imagined only by the technologically fluent" (27).

Despite this, Cameron Tonkinwise argues that the design imagination directed toward shaping alternative futures is itself shaped in spaces of Western technoscientific privilege. As such, CSD only reinstates the systems of power that it critiques. According to Tonkinwise,

> Speculative (Critical) Design has to date been irresponsibly lacking in diversity. This is not just a political correctness problem, though it is clearly part of ongoing normalizations of North Atlantic late capitalism that marginalize the different lifestyles and values of peoples in the Global South and East Asia. It is an epistemological error when Speculative (Critical) Designers at the Royal College of Art, for instance, imagine what they believe to be dystopian scenarios in a distant future, when in fact people in other parts of the world are already living versions of those lifestyles. (n.d.)

Likewise, Luiza Prado and Pedro J. S. Vieira de Oliveira contend that "as a discipline born and established in Europe—and hence within the paradigm of modernity, progress, and development—design produces and perpetuates this epistemological and technological regime of domination" (2016, 30). The challenge for Prado and de Oliveira, as well as for Tonkinwise and a number of others, is to forge genuine interstitial spaces for the imagination within the dominant regimes of Western design (Kalantidou and Fry 2014). Unfortunately, what the reigning paradigms of speculative and critical design reinforce are forms of the technoscientific imagination that only offer new opportunities for neoliberal capture instead of genuine alternatives to it.

Our conception of governmental design contributes to this conversation in potentially interesting ways. In order to intervene in the design of the molecular animation apparatus, it is not sufficient to redesign the technologies on which it depends. This does not necessarily change the apparatus; it may only serve to intensify its hold. The apparatus does not lie in any of the technical materials—in the code, in the media platforms, in the datasets, or in the interface design. The apparatus consists in the quality of relations shared among the heterogeneities assembled. This is a qualitative field that is itself enmeshed in a much wider eco-social fabric that has emerged over time. This is in large part what *Molecular Capture* has attempted to show: that the design of the molecular animation apparatus has taken well over a century to develop and perfect. Earlier prototypes, such as Jean Comandon's film about syphilis and Alexis Carrel's time-based cellular experiments, were integral stages in the design process.

If one were to imagine alternative perceptual spaces, then one would have to intervene in the relations that have come to define the apparatus. But these relations are deeply woven into the histories and practices of science and media culture in the West. To reimagine these relations entails intervening in what governance has worked hard to normalize; it means disrupting the "communicative bonds" that government has established in its centuries-old design practice. This is a very different mode of critical and speculative design practice than what tends to be peddled in art galleries and in design schools. It is one that heeds the call of designers and theorists who demand that speculative interventions into design be attentive to historical, philosophical, and genealogical methods of analysis. This is a mode of design called upon to intervene in a relational field that is at once historical, economic, normative, political, as well as technological. But more than anything, this is a mode of design intervention sensitive to the fact that the molecular animation apparatus has been in the making since at least the late nineteenth century.

In this context, we might recall that for Whitehead, the procurement of intensive contrast in experience may very well entail admitting into feeling a pattern of relation that seems remote, distant, or wholly irrelevant to what matters to humans in the present ("there are no preconstituted particulars" [PR, 105]). And in our case, it might mean entertaining potentials that seem irrelevant now, that seem anachronistic or even impossible given the current climate of data visualization and technoscientific research, but that nevertheless generate heterogeneities in experience incapable of being homogenized and subsumed under the dominant forms of social infection. This might mean entertaining the penumbra of potentials that lurked within the formative stages of apparatus design in life science research, but were nevertheless dismissed (not unlike the entertainment of alternatives for the Battle of Waterloo in Whitehead [PR, 185]). For example, what would it mean for the cinema to be used otherwise in early twentieth-century science? Not as a "portal" onto another world that permits the perceptual spaces of subjects to fuse with the temporally unfolding natural world, but as medium for vivifying radical heterogeneity? What would it mean, in other words, to use the cinematograph in science to express the alterity of living systems? Would this have led to a very different present?

This is the space of scientific research that Jean Painlevé (1902–89) opened up with his experimental science cinema, despite the fact that it always remained in the shadows of both art and science (Painlevé 2000).[4] Like Comandon and Carrel, Painlevé thought that the cinema was a research tool (often used in conjunction with a microscope) (162). However, what distinguishes Painlevé's work is that it refuses to give the viewer the satisfaction of recognition she desires. In the numerous "art–science" films Painlevé made between the 1920s and the 1980s (there are over two hundred)—such as *The Hermit Crab* (1931), *The Seahorse* (1934), *The Vampire* (1945), *The Love Life of the Octopus* (1967), and *Acera, or The Witches' Dance* (1972)—he never fails to unsettle our preconceptions of the biological world and force us to confront our own epistemic limitations.

Figure 30. *The Love Life of the Octopus.* Directed by Jean Painlevé, 1967.

For example, in *The Love Life of the Octopus,* our recognition of the "sleepy mollusk" is undermined by shots of its sex life (sex with sixteen tentacles) or close-ups of a suction cup on the octopus's tentacle, which makes a recognizable feature of its physiology nearly unidentifiable. Painlevé's camera delights in exposing how the entire range of human identification, from our austere molecular-perceptions to our aesthetic amusements—perfectly expressed by the shots of the erect seahorse in *The Seahorse*—are uprooted through contrasts that express life's resistance to representation: "There is an infinite field of magnificent and continual joys," explains Painlevé, "that prevents us from completely elucidating the mystery or the miracle" of life (2000, 123).

In other moments, Painlevé is less concerned with illustrating life's resistance to representation through contrasts and more interested in contemplating the impossibility of establishing any reference point at all. Instead of flooding the visual field with irreducible oppositions, he contemplates the helplessness that human cognition experiences before the vastness of "life itself." In these instances, life shows itself to be moving, creeping, and crawling without a cause or relation to any noticeable context. Such deterritorialized movement makes classification impossible. A scene from *Acera, or The Witches' Dance* makes this vivid: set to a

Figure 31. *The Seahorse.* Directed by Jean Painlevé, 1934.

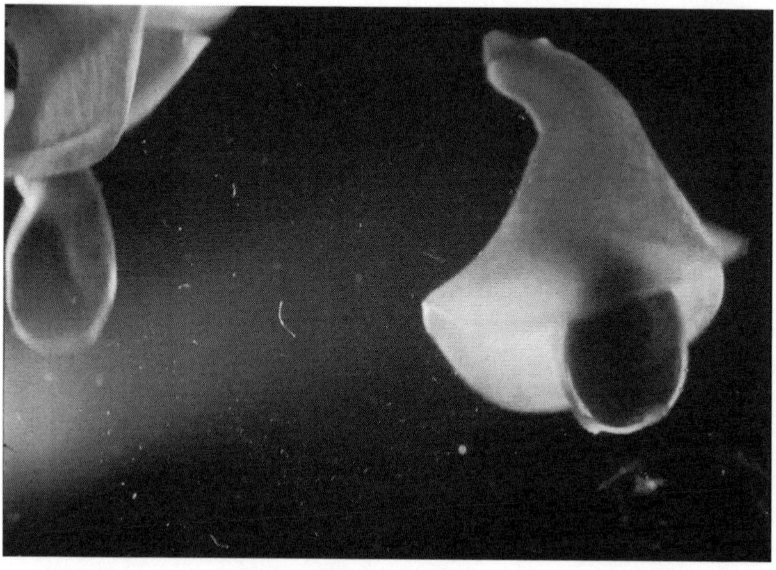

Figure 32. *Acera, or The Witches' Dance.* Directed by Jean Painlevé, 1972.

score by Pierre Jensen, mollusks weightlessly rise and fall in an underwater choreography. However, the viewer is left with little to secure this visual reference. The mollusks could just as easily be flying mushrooms. Are these magnificent creatures underwater or out of water? Flying or swimming? Plant or animal? Organic or inorganic? None of these questions are answered definitively. Painlevé lures the viewer into marveling at images emptied of reference; and left in the place where representation once had a firm hold is a suspension of judgment and determinacy (Cahill 2019; Cahill 2013).

Not only were Painlevé's works dismissed by scientists but they were also on the fringes of the avant-garde, since they did not strictly adhere to avant-gardist principles; and what's more, they were far outside of what could be considered popular cinema. As André Bazin writes about Painlevé, "It is not certain, unfortunately, that this startling cinematic truth [that the cinema can be both artistic and scientific] can be widely accepted. It harbors too much potential scandal at a cost to current notions of art and science" (2000, 147). What Painlevé's work proposes (although it was not entertained by many) is an alternative set of perceptual patterns or relations for the actual world of cinema and science. These are "potentials for definiteness" in which perceptual subjects feel their relation to the biological world in terms of resistance and hesitation instead of optimization and domination. These are patterns of perceptual relation where there is no possibility for visual conquest. This is a way of organizing the actual world that is very different from the one we have encountered in our genealogy of molecular animation.

What I am suggesting in conclusion is that Painlevé's cinematic experiments may very well offer us unentertained propositions for the world. And these experiments might then be leveraged to construct propositional spaces capable of dismantling apparatuses that took well over a century to design. In this frame, his cinema is important in at least two ways: on the one hand, it proposes a perceptual logic that introduces irreducible heterogeneity into the spaces of science cinema—not unlike Thomas LaMarre's conception of "animetism" (2009, chapter 1). Painlevé's cinema can therefore be read as part of a genealogy of animatism where distance, hesitation, interruption, and disorientation are primary. This is in opposition to the (hyper)cinematism of microcinematography that attempts to create a smooth and uninterrupted communication between scientific perception and cellular development. On the other hand, Painlevé's cinema is situated at an essential moment in the history of apparatus design: right when visual design and information theory were taking shape and the merger of perception and information was just being proposed. Painlevé's work figures like a possible world for the actual world of perceptual design. It is a world that could have been and signifies that a different genealogy for perceptual subjectivization was possible, but never actualized. The restraint and minimalism displayed in Iwasa's *Molecular Motor Struts Like a Drunken Sailor* might very well be a part of this nonactualized genealogy. Its clear differentiation between planes of the image (foreground and background) and invitation to hesitate before obfuscating their difference actualizes a potential for visuality whose relevance for contemporary economies of scientific perception is only now coming into focus.

Painlevé's cinema offers a virtual genealogy that may become relevant to the design of propositions in the present and future. How these propositions are felt and what critical work they do is not entirely clear at this point and is for another book project. What is clear is that intervening in the tight grip that the molecular animation apparatus has on the perceptual economies of the life sciences will be at least as laborious as understanding its design, and animations like Iwasa's may very well be a part of its redesign. But what *Molecular Capture* has managed to show is that the stakes could not be higher: the future of knowledge practices in the life sciences hangs in the balance. The hope is that the elements for such an intervention have been sufficiently sketched and are ready to be taken aboard the next flight of the speculative imagination.

ACKNOWLEDGMENTS

Molecular Capture, like the apparatuses it theorizes, was a collaborative effort from the very beginning: it drew numerous individuals, histories, and institutions together in order to become the book that it is. The project started as a PhD dissertation in the Department of Comparative Literature, Cinema, and Media at the University of Washington under the directorship of Phillip Thurtle. The dissertation was a very different project from *Molecular Capture,* and it is thanks to the careful guidance of my committee (Phillip Thurtle, James Tweedie, and Robert Mitchell) that I could see that a different (if related) book could be developed from the dissertation. The institutional support of the Comparative History of Ideas Department (then Program) provided invaluable intellectual community and financial support during the intense period of dissertation completion.

I owe an especially heartfelt thank you to Phillip Thurtle. Phillip was not only my PhD advisor and intellectual mentor but he also became a dear friend in the process. This meant that he counseled, challenged, and encouraged me during the stressful years of dissertating, but he also was a true friend during difficult life transitions. Phillip has remained an important mentor to me. He has read numerous versions of chapters 2 and 3, and I am tremendously grateful for his time and care in giving me thoughtful feedback. Phillip encouraged me to be unapologetic about bringing unconventional materials and genealogies together, especially ones that cut across media, science, and philosophy.

Much of the theoretical and intellectual backbone for this project is indebted to the work of Deleuze, Foucault, and Whitehead. For Deleuze and Foucault in particular, I owe a great debt to the Deleuze Studies community for guiding me through various stages of intellectual development: from helping me to see past naive misreadings and youthful embellishments to introducing me to a community of scholars receptive to reading Deleuze and Foucault alongside critical studies of media, science, and design. The Whitehead community also played an active role in shaping the theoretical direction of the book project. A special thanks goes out to the Whitehead Research Project at the Center for Process Studies within the Claremont School of Theology for inviting me to present early versions of chapter 5 and the Postscript. I am also especially grateful to the Whitehead reading group for patiently listening to excerpts of the book. Nicholas Gaskill and Michael Halewood have also been key interlocutors in working through some of Whitehead's more difficult work on speculative pragmatics.

During the formative stages of writing and editing the chapters on theoretical biology, I was a faculty fellow at the Amsterdam Institute for Advanced Studies (IAS). I am very grateful to the director of the IAS, Peter Sloot, for taking a chance on a scholar trained in the theoretical humanities to work alongside theoretical scientists from biology, computer science, physics, and economics. My understanding of experimental and theoretical biology was also shaped by long conversations with friends and partners in the Ontogenetics Process Group, whose members include Stuart Kauffman, Sha Xin Wei, Phillip Thurtle, Erin Espelie, Cary Wolfe, Michael Epperson, Gaymon Bennett, Giuseppe Longo, and Helga Wilde-Dimiris. Their patience, friendship, and scrupulous feedback have been invaluable to developing robust frameworks for integrating biology and philosophy. I also received valuable feedback on sections of chapters 3 and 4 during a keynote address Szymon Wrobel invited me to give as a part of the Technology and Socialization Conference at the University of Warsaw.

While *Molecular Capture* is not a book on design per se, design theory and philosophy are nevertheless everywhere present in it. For this, I have to thank those I have worked with at the Laboratory for Expanded

Design (LxD) over the past several years. The LxD's experimental engagements with design, philosophy, and performance have transformed how I think across theoretical and social practice, and I don't think I can ever go back. Steven Loo, Oliver Feltham, Mike Waller, Barbara Formis, Sha Xin Wei, Duncan Fairfax, Stacey Moran, Michael Hornblow, Tony Fry, Anne-Marie Willis, and others have all been a part of this transformative work at the LxD. A special thanks goes out to Tony Fry, whose work and now friendship have been especially important to how I think about the relation between philosophical critique and ontological design. Ontological design is at the heart of the conception of the apparatuses I theorize in this book (even if it is not always stated) and will continue to propel my work forward. In this context, a special thanks goes out to Iris van der Tuin and Maaike Bleeker for inviting me to give a draft of the Postscript on ontological design to the Transmission in Motion lecture series at Utrecht University. Thanks also to Vera Bühlmann at TU Wien for inviting me to lecture on design and governmentality to the architecture school. I also have to thank my partners at NON+ Design in Amsterdam—Conny Groenewegen, Maurizio Montalti, and Stacey Moran—for their support and patience with me as I continue to insist that a speculative design proposition (Whiteheadian style) can be built and experienced.

Molecular Capture is fundamentally an experiment that puts its trust in the possibility that transdisciplinary encounters with biology, philosophy, art, and design can build propositions for thinking and feeling the world differently. This trust is very much borne out of the work that I have the privilege of doing as founding director of the Center for Philosophical Technologies (CPT) at Arizona State University (ASU). I have no words to thank those inspired and crazy souls who work with me to imagine speculative encounters where the technicities of philosophy and the conceptual architectures of technology meet. I am especially indebted to those folks with whom I work every day at the CPT—especially Silvia Nerreti and Stacey Moran—as well as members of the CPT Faculty Board, including Ron Broglio, Gaymon Bennett, Cala Coats, and Grisha Coleman, and the numerous fellows the CPT has hosted over the years. CPT graduate students have also played a significant role in steering the direction of this work, and so my sincerest gratitude goes out to Jonathan Bratt, Garrett Johnson, Zach Thomas, and

Celina Osuna. Finally, the critical and speculative thought experiments that take place in this book, and are carried out at the CPT, would not be possible without the encouragement and investment of the ASU administration: President Michael Crow, Executive Vice President Sethuraman "Panch" Panchanathan, Dean Stephen Tepper, and Senior Associate Dean Sandra Stauffer.

I also owe a great deal of gratitude to numerous members of the wider ASU and academic community who have supported me during intense periods of writing and editing. The faculty and staff in my home department, the School of Arts, Media, and Engineering, have offered tremendous encouragement and have been a constant source of inspiration. In particular, I have to thank my former director, Sha Xin Wei, for both his mentorship and friendship. He was an inspired interlocutor during the entire writing process, and I cannot thank him enough for his support. Also at ASU, my dear friend and colleague, Ron Broglio, could not have been a better friend, mentor, and cheerleader during the many ups and downs I experienced writing and editing this long and unruly manuscript. Ron is the original and only true field marshal (the "Molecular Revolution" is next!). Likewise, Cary Wolfe has mentored me from the very beginning of this project: from late in the dissertation through peer review to completion of the manuscript, Cary has been uncompromising in the feedback and encouragement he has offered me. And at the University of Minnesota Press, Doug Armato's incisive reading, judgment, and advice are unparalleled and made it possible for *Molecular Capture* to reach its potential. Zenyse Miller has also been a tremendous resource throughout the production phases of this project. Finally, I greatly appreciate the help of Jennifer Quincey's keen editorial sense during the final stages of manuscript preparation.

If Whitehead teaches us anything, then it's that there is no such thing as a work in isolation. A work stands or falls based on its support network. This could not be truer than here. My parents, Jim and Patty, and my sister, Meredith, have given me nothing but love, support, and encouragement, even when there seemed like no end was in sight. Thank you for this. I dedicate this book to the intelligent and inspired women whom I share every day with: my wife, Stacey, and my daughters, Fiona and Ivy. They are the reason this book made it to press. Without Stacey's unflagging love, patience, and insight (which are everywhere present in these pages), and without Fiona and Ivy's humor, razzing, and support, I could not imagine this work would have been completed. In particular, Stacey's patient reading and brutally honest feedback during morning coffee and evening wine has made this book what it is. Too many others have sacrificed, and too much encouragement and advice have been given, for my name alone to appear on the book. This book is for you.

NOTES

INTRODUCTION

1. I am using media philosophy and media theory here as a single field of study, when in reality they have their own genealogies, interests, and problems. Media theory tends to have roots in cinema and media studies, while media philosophy owes much of its discursive and disciplinary formation to theoretical debates concerning broader problems of "mediation" in late twentieth-century Germany (see Hartmann 2000; Kittler 1999). Media philosophy asks foundational questions about the metaphysical and epistemological conditions of mediation, where media theory tends to concern itself with specific media and their effects. In many ways, this book is indebted to both histories and methods: it investigates a specific media practice and its effects and draws on a philosophy of mediation derived from Alfred North Whitehead and Michel Foucault to anchor its understanding of molecular animation. Nevertheless, the philosophical framing of animation looms large in the book (it is the precondition for engaging with animation), and for this reason *Molecular Capture* sits most comfortably alongside other works of media philosophy.

2. On the relation between problems and concepts in Deleuze's work, see Smith 2012.

3. Stengers explains that "the mathematician's trust is inseparable from a commitment not to mutilate the problem in order to solve it and to take its demands fully into account. Yet it implies a certain deliberate amnesia with regard to the obviousness of obstacles, an active indetermination of what the terms of the problem 'mean'" (2011b, 15).

4. See Gaskill and Nocek 2014.

5. For all Whitehead citations in this book, I follow the style of citation used in the Whitehead scholarship. Please see the list of abbreviations at the beginning of the bibliography for a complete list of all Whitehead citations.

6. On the connection between the cyclical nature of speculative thought and learning, see Allen 2012.

7. On the mechanistic paradigms that dominate epistemology in the life sciences, see Henning and Scarfe 2013.

8. See Landecker 2007.

9. See, for example, Mitchell 2011 and Kember and Zylinska 2012, as well as Levitt 2017. There are a number of other important works in media theory that underscore the affective and processual nature of media and mediation, if not necessarily their "vitality." In this context, see, for instance, Hansen 2004 and Sha 2013.

10. See Nocek 2014b.

11. In particular, chapter 1 returns to Lev Manovich's polemical typology of new-media objects (2001) and software metamediation (2013). It also engages with debates in cinema and media studies over the ontological status of animation and live-action film in the era of digital cinema. LaMarre is especially helpful in sorting out many of the issues surrounding the nature and composition of the animate image. These debates help to situate molecular animation as a media object.

12. For Han, a critical stance toward digital media culture calls for reframing Foucauldian notions of governmentality in terms of a distinct brand of Hegelianism capable of countering the positivity, immediacy, and visibility of today's media society with a fresh understanding of negativity, distance, and invisibility. See Han 2017a.

1. MOLECULAR ENTERTAINMENT

1. Natasha Myers briefly discusses the relation between *The Inner Life of the Cell* and *Fantastic Voyage* in *Rendering Life Molecular.* She contends that the body politics of the films are very different: "If Bolinsky narrates molecules in *The Inner Life of the Cell* as hard-working neo-liberal subjects, then *Fantastic Voyage* dramatizes physiological and molecular forces through the violent and sexualized imagery of the Cold War" (2015, 186).

2. See Lorraine Daston and Peter Galison's *Objectivity* for their analysis of scientific visuality in the context of the construction and circulation of scientific atlases. In particular, they argue that with the emergence of digital visualization tools and online atlases in the twenty-first century, accurate representation

is no longer the abiding concern of scientific imagists. "In this corner of science," Daston and Galison insist, "the representation of the real—the use of images to finally get nature straight—may be coming to a close" (2007, 392). It is for this reason that scientific images are now positioned as art. See chapter 7, "Representation to Presentation."

3. It is important to note that one of the most well-known molecular animations, *The Inner Life of the Cell,* was first screened at the 2006 SIGGRAPH in Boston and presented to a media arts and science audience and not to molecular biologists.

4. See Daston and Galison 2007, chapter 7.

5. Using ethnographic methods, Myers depicts the work of protein modelers. She portrays modeling as incredibly technical and time-consuming work that is really only meant for the eyes of other scientists. 3D animation, on the other hand, has many of the formal characteristics of a Hollywood feature film and it is gaining the attention of diverse audiences.

6. In *Rendering Life Molecular,* Myers discusses the praise and criticism that molecular animation has received. She frames her discussion in terms of the spirited conversation that broke out after David Bolinsky's TED Talk. I weigh in on this commentary and offer my own insight into the scientific legitimacy of computer animation in later sections of the book.

7. MIMs were first introduced by Kurt W. Kohn in 1999. The map has become an essential tool that allows scientists to visualize a wide variety of molecular interactions. Since its introduction, the notation on MIMs has become increasingly standardized. Please see Kohn 1999 and the Systems Biology Graphical Notation consortium (http://www.sbgn.org/), which standardizes graphical notation.

8. It is important to note that this presentation of molecular data conforms to the mechanistic conception of molecular interaction that the many scientists whom Myers interviews affirm but continually subvert. Myers argues that molecular modelers enact a more vital, dynamic, and embodied understanding of the molecular world, even though they continue to speak of it mechanically. See Myers 2015.

9. In *Modest-Witness@Second-Millennium.FemaleMan-Meets-OncoMouse: Feminism and Technoscience,* Donna Haraway draws on Simon Schaffer and Steven Shapin's notion of the "modest witness" and argues that modesty "is the virtue that guarantees that the modest witness is the legitimate and authorized ventriloquist for the object world, adding nothing from his mere opinions, from his biasing embodiment. And so he is endowed with the remarkable power to establish the facts. He bears witness: he is objective; he guarantees the clarity and purity of objects. His subjectivity is his objectivity. His narratives

have a magical power—they lose all trace of their history as stories, as products of partisan projects, as contestable representations, or as constructed documents in their potent capacity to define the facts. The narratives become clear mirrors, fully magical mirrors, without once appealing to the transcendental or the magical" (1997, 24).

10. As McGill (2008) notes in relation to molecular animation, there are some seventeen orders of time, and molecular animation is directly confronting the challenge of modeling these orders of time.

11. See http://multimedia.mcb.harvard.edu/; http://www.cellimagelibrary.org/; http://eowilsonfoundation.org/.

12. See, for example, the molecular animation workshop hosted by the Resource for Biocomputing, Visualization and Informatics (RBVI) and the National Center for Macromolecular Imaging (NCMI): http://www.ncbi.nlm.nih.gov/.

13. See Janet Iwasa's NSF-funded project, Molecular Flipbook, which provides biologists with "intuitive" software for building animations: https://www.molecularflipbook.org/.

14. For important work on this topic, see Flaxman 2012.

15. A number of theorists of early internet culture have confused Deleuze's work on the virtual with cyberspace. For example, Neil Spiller once wrote that *A Thousand Plateaus* is the "philosophical bible of the cyber-evangelists" (2002, 96). Also see Lévy 1998. For a full discussion of the misappropriation of Deleuze's concepts for the discourse on cyberspace, see Marks 2006.

16. I have to thank Gregory Flaxman for helping me see the way in which Deleuze's transcendental philosophy is crucial for understanding his cinema books.

17. As will become clear throughout this chapter, I am indebted to LaMarre's careful articulation of cinematic and animetic tendencies of the image. LaMarre's conception of tendency is also indebted to Félix Guattari's notion of the abstract machine (LaMarre 2009).

18. To see the rise of animation within film and media studies, one only has to note the success of *Animation: An Interdisciplinary Journal,* which was founded in 2006, as well as the growing number of edited collections on animation: *Robot Ghosts and Wired Dreams: Japanese Science Fiction from Origins to Anime* (2007), *Pervasive Animation* (2013), *Animating Film Theory* (2014), and *Animating Biophilosophy* (2014). Also noteworthy is the rise in the publication of monographs on animation in recent years (LaMarre 2009; Azuma 2009; Buchan 2010; Steinberg 2012; LaMarre 2018).

19. See, for instance, Manovich 2001; Andrew 2010; Rodowick 2007; Elsaesser 1998; Hansen 2004. In more recent years, Gregory Flaxman has called

upon us to "reclaim our faith" in cinema in the digital era because of its fundamental relation to off-screen space. In other words, what is essential to the cinema is not the "art of the index" but rather the production of outsides (Flaxman 2012).

20. See, for instance, Manovich and Kratky 2005 and Pisters 2012.

21. On this question, see Buchan 2013; Beckman 2014a.

22. See especially Beckman 2014b; Mihailova and Mackay 2014.

23. In *The World Viewed,* Stanley Cavell writes that "there is one whole region of film which seems to satisfy my concerns with understanding the special powers of film but which explicitly has nothing to do with projections of the real world—the region of animated cartoons. . . . If this region of film counters my insistence upon the projection of reality as essential to the medium of the movies, then it counters it completely" (1979, 167; cited in Buchan 2013, 4).

24. In more recent work, Tom Gunning challenges the quick dismissal of animation's relation to photography.

25. Not only has Manovich been attacked by cinema and media studies scholars but he has also received pointed criticism from the political Left, especially for his lack of engagement with the social and political realities of digital media. See, for instance, Holmes n.d. For a careful assessment of these and other political critiques of Manovich, see Galloway 2011.

26. On the epistemic mediation of software, and its distinctly Kantian tone, see https://redassiniboineresearchunit.wordpress.com/.

27. Manovich actually revises this statement somewhat in *Software Takes Command* by noting that it idealizes computer science in a way that he would no longer endorse (2013, 10).

28. See also Manovich n.d.

29. I do not want to suggest that *all* molecular animations produce numerical images. This fails to appreciate the wide range of possibilities for animation to engage with the molecular and cellular world. Wellmann (2017a) thinks about animation in a much broader way than I am here, and as such, she is able to expand the possibilities for animation to engage biology. Once we arm ourselves with a more expansive conception of animation—indeed, one that traverses live-action and manual construction—then, in principle, it should be easier to imagine the variety of material possibilities for molecular animation. However, one of the central theses of this book is that many of these other possibilities—including microcinematography and other time-based techniques—are also captured by the same historical and political forces as the computerized animation of molecules. As such, conceiving of alternatives is a more conceptually and genealogically difficult task than redesigning or reconceiving the technology. See the Postscript of this book.

30. For example, the Double-Stranded DNA (dsDNA) Kit, as well as a variety of other kits, for mMaya automates simulation, see https://clarafi.com/tools/mmaya/. See chapter 3 for my discussion of mMaya.

31. See Thurtle 2014.

32. See http://www.ribosomestudio.com/.

33. See https://www.nytimes.com/.

34. See https://www.ted.com/.

35. YouTube comments cited in Myers.

36. While Myers never uses naive realism to characterize computer animation skeptics, I do think her critique raises the question of a kind of naivete, which is what I am attempting to bring out.

37. Scientists (and philosophers) are well aware of the differences between the photographic apparatus and the microscopic apparatus.

38. On diffraction in classical and quantum physics, see Plotnitsky 2013. I have to thank Stacey Moran and her expansive dissertation, *Feminist Materials: Quantum Physics and Critical Writing Practices for New Material Feminism,* for helping me to understand the importance of quantum diffraction for STS and humanities scholars.

39. See also Allchin 1999; Campbell 1974. According to Allchin, what Hacking demonstrates with his theory of microscopic visual epistemology is what Campbell calls vicarious selection, essentially an "indirect, or secondary but not necessarily unreliable means of probing phenomena that are otherwise difficult to observe." In short, "the microscopic image is a vicarious selector of a specimen" (Allchin 1999, 288).

40. Also see Morana Alač's discussion of shearing (2011, 125, 129–30).

41. As Hacking argues in his essay "On Not Being a Pragmatist: Eight Reasons and a Cause," his version of pragmatism is less committed to classical genealogies of pragmatism (Pierce, James, Dewey, et al.) and much more interested in drawing on diverse sources (from Popper to Foucault) not often associated with pragmatism as such. Nevertheless, Hacking is clear in *Representing and Intervening* that "John Dewey's jeers at the 'spectator theory of knowledge' are equally germane for the spectator theory of microscopy" (1983, 189).

42. It is important to distinguish Hacking's work from "theory realism." For instance, one can believe in electrons—that is to say, believe they "exist"—even if there is as of yet no theory of electrons (Hacking 1983, 27). One can be an entity realist without being a theory realist (Sankey 2012, 33–34).

43. The teleological critique of Manovich's work is by now familiar to film and media scholars who argue that his work on animation is too deterministic: the valorization of digital animation leaves puppetry, stop-motion, and other expressive potentials of animation unacknowledged (Buchan 2014; Beckman 2014b).

44. Here's Manovich: "There is no such thing as 'digital media.' There is only software, the 'properties' of digital media are defined by the particular software as opposed to solely being contained in the actual content (i.e., inside digital files)" (2013, 152).

45. What is also important for apparatus theory is that technical materials are ideologically determined. See, for instance, Baudry 1974–75; Comolli 1985.

46. LaMarre writes that "even though I associate compositing with cel animation, and camera mobility with live-action cinema, these are two tendencies of the moving image. Cinema can use techniques of compositing" (2009, 124).

2. VISUALITY AND EXPERIMENTAL KNOWLEDGE PRACTICES

1. Because histories are always partial and limited, this history does not discount, exclude, or in any way reduce other ways of telling the history of visual epistemology in the biosciences, especially non-Western histories. Rather, this genealogy illuminates a certain range of possibilities and limitations for experimental knowledge practices in the Western life sciences.

2. Note that Keller and Grontkowski's claim about scientific objectivity's origins in Descartes could be challenged by Lorraine Daston and Peter Galison's (2007) history of "objectivity" in Western science.

3. The relation between theory and observation in scientific reasoning has been debated extensively in the analytical philosophy of science. See, for instance, Hempel 1970; Duhem 1991; Suppe 1977.

4. There are many distinctions to be made with "theory-ladenness": for example, between perceptual and semantic theory-ladenness (Kuhn 1962) as well as between theory-ladenness in its weak (Popper) and strong (Feyerabend and Kuhn) versions.

5. There is an important note to be made about the term "excess" and how I am using it in this chapter. Insofar as the "seeable" and the "knowable" are inherently linked in this genealogy, I am using "excess" to demarcate what is "unseen" and "unknown." This does not mean that excess lacks existence; on the contrary, it means that excess has a different onto-epistemological status, but it is one that is not unrelated to what is known and seen. The onto-epistemological status of this excess is alluded to when I discuss Deleuze on the virtual at the end of the section on classical genetics. (Note: this question is perhaps best treated by Deleuze in his books on Foucault and Leibniz, where outsides and insides are folded in ever more complex ways in the history of subjectivation [Deleuze 1988, 1993]). Given this, there are certainly resonances between what I am

claiming about the visible and the knowable and the "seeable" and the "sayable" in Foucault's *The Birth of the Clinic.* There he shows how the largely unconscious articulation of seeing and saying in the nineteenth century is what structures possibilities for knowing—evidenced in the medical gaze, and so on. Despite these connections, what I am claiming here is somewhat different: what is known about life's main functions—heredity and development, for instance—is made possible by what is visible via technologies of perception; and yet, what is *literally* invisible (and hence, unknown) continues to structure (virtually) the very meaning of what is seen and known. For instance, the location of life's most essential features is outside of what can be seen (literally), despite the fact that it gives meaning to all that is seen; the same can be said for time in the postmolecular era.

6. Although cellular observations in animals proved to be more difficult due to the fact that animal cells tended to be soft and low in contrast, all of which made observation and discernment of microstructure difficult (Coleman 1977, 24).

7. For Coleman, the brilliance of Schleiden and Schwann was their ability to tie botanical and zoological observations together into a coherent theory, resolving the disparate foundations of plant and animal life. According to the official story, Schwann had a conversation with Schleiden about his microscopic observations, and Schwann was "struck by the resemblance of this important body, that is, the plant cell nucleus, with the body which he [Schwann] had already observed in animal tissues" (Schwann, quoted in Coleman 1977, 26). He recalled seeing a similar structure in the cells of a notochord, and this convinced both of them that the cell was what spatially inscribed life itself.

8. For example, Lorenz Oken (1779–1851), an ardent supporter of *Naturphilosophie,* made claims that resembled Schleiden and Schwann's, especially regarding the generation of life. For Oken, the diversity of living forms was made possible by a more primordial unifying substance—an "undifferentiated mucouslike fluid" (Coleman 1977, 25). This unifying element was not unlike the generative principle holding plant and animal life together in cell theory, but Oken's theories were derived from "metaphysics and gratuitous assumptions," where the cell theory, at least according to its promoters, was the result of the careful empirical investigations of accomplished microscopists (25–26). Unlike the speculations of vitalists, cell theory was supposedly founded upon the secure foundations of observation. It was through microscopic observation, and not metaphysical speculation, that Schwann and Schleiden devised cell theory.

9. As Coleman explains, "The age-old persuasion of the continuity of life took on new and more concrete meaning, for in the cell could be seen a

self-perpetuating structural and functional unit common to all living things. These discoveries, of course, contradicted and destroyed the idea of cell precipitation in a formative and nonliving cytoblastema" (1977, 30).

10. See "Three Centuries of Sperm Research" in *Sperm Biology: An Evolutionary Perspective* (18–19) for a helpful description of the revolution Hertwig and Fol's work brought to evolutionary biology.

11. For a helpful reference, see Rheinberger, Müller-Wille, and Robert Meunier's "Gene" contribution to *The Stanford Encyclopedia of Philosophy.*

12. This distinction was introduced in Johannsen 1909 and developed more fully in Johannsen 1911.

13. In particular, Johannsen designed pure-line breeding experiments on the princess bean *(Phaseolus vulgaris)* and barley *(Hordeum vulgare).* He raised self-fertilizing plants and studied the characteristics of each generation of seed. Based on the observed characteristics of the different generations of seeds (their physical dimensions, etc.), he was able to place the pure lines into groups. Availing himself of statistical analysis, he discovered that these groups were better determinants of the characteristics a seed would produce than the characteristics of the mother. What this meant is that there had to be some grouping, or "type," which was distinct from what were assumed to be the heritable characteristics that organisms transmit to their offspring. By 1905, Johannsen called this type the "genotype," which was the organism's hereditary information, and distinguished it from the "phenotype," which indexed the actual observed characteristics of the organism. According to Johannsen, the genotype hypothesis disproved the "transmission" theory of heredity (Müller-Wille and Rheinberger 2012, 139). Where the latter theory postulated that parents transmitted their characteristics to offspring (Darwin's pangenesis is an example of this [Holterhoff 2014]), the former asserted that "the qualities of both ancestor and descendant are in quite the same manner determined by the nature of the sexual substances—i.e., the 'gametes'—from which they have developed" (Johannsen 1911, 130).

14. For this reason, Müller-Wille and Rheinberger assert that Johannsen's genotype is closer to an atom or a molecule than it is to a living substance, though Johannsen refused to assign a chemical makeup to it (2012, 140). In fact, he even claimed that "as to the nature of 'genes' it is as of yet no value to propose any hypothesis" (Johannsen 1909, 124).

15. Correns, for his part, developed a unique experimental regime where research questions were derived from the plant variety on which he experimented. This is in marked contrast to the "model organism" approach to genetic experimentation that we come across in Thomas Hunt Morgan, for example. Rheinberger notes that after Correns's death, his biographer, Emmy Stein, went through his lab notes and found "no less than 340 genera with whose species

or varieties Correns had experimented from the late 1880s to the early 1930s" (Rheinberger 2015, 54). With such a baffling number of experiments, and so many of them happening concurrently, it's difficult to measure the true impact and reach of his work. This is especially true given that the annals of science generally overlook his contribution to early genetics.

16. Müller-Wille and Rheinberger (2012, 145) discuss the mechanism responsible in terms of "triggered gene actions." In this context, also see Keller's (2002, chapter 4) interesting discussion of gene action.

17. In later work, Correns would experiment on variegation (which is the occurrence of different coloring in plant leaves) in *Mirabilis jalapa* crosses and come to the conclusion that there had to be cytoplasmic inheritance as well. Put generally, Correns mapped a distinctive cellular architecture based upon his experiments with plant hybrids. In his blueprint, dispositions were the unobservable motors of inheritance and development—although the latter required a yet-to-be-identified cytoplasmic mechanism that would allow the dispositions "to unfold in the right place"—that were only "knowable" through visible expressions in characters.

18. Subsequent researchers, including Morgan's student Herman J. Muller, became convinced that they could hunt down the physicochemical gene. Of particular note were his experiments with X-rays that induced Mendelian mutations in *Drosophila* (Müller-Wille and Rheinberger 2012, 151–52). But even with these experiments in hand, which seemed to suggest that the gene had a chemical makeup that could be altered through radiation, Muller admitted in 1950 that "the real core of gene theory still appears to lie in the deep unknown. That is, we have as yet no actual knowledge of the mechanism underlying that unique property which makes a gene a gene—its ability to cause the synthesis of another structure like itself, [in] which even the mutations of the original gene are copied. [We] do not know of such things yet in chemistry" (Muller, cited in Müller-Wille and Rheinberger, 152).

19. There is an interesting parallel that could be made between what I am suggesting about the era of classical genetics and Richard Doyle's claim about the "vanishing point" of life in Foucault's *The Order of Things*. As Doyle writes in relation to Foucault, "Whereas in the regime of natural history living beings were compared on the basis of a taxonomy that could be gleaned from a single plane, the visible surface, 'life' invisibly comes into its secret existence by introducing or injecting depth into the invisible, interior bodies of organisms" (1997, 10).

20. Given the way that Watson obtains the photo, Watson and Crick's brief mention of the "workers" at King's College suggests that they view themselves as lone visionaries.

21. To get a sense of how difficult it is to form crystals, see the documentary *Naturally Obsessed: The Making of A Scientist* (2009). Myers also offers a helpful analysis of the film in *Rendering Life Molecular.*

22. See Myers's analysis of crystallization in chapter 1 of *Rendering Life Molecular.*

23. See https://www.theguardian.com/science/2015/jun/23/sexism-in-science-did-watson-and-crick-really-steal-rosalind-franklins-data.

24. See "Linus Pauling's Triple DNA Helix Model, 3D Animation with Basic Narration" at https://www.dnalc.org/.

25. See "Linus Pauling and the Race for DNA" at http://scarc.library.oregonstate.edu/.

26. Note that Alex Wild updated his piece after Curry's criticism.

27. Hannah Landecker has written at length about how microcinematography in the early twentieth century functioned like a portal into the world of the *temporally* developing cell (2011, 2007). But as she notes in a co-authored piece with Christopher Kelty (2006), this work was on the margins of knowledge practices in the life sciences. Cytological investigations largely gave way to genetics research and then to molecular genetics. See my discussion of microcinematography in chapters 2 and 7.

28. Helmut Ruska, Ernst Ruska's brother, was a little-known but crucial player in the early deployment of the electron microscope in the biomedical sciences. For example, he used the EM to study viruses, parasites, and bacteriophages and became known in Europe for producing the first published bacteriophage micrographs, in 1940 (Ackermann 2011).

29. Of course, many modifications have been made to the TEM and SEM since their inception. For example, Somnath Chatterjee and his colleagues (2012, 892) review these developments and point to cryo-electron microscopy (cryo-EM) as a significant advancement in TEM technology. With the cryo-EM the sample is now imaged in cryogenic temperatures, which means that there is no need to stain and fix it. This is a crucial advancement over traditional TEM and X-ray crystallographic techniques, since they both damage the specimen. And recently, scientists using cryo-EM have obtained images with a resolution fine enough to compete with and perhaps surpass X-ray crystallography (Campbell et al. 2015). Chatterjee and his colleagues note other advancements in TEM, such as the Environmental TEM (ETEM), which permits imaging of nanostructures in situ instead of within the high-vacuum conditions inside the microscope column (traditional TEM). They also cite developments in SEM technology, and in particular, Environmental SEM (ESEM), which permits scientists to obtain micrographs of specimens using the SEM, but within their "wet" state, instead of being cleaned and made electronically conductive (see Bogner et al. 2007).

30. Curry explains that these images give us no indication of their function. And as Somnath Chatterjee and his colleagues (2012) note, this ought to be the goal of electron microscopy.

31. See Keller's discussion of the "vital barrier" in life science imaging in *Making Sense of Life* (2002, chapter 7).

32. Of course, this does not mean that moving images did not have an important role to play in educational settings, and so on. For a more detailed analysis of this, see Gaycken 2015, as well as Hentschel 2014 and Ostherr 2013.

33. Orlando and his colleagues explain that "in many cases it is not even known how many processes underlie the measured signal or what their respective time scales are. These questions are particularly relevant to the field of developmental biology. Developmental studies focus on systems such as animal embryos or plant organs in which processes such as growth, segmentation and differentiation can all take place simultaneously, but on different time scales, complicating the interpretation of expression data" (2010, 1).

3. A FEELING FOR THEORETICAL BIOLOGY

1. I follow parts of Evelyn Fox Keller's careful assessment of D'Arcy Thompson and Alan Turing in *Making Sense of Life* (chapter 2), although Stephen J. Gould offers a useful supplement to her analysis.

2. In fact, "all but three or four of its seventeen chapters," writes Gould, "are devoted to the citation of correspondences between organic forms and the shapes that physical forces produce in acting upon non-living material of similar size, density, viscosity, and rigidity. From these correspondences, D'Arcy Thompson inferred that organic forms had been fashioned in the same way—by the direct action of physical forces" (1971, 240).

3. Some have argued that there was a split between embryologists and geneticists during this period. However, the reality of this split, according to some historians, seems exaggerated. See Keller, *Making Sense of Life* (especially chapter 4), Staffan Müller-Wille and Hans-Jörg Rheinberger, *A Cultural History of Heredity*, and Lily E. Kay, *Who Wrote the Book of Life?*

4. I have to thank Giuseppe Longo for pointing out to me the important difference between Turing's work on the Discrete-State Machine in 1950 and his model of morphogenesis in 1952. This conversation occurred at the Amsterdam Institute for Advanced Study's Life beyond the Algorithm symposium: http://ias.una.nl. Longo's paper is available at https://www.di.ens.fr.

5. As Keller explains, C. H. Waddington cannot be considered a representative figure in twentieth-century biology by any stretch of the imagination.

Not only did he "straddle" the fields of genetics and embryology but he was also a founding member of the Cambridge-based Theoretical Biology Club, which I discuss briefly at the end of chapter 7 (Keller 2002, 104). What's more, Waddington endorsed Whitehead's philosophy of organism, a very unpopular position among theoretical biologists at that time (Peterson 2011).

6. Although on this point, I think Thurtle is much better than Keller, mainly because he has a more genealogically robust way of tracing regulation back to Bateson and Lewis, essentially suggesting that it comes out of classical genetics and not Waddington's epigenetics (Thurtle 2018).

7. See M. J. Apter and Lewis Wolpert's lengthy discussion and critique of Dancoff and Quastler's information-based biology. As they put it: "Essentially their [Dancoff and Quastler's] method is to take the object whose information content they wish to measure, along with enough building stones to build another object identical with this one, and then write a list of instructions specifying how the second object is to be built from these building stones" (1965, 248). Among the many problems with this proposal is "that the information figure estimated is so arbitrary as to make the values obtained meaningless. This method of measuring information overlooks the fact that if we were going to build one object identical with another one, we would need to know not only what sort of building blocks had to be put in each position, but we would also need to know where each position was" (249).

8. Of course, this did not stop the wider mechanisms of military and industrial power from absorbing biology into the discourse of automated communication. And this also does not mean that biology was not eventually informationalized, as I argue in later sections of this book. What it does mean is that the informationalization of biology was not readily accepted by working molecular biologists.

9. I am not claiming that Thompson and Turing's work in mathematical biology is equivalent to the information-based research in cybernetic biology. Nor am I asserting that Thompson and Turing are working with equivalent models. Rather, I am pointing out that this theoretical work in biology experienced similar forms of resistance in the mainstream culture of experimental biology, despite its (possible) relevance. This is not unlike what Keller argues in *Making Sense of Life,* although this point is sometimes obscured.

10. As the plant physiologist and molecular biologist Anthony Trewavas writes, "The understanding of systems has had enormous impact on what are loosely regarded as human sciences, including economics, sociology, psychology, and medicine. Systems biology has generated revolutions in ecology, population biology, and evolutionary studies and is slowly making inroads into biochemistry, development, genetics, and whole-plant biology. But it is only very

recently that molecular biology has adopted a systems approach. The enormous growth in genomics now makes this possible" (2006, 2420).

11. This claim may be true for molecular biology, strictly speaking. But Karl Ludwig von Bertalanffy, for instance, one of the founding fathers of general systems theory, made notable contributions to the biology of the organism as well as to cybernetics, philosophy, and psychiatry, among other disciplines. However, within the dominant paradigms of molecular biology, von Bertalanffy's work was largely passed over. See von Bertalanffy's *General System Theory: Foundations, Development, Application.*

12. Charles F. Stevens (2004) offers a comprehensive overview of the differences between molecular and systems biology. And while these two epistemological cultures did not generally overlap in the twentieth century, there are some notable exceptions; F. J. Bruggeman and Hans Westerhoff (2007) note these.

13. In making the case for system-level analysis, Bruggeman and Westerhoff argue that sequencing an entire genome gives us very little when it comes to understanding how the components of the system function. Hence, a system-level analysis is crucial for understanding genome-wide function.

14. See Daston and Galison's interesting discussion of the emergence of an engineering-based ethos in scientific practice in their *Objectivity,* chapter 7.

15. See Cristian Calude and Giuseppe Longo's "The Deluge of Spurious Correlation in Big Data" for an insightful and critical analysis of the use of correlations in data-driven science.

16. http://www.siliconcell.net/.

17. See "The iGEM Competition: Synthetic Biologists and Molecular Animation" at https://blogs.scientificamerican.com/.

18. https://www.rcsb.org/.

19. Gaël McGill describes the difficulty of this process. Essentially, there is one of two ways to get molecular structures into Maya: either find existing MEL scripts online—most famous is Tom Doeden's PDBreader.mel—or use a different molecular graphics application, such as Chimera. You then build a representation in those programs and export the geometry to a file format that is recognizable to Maya.

20. See, for example, Autodesk comments on MEL and Python: https://www.reddit.com/r/Maya/.

21. http://biopython.org/.

22. Sharpe and his colleagues note that MEL++ is available for projects that require OOP; there is also PyMEL, which translates MEL commands into Python functions.

23. It is important to point out that this is an inexact method. One scientist-animator's cartoon is another scientist-animator's detailed structure.

24. Another, and perhaps more provocative, way to put this would be to say that these models help researchers gain a "feeling" for the molecular world. As Myers demonstrates in *Rendering Life Molecular,* biologists perceive and think with their models.

4. ECO-SOCIAL MEDIA

1. See Bolinsky 2007.

2. Similar to what Jonathan Crary notes in *Techniques of the Observer,* this visual logic "relocate[es] vision to a plane severed from a human observer," so that "most of the historically important functions of the human eye are being supplanted by practices in which visual images no longer have any reference to the position of an observer in a 'real,' optically perceived world. If these images can be said to refer to anything, it is to millions of bits of electronic mathematical data" (1990, 1–2). See chapter 6 for a more detailed discussion of Crary's work.

3. What I am arguing has certain resonances with Peter Haas's (1992) analysis of epistemic community as defined by those who share a set of normative commitments. While I acknowledge the advantages of this notion in certain circumstances, and that it has been influential in certain STS circles, I am more interested in the formation of an assemblage of entities and actors without the conscious consent of those individuals who make it up.

4. The pragmatic reading of Whitehead's metaphysics has gained traction in the Anglophone world after the translation of Stengers's *Penser avec Whitehead* into English. One of the many advantages of this reading is that it takes Whitehead's speculative method, and not just his concepts, seriously. And in doing so, pragmatic readers of Whitehead tend to insist that his notions are propositions whose efficacy is determined by what they do and what they make matter.

5. "'Actuality,'" Whitehead explains, "is the decision amid 'potentiality'" (PR 43). On the notion of the decision in Whitehead, see chapter 2 in Shaviro (2009).

6. Whitehead is also and not unintentionally overturning Kant's claim that the subject constitutes its objects. For Whitehead, the feeling subject (or feeler) does not preexist the feeling; it's just the opposite: "The feelings aim at the feeler" (PR, 222). Whitehead reverses Kantian priorities: "For Kant, the world emerges from the subject; for the philosophy of organism, the subject emerges from the world" (88).

7. Whitehead explains that "eternal objects of the objective species are the mathematical Platonic forms. They concern the world as a medium" (291).

8. See Elizabeth Kraus's (1998, 30) discussion of the adverbial nature of eternal objects.

9. Please see Kraus's (1998, 28–35) excellent discussion of the realm of eternal objects.

10. Whitehead will argue that God is "where" eternal objects are located. But God is also an actual entity. See my discussion of Whitehead's God in chapter 5.

11. In *Modes of Thought,* Whitehead characterizes this complexity in terms of self-enjoyment and concern. The actual entity is (1) a precognitive process of vivid and immediate "self-creation" qua self-enjoyment that must "be understood without reference to any other concurrent occasions" (MT, 151); and it is also (2) "an activity of concern, in the Quaker sense of that term. . . . The occasion is concerned, in the way of feeling and aim, with things that in their own essence lie beyond it" (167; see also Shaviro 2014). In the language of *Process and Reality,* an occasion's evaluations are at once tied to the singular and immediate self-enjoyment of a datum and also to the concern it feels for the world that lies beyond it and that it will affect.

12. See, for instance, Morton 2007.

13. Harman (2002) contends that Whitehead is one of the forgotten twentieth-century thinkers of metaphysical reality outside of the confines of human thought and perception (correlationism). In fact, Harman contends that both Heidegger and Whitehead are aligned in their shared goal to undermine substance-based ontologies, but it is precisely for this reason that they are both opposed to Harman's own object-oriented metaphysics (227). Levi Bryant makes similar arguments about Whitehead's ability to think noncorrelational reality, but disagrees with Whitehead concerning his distinction between occasions and societies (2011, 233–34). Likewise, in *The Universe of Things,* Shaviro places Whitehead squarely in the middle of speculative realist debates but draws on Whitehead's aesthetics to move past certain dead ends in speculative realism and object-oriented ontology.

14. Whitehead explains that "it will also be noticed that this ideal of speculative philosophy has its rational side and its empirical side. The rational side is expressed by the terms 'coherent' and 'logical.' The empirical side is expressed by the terms 'applicable' and 'adequate'" (3).

15. Gilbert Simondon's critique of hylomorphism is an important partner in this conversation about the organization of technical individuals. See Simondon's *L'Individuation psychique et collective* and *On the Mode of Existence of Technical Objects,* as well as Muriel Combes's superb elaboration of his philosophy in *Gilbert Simondon and the Philosophy of the Transindividual.*

16. I want to be careful not to inappropriately apply Whitehead's concept of the common element of form to molecular animation. For this reason, I will cautiously say that the common element of form helps vivify aspects of

the relations that hold among elements of a media ensemble. If I emphasize that it is a "logic" or a "pattern of relations" that is the common element, then it is because in Whitehead the common form—a complex eternal object—does not depend on any epoch for its existence—it is eternal. It is entirely unclear to me whether the pattern transcends any particular epoch. This would mean, for example, that this is a pattern that does not depend upon actual vision and could be an element of form in a sonic society, or any other society of occasions.

5. GOVERNING THE SOCIAL

1. I have to thank Erin Manning and Brian Massumi for pushing back on my formulation of Whitehead's generic philosophy. I am now less inclined to see Whitehead's project as lacking specificity or even unable to address historical and political realities as such. However, I still think that Whitehead's thought needs to be transformed in specific ways to account for the historical formation of human societies and the systems of power that circulate within them.

2. See Alexander Galloway's week-long seminar published in a pamphlet titled "French Theory Today: An Introduction to Possible Futures" at http://cultureandcommunication.org/. Also see Galloway's response to Graham Harman on the blog *An und für sich* at http://itself.blog/.

3. For Foucault's relation to other theoretical movements (such as structuralism, psychoanalysis, and Marxism) and thinkers, see Nealon 2008; Dreyfus and Rabinow 1982; Han 2002.

4. There are few but important places in Foucault's work that provide clear and precise characterizations of philosophical concepts. "The Subject and Power" is one of these important places. As Beatrice Han notes, this is likely because Foucault is constantly working between two interdependent levels of analysis: he provides "detailed analyses of concrete techniques where a specific power and a specific knowledge are articulated" alongside more general conceptualizations of "'power,' 'knowledge,' and 'truth'" (2002, 108).

5. In *The Birth of Biopolitics,* Foucault explains that "the term itself, power, does no more than designate a domain of relations which are entirely still to be analyzed, and what I have proposed to call governmentality, that is to say, the way in which one conducts the conduct of men, is no more than a proposed analytical grid for these relations of power" (2008, 186).

6. The relation between Whitehead and autopoietic systems theory has not been adequately theorized, although the self-organization of sociality suggests that there are interesting connections to be made. For a careful introduction

to autopoiesis and to how it aligns with deconstructive, posthumanist, and systems theoretical genealogies, see Wolfe 2010.

7. Also see Whitehead's discussion of persuasion and force in *Adventure of Ideas.* He explains that "civilization is the maintenance of social order, by its own inherent persuasiveness as embodying the nobler alternative. The recourse to force, however unavoidable, is a disclosure of the failure of civilization" (AI, 83).

8. Some scholars have argued that Foucault's late work on the "care of the self" is just such an intervention into neoliberal forms of self-governance—qua neoliberal *homo œconomicus.* Indeed, the turn toward subjectivity and ethics in the last two volumes of the *History of Sexuality* are helpful in spelling out Foucault's relation to and distance from neoliberal forms of subjectivity. Yet how precisely Foucault's late work on the care of the self functions as a mode of resistance is by no means clear, and it leaves scholars to ponder what he might have had in mind. See, for example, Dilts 2011.

9. The question of the proposition bears directly on Whitehead's larger conversation about symbolic reference as the intersection of causal efficacy and presentational immediacy. In his lectures titled *Symbolism: Its Meaning and Effect,* he notes that symbolic reference always risks being wrong (S 80).

10. See my longer discussion of Whitehead's conception of error and imaginative freedom in Nocek 2017.

11. I have to thank Steven Meyer for pointing out the different occasions for Whitehead's lectures. The 1926 Lowell Lectures, *Religion in the Making,* were delivered primarily to a Christian audience at King's Chapel in Boston. At this time, Whitehead wanted to appease his audience as much as possible. By contrast, the 1927–28 Gifford Lectures, titled *Process and Reality,* were delivered at the University of Edinburgh, and Whitehead did very little to accommodate his audience.

12. In *The Metaphysics of Experience,* Elizabeth Kraus insists that Whitehead's conception of God remains at the blueprint stage at best, so we cannot expect for all of the difficulties associated with it to be neatly resolved. "The details" of the dialectic of God and World that unfold in the last chapter of *Process and Reality* "are not worked out; the language is more evocative than precise; more is concealed than is revealed," even though the dialectic is rich in poetry and suggestiveness (1998, 168).

13. In other work, Steven Shaviro experiments with the proposition that Whitehead's God shares certain features with Deleuze and Deleuze and Guattari's notion of the exclusive and inclusive use of the disjunctive synthesis, and then uses this connection to illustrate a structural parallelism between the function of God and the "Body without Organs" (2009, 99–142). Donald

Sherburne (1967), on the other hand, goes so far as to argue that there is nothing worth saving in Whitehead's God, and in this spirit, even contends that his system is internally incoherent with the concept of God. Likewise, Judith Jones worries about God, though seems more sympathetic to the role that God plays in the system and is not convinced that there is no way to eliminate it without causing problems to the rest of the system. As she writes, "This function of the divinity is not without problems, which is why I have attempted to develop a view without direct recourse to the status of God, except as that element of the universe in virtue of which each entity entertains an ideal of itself in its subjective aim. Subjective aim might be derivable from the general nature of the universe itself, without recourse to a primordial actual entity, though how this is to be done without stretching the ontological principle to the point of unintelligibility is unclear to me. The grounds of order and the unique possibilities of intense order proper to a particular atomic standpoint do not in principle point to the necessity for any singular entity, however" (1998, 127).

14. There is an important way in which these two tendencies of the image map onto Deleuze's notion of the action-image and the time-image. Indeed, the crisis in the action-image (cause and effect and goal-oriented action in classical Hollywood cinema) forecloses the possibility of connecting to actions in a linear, causal chain, and so on, and paves the way for the time-image, which introduces rupture, discord, confusion, and hesitation (Deleuze 1989; LaMarre 2009, 197–200).

15. For instance, Quentin Meillassoux notes in an interview with Rick Dolphijn and Iris van der Tuin, "Concerning Foucault I will simply answer as follows. His investigation focuses on past dispositives of knowledge-power, and eventually on dispositives that are contemporary to him. He can bring us nothing in regards to the disqualification of strong correlationism, because the disqualification is situated at a level that his research does not address, but rather presupposes. Indeed, I examine how correlationism, from its point of departure in the Cogito, has come to dominate all of modern thought, including the most resolutely anti-Cartesian: the 'great confinement' was not that of the fools in the asylum, but that of the philosophers in the Correlate—and this also applies to Foucault. Indeed, Foucault does not say anything that would embarrass a correlationist, as all his comments can easily be considered as a discourse-correlated-to-the-point-of-view-of-our-time, and rigorously dependent on it. This is a typical thesis of some correlationist relativism: we are trapped in our time, not in Hegelian terms, but rather in a Heideggerian fashion—that is to say in the modality of knowledge-power that always already dominates us. . . . I am not at all hostile to Foucault's thesis, even though in my view he thinks within a historicist ontology that remains unthought in its deep nature" (2012, 77).

16. This would be my counter to scholars in critical animal studies who critique Foucault's thought for excluding human–nonhuman relations. See, for instance, Shukin 2009. For thinkers of biopower, Shukin argues, "animality functions predominantly as a metaphor for that corporeal part of 'man' that becomes subject to biopolitical calculation" (2009, 9–10). On the inclusion of animals and plants within the purview of biopower, see Wolfe 2012; Nealon 2015.

17. Shaviro (2014, chapter 1) suggests a similar aesthetic reading of Emmanuel Levinas and Graham Harman.

18. In her lecture "Another Science Is Possible! A Plea for Slow Science," Stengers discusses the case of the scientist Barbara van Dyke, who was sacked from the Katholieke Universiteit Leuven for entertaining a proposition for agricultural production that was nonconformal with the values of fast technoscience. Van Dyke was publicly dismissed and accused of being an "enemy of science" (Stengers 2011a; Nocek 2017).

19. This is why Stengers writes that her notion of cosmopolitics "must therefore be distinguished here from any particular cosmos, or world, as a particular tradition may conceive of it. . . . In the term cosmopolitical, cosmos refers to the unknown constituted by these multiple, divergent worlds, and to the articulations of which they could eventually be capable, as opposed to the temptation of a peace intended to be final, ecumenical" (2005, 995).

6. THE ANIMATION APPARATUS

1. In this context, Crary hones in on how the optical sciences, as well as many of the artistic and philosophical discourses of the time, were marshaled to normalize this nineteenth-century observer-subject/consumer (1990, 68–96).

2. This is closely linked to Crary's crucial observation in *The Suspensions of Perception* that the rise of the entertainment industry was largely a consequence of the mass production of images as measurable commodities and the design of subjects who could consume these images.

3. For Kracauer, "the industrial body of the Tiller Girls," Hito Steyerl explains, "is abstract, artificial, and alienated. Precisely because of this, it breaks with the traditional, and at the same time, radically imbued ideologies of origin, belonging, as well as with the idea of a natural, collective body created by genetics, race, and common culture" (2012, 181).

4. Marx's theorization of the commodity form in terms of animism and magic has been noted by many theorists. See especially Christopher Bracken's *Magical Criticism: The Recourse of Savage Philosophy.* In this work, Bracken

examines the double life of the commodity in animist terms—commodities "speak" to one another (see Bracken 2007, 155–65).

5. See Steven Shaviro's blog post, "The Cinematic Mode of Production," at http://www.shaviro.com/.

6. See Beller's (2006, chapter 4) detailed discussion of the falling rate of profit.

7. See Lambert's (2015) discussion of the collective in ordoliberalism.

8. For a more detailed account of the rise of neoliberalism as a theory and practice, see Brown 2019; Dardot and Laval 2017.

9. Both essays, "Economics and Knowledge" (1935) and "The Use of Knowledge in Society" (1945), are reprinted in the collection of essays *Individualism and Economic Order* (1948).

10. Adam Greenfield discusses how Google Glass was a failure, largely because the hands-free, head-mounted AR technology had not been perfected by the time it went to market. However, this does not mean that better designs are not already close to being market ready (Greenfield 2017, 73–84).

11. See Bo Begole's post, "Omnipresence and the Coming Age of 'Remote Reality,'" at http://venturebeat.com/.

12. Note that Sha's examination of responsive media develops a way to engage these media as sites of resistance to forms of political and economic capture through "ethico-aesthetic" practices indebted to Deleuze, Deleuze and Guattari, and Whitehead.

13. This is what Manovich's taxonomy of software cannot appreciate with its largely apolitical reading of new media (Galloway 2012).

14. See Neidich 2013. Also see Moulier-Boutang 2011, as well as Lazzarato's (2012a) conception of noo-politics.

15. There is a growing body of research at the intersection of entrepreneurship and neuroscience, also known as "neuroentrepreneurship." In addition to Day, Boardman, and Krueger 2017 and Zald and his colleagues' (2008) work on dopamine receptors, see Lawrence et al. 2008.

16. It's worth noting that while copyediting this manuscript during the COVID-19 pandemic, the public consumption of molecular animations has spiked dramatically. In just three months, Bolinsky's and Berry's TED talks have each received over one million new views, as has *The Inner Life of the Cell* animation. My suspicion is this is largely because computer graphics and animations of COVID-19 have become essential tools for both communicating important information about the virus to the general public and for sensationalizing mass tragedy on cable news and social media. As of June 2020, the computer animation, *COVID-19 Animation: What Happens If You Get Coronavirus?* has received close to 15,000,000 views. See https://www.youtube.com/watch?v=5DGwOJXSxqg.

17. See http://www.digizyme.com/ and https://axs3d.com/.

18. See "What Is a Dispositif?—Part 1" at http://jcrt.org/religioustheory/.

19. Both Lambert and Pasquinelli note that Canguilhem is cautious not to overextend the notion of biological normalization into the social, for it risks becoming a kind of "organicism" that "sloppily" applies biological notions to the political (see Pasquinelli 2015, 85). What's more, in Foucault the normalizing effects of the neoliberal biopolitical dispositif does not eclipse the *mechanisms* of discipline, it just reconfigures them.

20. Pasquinelli explains, "For Goldstein, in general, the normative power is the ability of each organism (and specifically of the human brain) to invent, modify and destroy its own norms, internal and external habits, rules and behaviors, in order to adapt better to its own Umwelt (or surrounding environment), particularly in cases of illness and traumatic incidents and in those conditions that challenge the survival and unity of the organism. In Goldstein, the normative proceeds from the constant antagonism of the organism against its environment" (2015, 83).

21. For a thorough genealogy of the dispositif, from Goldstein through Canguilhem to Foucault, see Pasquinelli 2015.

22. We should be careful here not to assume that all modes of social togetherness in Whitehead are apparatuses. This is very clearly not the case.

23. It is interesting to consider whether there are elements in the actual world essential to the reproduction of this form. In a strict Whiteheadian perspective, none of them would be essential, since forms of definiteness do not depend on any particular actuality (or epoch) for their existence. Molecular animation is a potential exemplification of this form, and there are many others, including music videos, blockbuster films, etc., but none of them are essential. What's more, this logic is not vision dependent (an evolutionary contingency) and can in principle be exemplified in other societies. However, such a strict Whiteheadian reading may only go so far, and may need to be modified, since it is not at all clear whether we could say this logic or pattern is a quality of relation abstracted from the emergence of any particular epoch. Although this brings up much deeper questions about the "eternality" of eternal objects, and whether we might turn to Deleuze's potentiality for help.

7. EPISTEMIC CAPTURE

1. In an interview, Foucault notes that "I set out from a problem expressed in the terms current today and I try to work out its genealogy. Genealogy means that I begin my analysis from a question posed in the present" (Kritzman 1988, 262).

2. Oliver Gaycken (2015, 20) notes that the original program no longer exists, and these are the titles of the films based on reviews and Urban's catalogue.

3. This is not to say that there were not a number of scientists working with cinematography in these years. There is a growing body of research in early cinema that examines how a number of early twentieth-century scientists were using film, and their experiments were essential to the development of the medium. For instance, Auguste Lumière was a scientist who made contributions to medical biology, pharmacology, and physiology; Étienne Jules Marey was a physiologist (Cartwright 1995).

4. Curtis explains that, later in his career, Bergson embraces other potentials for the cinema: "Bergson recognized that cinema had the capacity to provide both analysis and synthesis, that it had an inherently ambiguous character precisely because it is both still and moving. Indeed, Bergson's invocation of the cinematographic apparatus was never an outright condemnation of the cinema but instead a way of describing a tendency in our habitual, everyday way of thinking that science had codified into experimental method" (2015, 36).

5. Gaycken details Pathé's role in the promotion and dissemination of Comandon's work in order to create new markets.

6. The language of eugenics and even a kind of transhumanism is unmistakable here, though Landecker does not take her analysis in this direction.

7. Landecker discusses these new forms of mediated cellular knowledge in the work of Michael Abercrombie, Warren Lewis, Charles Pomerat, and Marcel Bessis.

8. I do not want to claim that Landecker is not interested in these questions. She writes, for example, "It follows that the microcinematograph extends our perception into the world of the very slow, the very small, and thus modifies our experience of cellular life and mediates the very possibility of its study" (2011, 385). What I am suggesting is that this question is not examined in depth, and there is no serious discussion of political and economic power in the context of visual media in the twentieth century. She alludes to Philip Pauly's notion of "biological modernism" (1994) and cites Doane's (2002) work on cinematic modernity, although the political stakes of the visual in biology are not examined. What I am interested in here might be seen as an extension of this project.

9. See Curtis's (2015) detailed discussion of "enlightened" modes of viewing, in chapters 2 and 3 in particular. This also fits into wider conversations in the history of the science about training objective perceivers. See Daston and Galison 2007.

10. There are other labeling options that Gaycken discusses. In particular, Miriam Hansen's "vernacular science" is an option, but this distorts the fact that the layperson is not creating the images (Gaycken 2015, 8).

11. For example, see Cary Wolfe's important critique of Roberto Esposito's "affirmative biopolitics" as a form of political vitalism, which functions as an onto-theology propping up Christian ideas (2012, 59). Also see Alexander Galloway and David M. Berry's discussion of a particular style of Deleuzian thought, especially the "Google Deleuzians."

12. As Huxley writes in "Eugenics and Society," "We must therefore concentrate on producing a single equalized environment; and this clearly should be one as favourable as possible to the expression of the genetic qualities that we think desirable. Equally clearly, this should include the following items. A marked raising of the standard of diet for the great majority of the population, until all should be provided both with adequate calories and adequate accessory factors; provision of facilities for healthy exercise and recreation; and upward equalization of educational opportunity" (1936, 25).

13. See "Biology and the Bauhaus" at https://www.tate.org.uk.

14. See Christina Cogdell's *Eugenic Design: Streamlining America in the 1930s* for her analysis of the tight coupling between eugenics programs and industrial design in the early twentieth century. She writes that "by establishing the precedent set by modern architects . . . of basing architectural and design theories upon evolutionary ideas, I propose the 'popular' style of streamlining and its relation to eugenics as a logical extension of 'high' modernist architectural theories" (2004, 6).

15. In the foreword to *Vision in Motion,* Moholy-Nagy writes: "This book is written for artists and the layman, for everyone interested in his relationship to our existing civilization. It is an extension of my previous book, 'The New Vision.' But while 'The New Vision' gave mainly particulars about educational methods of the old Bauhaus, 'Vision in Motion' concentrates on the work of the Institute of Design, Chicago, and presents a broader, more general view of the interrelatedness of art and life. . . . Thus this book is an attempt to add to the political-social a *biological 'bill of rights.'* . . . It proposes that new tools and technologies cause social change; that they shift ways of production, possessions, wealth, and power. . . . The emotional prejudice—or inertia—is the greatest hindrance to necessary adjustments and social reforms. The remedy is to add to our intellectual literacy and emotional literacy, an education of the senses" (1947, foreword).

POSTSCRIPT

1. In her short and provocative book *Medium Design,* Keller Easterling proposes a notion of medium design that resonates with apparatus design. Here's Easterling: "But in any context, large or small, designing the medium is

managing the potentials and relationships between objects, the activity or disposition immanent in their organization. The disposition of any organization makes some things possible and some things impossible. Like a growth medium, it determines what will live or die. Like an operating system, it sets the rules of the game that link and activate the components of an organization" (2018, 9).

2. In Flusser's short essay on the word "design," he writes that "[t]he word occurs in contexts associated with cunning and deceit. A designer is a cunning plotter laying his traps" (1999, 17). And then just below, he adds that "to deceive nature by means of technology, to replace what is natural with what is artificial and build a machine out of which there comes a god who is ourselves. In short: The design behind all culture has to be deceptive (artful?) enough to turn mere mammals conditioned by nature into free artists" (19).

3. For my full discussion of critical and speculative design, see Nocek 2019.

4. Jean Painlevé's cinematic "science experiments," of which there are more than two hundred between 1927 and 1982, are almost as absent from film history cannons as they are from the annals of scientific imaging. For André Bazin, Painlevé's films are as anomalous as they are brilliant (2000, 147). They occupy a strange middle ground between science and art and are largely unacceptable to histories of either one: too scientific to be art, and too artistic to be science. Painlevé scholars have noted that he was a gifted biologist and that he could have had an illustrious scientific career, having coauthored a paper in 1923 on the color-staining of glandular cells in hironomid larvae, which was presented at the Académie des Sciences, and made the following headlines: "Twenty-Year-Old Son of Statesman Paul Painlevé, Youngest Researcher Ever to Present a Paper to the Academy." Even with such an early accomplishment, Painlevé found himself frustrated with the stifling intellectual space of academia (Berg 2000, 12). Thus instead of seeking the company of other scientists, Painlevé sought friendship and inspiration from avant-garde artists in France, many of whom were deeply influenced by surrealism. Despite the many friendships with artists (especially surrealists) that Painlevé forged, he never officially joined any of the so-called surrealist camps.

VIDEOGRAPHY

Acera, ou Le bal des sorcières / Acera, or The Witches' Dance. Directed by Jean Painlevé, 1972.

Cell Membrane. Directed and produced by Ribosome Studio: Oberrieden, Switzerland, 2018.

Clathrin Mediated Endocytosis. Directed by Janet Iwasa and Tomas Kirchhausen, 2012.

A Day in the Life of a Motor Protein. Directed and produced by Casper Hoogenraad et al. Utrecht: Hoogenraad Lab, Utrecht University, 2013.

The Inner Life of the Cell. Directed and Produced by XVIVO Scientific Animation: Hartford, Conn., 2006.

Kinesin Motor Protein. Directed by John Liebler. Guilford, Conn.: Art of the Cell; Medical and Scientific 3D Animation and Illustration, 2014.

Le bernard l'ermite / The Hermit Crab. Directed by Jean Painlevé, 1929.

Le vampire / The Vampire. Directed by Jean Painlevé, 1945.

L'hippocampe / The Seahorse. Directed by Jean Painlevé, 1934.

Man with a Movie Camera. Directed by Dziga Vertov, 1929.

The Molecular Interactions of the MAPK Pathway. Directed by Ribosome Studio: Oberrieden, Switzerland, 2018.

Molecular Motor Struts Like a Drunken Sailor. Directed by Janet Iwasa. Produced by Reck-Peterson Group, 2012.

Mours de la pieuvre / Love Life of the Octopus. Directed by Jean Painlevé, 1965.

Naturally Obsessed: The Making of a Scientist. Directed by Carole Rifkind and Richard A. Rifkind, 2009.

Our Secret Universe: The Secret Life of the Cell. Directed by Mike Davis. Los Angeles: Wide-Eyed Entertainment, 2012.

Plasma Membrane. Directed and produced by Ribosome Studio: Oberrieden, Switzerland, 2018.

Protein Packing. Directed and produced by XVIVO Scientific Animation: Hartford, Conn., 2013.

The Secret Life of Levodopa. Directed and Produced by Ribosome Studio: Oberrieden, Switzerland, 2018.

Strike. Directed by Sergei Eisenstein, 1925.

X Inactivation and Epigenetics. Directed by Etsuko Uno and Drew Berry. Produced by Walter and Eliza Hall. Melbourne: WEHImovies, Walter and Eliza Hall Institute of Medical Research, 2012.

BIBLIOGRAPHY

ABBREVIATIONS OF WHITEHEAD TITLES

AI *Adventures of Ideas*
CN *The Concept of Nature*
MT *Modes of Thought*
PR *Process and Reality: An Essay in Cosmology*
RM *Religion in the Making*
S *Symbolism: Its Meaning and Effect*
SMW *Science and the Modern World*
SP *Science and Philosophy*

Ackermann, Hans-W. 2011. "The First Phage Electron Micrographs." *Bacteriophage* 4, no. 1: 225–27.

Agamben, Giorgio. 2009. *What Is an Apparatus?* Trans. David Kishik and Stefan Pedatella. Palo Alto, Calif.: Stanford University Press.

Ahn, J., and J. Lee. 2008. "X Chromosome: X Inactivation." *Nature Education* 1, no. 1: 24.

Alač, Morana. 2011. *Handling Digital Brains: A Laboratory Study of Multimodal Semiotic Interaction in the Age of Computers.* Cambridge, Mass.: MIT Press.

Alberghina, Lilia, and Hans Westerhoff, eds. 2005. *Systems Biology: Definitions and Perspectives.* Berlin: Springer.

Allchin, Douglas. 1999. "Do We See through a Social Microscope? Credibility as a Vicarious Selector." *Philosophy of Science* 60 (Proceedings): S287–S298.

Allen, George. 2012. *Modes of Learning: Whitehead's Metaphysics and the Stages of Education.* Albany: State University of New York Press.

Andrew, Dudley. 2010. *What Cinema Is! Bazin's Quest and Its Charge.* Sussex: Wiley and Blackwell.

Apter, Michael J., and Lewis Wolpert. 1965. "Cybernetics and Development." *Journal of Theoretical Biology* 8:244–57.

Armstrong, Rachel. 2011. "How Protocells Can Make 'Stuff' Much More Interesting." *AD: Protocell Architecture* 81, no. 2: 68–77.

Auffray, Charles, Sandrine Imbeaud, Magali Roux-Rouquié, and Leroy Hood. 2003. "From Functional Genomics to Systems Biology: Concepts and Practices." *Comptes Rendus Biologies* 326, nos. 10–11: 879–92.

Avery, Oswald T., Colin M. MacLeod, and Maclyn McCarty. 1944. "Studies on the Chemical Nature of the Substance Inducing Transformation of Pneumococcal Types." *Journal of Experimental Medicine* 79:137–57.

Azuma, Hiroki. 2009. *Otaku: Japan's Database Animals.* Trans. Jonathan E. Abel and Shion Kono. Minneapolis: University of Minnesota Press.

Balaguera, Maria Angels de Luis, Adam P. Fisher, Natalie M. Clark, Maria Guadalupe Fernandez-Espinosa, Barbara K. Möller, Dolf Weijers, Jan U. Lohmann, Cranos Williams, Oscar Lorenzo, and Rosangela Sozzania. 2017. "Predicting Gene Regulatory Networks by Combining Spatial and Temporal Gene Expression Data in Arabidopsis Root Stem Cells." *Proceedings of the National Academy of Sciences* 114, no. 36: E7632–E7640.

Balázs, Béla. 1970. *Theory of the Film: Character and Growth of a New Art.* Trans. Edith Bone. New York: Dover.

Barad, Karen. 2007. *Meeting the Universe Halfway: Quantum Physics and the Entanglement of Matter and Meaning.* Durham, N.C.: Duke University Press.

Baudry, Jean. 1974–75. "Ideological Effects of the Basic Cinematographic Apparatus." *Film Quarterly* 28, no. 2: 39–47.

Bazin, André. 2000. "Accidental Beauty." In *Science Is Fiction: The Films of Jean Painlevé,* ed. Andy Masaki Bellows, Marina McDougall, and Brigitte Berg, 144–47. Trans. Jeanine Herman. Cambridge, Mass.: MIT Press.

Bazin, André. 2005. *What Is Cinema?* Trans. Hugh Gray. Berkeley: University of California Press.

Becker, Gary S. 1965. "A Theory of the Allocation of Time." *Economic Journal* 75, no. 299: 493–517.

Beckman, Karen, ed. 2014a. *Animating Film Theory.* Durham, N.C.: Duke University Press.

Beckman, Karen. 2014b. "Animating Film Theory: An Introduction." In *Animating Film Theory,* ed. Karen Beckman, 1–22. Durham, N.C.: Duke University Press.

Begole, Bo. 2017. "Responsive Media: Media Experiences in the Age of Thinking Machines." *APSIPA Transactions on Signal and Information Processing* 6:1–8.

Beller, Jonathan. 2006. *The Cinematic Mode of Production: Attention Economy and the Society of the Spectacle.* Hanover: University Press of New England.

Benjamin, Walter. 2008. "The Work of Art in the Age of Its Technological Reproducibility." In *The Work of Art in the Age of Its Technological Reproducibility, and Other Writings on Media,* ed. Michael W. Jennings, Brigid Doherty, and Thomas Y. Levin, 19–55. Cambridge, Mass.: Harvard University Press.

Berardi, Franco. 2009. *The Soul at Work: From Alienation to Autonomy.* Trans. Francesca Cadel and Giuseppina Mecchia. Cambridge, Mass.: MIT Press.

Berardi, Franco. 2011. *After the Future.* Ed. Gary Genosko and Nicholas Thoburn. Edinburgh: AK Press.

Berg, Brigitte. 2000. "Contradictory Forces: Jean Painlevé, 1902–1989." In *Science Is Fiction: The Films of Jean Painlevé,* ed. Andy Masaki Bellows, Marina McDougall, and Brigitte Berg, 2–47. Trans. Jeanine Herman. Cambridge, Mass.: MIT Press.

Berry, David M., and Alexander Galloway. 2015. "A Network Is a Network Is a Network: Reflections on the Computational and the Societies of Control." *Theory, Culture and Society* 33, no. 4: 151–72.

Birkhead, Tim R., and Robert Montgomerie. 2009. "Three Centuries of Sperm Research." In *Sperm Biology: An Evolutionary Perspective,* ed. Tim R. Birkhead, David J. Hosken, and Scott Pitnick, 1–42. Burlington, Mass: Elsevier.

Bleecker, Julian. 2009. "Design Fiction: A Short Essay on Design, Science, Fact and Fiction." *Near Future Laboratory* (blog). March 17, 2009. http://blog.nearfuturelaboratory.com/.

Boehnert, Joanna. 2018. *Design, Ecology, Politics: Towards the Ecocene.* London: Bloomsbury.

Bogner, A., P. H. Jouneau, G. Thollet, D. Basset, C. Gauthier. 2007. "History of Scanning Electron Microscopy Developments: Towards 'Wet-STEM' Imaging." *Micron* 38, no. 4: 390–401.

Bogue, Ronald. 2007. "The Art of the Possible." *Revue internationale de philosophie* 3, no. 241: 273–86.

Bolinsky, David. 2007. "Visualizing the Wonder of a Living Cell." Filmed July 2007 in New York, N.Y. TED video, 9:45. https://www.ted.com/.

Bolton, Christopher, Istvan Csicsery-Ronay Jr., and Takayuki Tatsumi, eds. 2007. *Robot Ghosts and Wired Dreams: Japanese Science Fiction from Origins to Anime.* Minneapolis: University of Minnesota Press.

Boogerd, Fred C., Frank J. Bruggeman, Jan-Hendrik S. Hofmeyr, and Hans V. Westerhoff, eds. 2007. *Systems Biology: Philosophical Foundations.* Amsterdam: Elsevier.

Boradkar, Prasad. 2010. *Designing Things: A Critical Introduction to the Culture of Objects.* Oxford: Berg.

Boy, Guy André. 2017. "Human-Centered Design of Complex Systems: An Experience-Based Approach." *Design Science* 3, no. e8: 1–23.

Bracken, Christopher. 2007. *Magical Criticism: The Recourse of Savage Philosophy.* Chicago: University of Chicago Press.

Bratton, Benjamin. 2015. *The Stack: On Software and Sovereignty.* Cambridge, Mass.: MIT Press.

Brenner, Sydney. 1999. "Theoretical Biology in the Third Millennium." *Philosophical Transactions of the Royal Society, London* B 354:1963–65.

Brown, Wendy. 2019. *In the Ruins of Neoliberalism: The Rise of Antidemocratic Politics in the West.* New York: Columbia University Press.

Bruggeman, F. J., and Hans Westerhoff. 2007. "The Nature of Systems Biology." *Trends in Molecular Biology* 15, no. 1: 45–50.

Bryant, Levi. 2011. *The Democracy of Objects.* Ann Arbor, Mich.: Open Humanities Press.

Buchan, Suzanne. 2010. *The Quay Brothers: Into a Metaphysical Playroom.* Minneapolis: University of Minnesota Press.

Buchan, Suzanne, ed. 2013. *Pervasive Animation.* New York: Routledge.

Buchan, Suzanne. 2014. "Animation, in Theory." In *Animating Film Theory,* ed. Karen Beckman, 111–27. Durham, N.C.: Duke University Press.

Cahill, James Leo. 2013. "Anthropomorphism and Its Vicissitudes: Reflections on Homme-Sick Cinema." In *Screening Nature: Cinema beyond the Human,* ed. Anat Pick and Guinevere Narraway, 73–90. New York: Berghahn Books.

Cahill, James Leo. 2019. *Zoological Surrealism: The Nonhuman Cinema of Jean Painlevé.* Minneapolis: University of Minnesota Press.

Cairns, Stephen, and Devisari Tunas, eds. 2017. *Future Cities Laboratory: Indicia 01.* Zurich: Lars Müller.

Calude, Cristian, and Giuseppe Longo. 2017. "The Deluge of Spurious Correlation in Big Data." *Foundations in Science* 22, no. 3: 595–612.

Campbell, Donald T. 1974. "Evolutionary Epistemology." In *The Philosophy of Karl Popper,* ed. P. A. Schilpp, 413–63. La Salle, Ill.: Open Court.

Campbell, Melody G., David Veesler, Anchi Cheng, Clinton S. Potter, and Bridget Carragher. 2015. "2.8 Å Resolution Reconstruction of the Thermoplasma Acidophilum 20S Proteasome Using Cryo-electron Microscopy." eLife 2015, no. 4: e06380.

Canguilhem, Georges. 2009. *Knowledge of Life.* Trans. Stefanos Geroulanos and Daniela Ginsburg. New York: Fordham University Press.

Carlson, Elof Axel. 2004. *Mendel's Legacy: The Origin of Classical Genetics.* Cold Spring Harbor, N.Y.: Cold Spring Harbor Laboratory Press.

Carrel, Alexis. 1931a. "The New Cytology." *Science* 73:297–303.

Carrel, Alexis. 1931b. "Physiological Time." *Science* 74:618–21.

Carrel, Alexis. 1938. "Introduction." In *Methods of Tissue Culture,* ed. Raymond C. Parker. New York: Paul B. Hoeber.

Carroll, Sean B. 2005. *Endless Forms Most Beautiful: The New Science of Evo Devo and the Making of the Animal Kingdom.* New York: W. W. Norton.

Cartwright, Lisa. 1995. *Screening the Body: Tracing Medicine's Visual Culture.* Minneapolis: University of Minnesota Press.

Cavell, Stanley. 1979. *The World Viewed: Reflections on the Ontology of Film.* Cambridge, Mass.: Harvard University Press.

Chang, Hsin-Yi, Chris Quintana, and Joseph S. Krajcik. 2010. "The Impact of Designing and Evaluating Molecular Animations on How Well Middle School Students Understand the Particulate Nature of Matter." *Science Education* 94:73–94.

Chapman, Brad, and Jeffrey Chang. 2000. "Biopython: Python Tools for Computational Biology." *ACM SIGBIO Newsletter* 20, no. 2: 15–19.

Chatterjee, Somnath, Anirban Roy, Aparna Laskar, and Snehasikta Swarnakar. 2012. "Electron Microscopy in the Perspective of Modern Biology: Ultravision and Ultradimension." In *Current Microscopy: Contributions to Advances in Science and Technology,* ed. A. Méndez-Vilas, 891–902. Badajoz, Spain: Formatex Research Center.

Cholodenko, Alan. Ed. 1991. *The Illusion of Life: Essays on Animation.* Sydney: Power Publications.

Cholodenko, Alan. 2008. "The Animation of Cinema." *Semiotic Review of Books* 18, no. 2: 1–10.

Christian, William. 1959. *An Interpretation of Whitehead's Metaphysics.* Westport, Conn.: Greenwood Press.

Clark, Andy. 2004. *Natural-Born Cyborgs: Minds, Technologies, and the Future of Human Intelligence.* Oxford: Oxford University Press.

Clark, Andy. 2007. "Re-inventing Ourselves: The Plasticity of Embodiment, Sensing, and Mind." *Journal of Medicine and Philosophy: A Forum for Bioethics and Philosophy of Medicine* 32, no. 3: 263–82.

Clark, Taylor. 2017. "How to Get Rich Playing Video Games Online: For the Stars of the Streaming Service Twitch, Success Means Working around the Clock." *New Yorker,* November 20, 2017.

Cobb, Matthew. 2015a. *Life's Greatest Secret: The Race to Crack the Genetic Code.* London: Profile Books.

Cobb, Matthew. 2015b. "Sexism in Science: Did Watson and Crick Really Steal Rosalind Franklin's Data?" *Guardian.* June 23, 2015. https://www.theguardian.com/.

Cogdell, Christina. 2004. *Eugenic Design: Streamlining America in the 1930s.* Philadelphia: University of Pennsylvania Press.

Coleman, William. 1977. *Biology in the Nineteenth Century: Problems of Form, Function, and Transformation.* Cambridge: Cambridge University Press.

Combes, Muriel. 2012. *Gilbert Simondon and the Philosophy of the Transindividual.* Trans. Thomas LaMarre. Cambridge, Mass.: MIT Press.

Comolli, Jean-Louis. 1985. "Technique and Ideology: Camera, Perspective, Depth of Field." In *Movies and Methods: An Anthology, Volume II,* ed. Bill Nichols, 40–57. Berkeley: University of California Press.

Conesa, Ana, and Ali Mortazavi. 2014. "The Common Ground of Genomics and Systems Biology." *BMC Systems Biology* 8, no. 2 (Supplement): S1.

Connolly, Patrick J. 2014. "Newton and God's Sensorium." *Intellectual History Review* 24, no. 2: 185–201.

Cooper, Melinda. 2008. *Life as Surplus: Biotechnology and Capitalism in the Neoliberal Era.* Seattle: University of Washington Press.

Cooper, Melinda, and Catherine Waldby. 2014. *Clinical Labor: Tissue Donors and Research Subjects in the Global Bioeconomy.* Durham, N.C.: Duke University Press.

Coopmans, Catelijne, Janet Vertesi, Michael E. Lynch, and Steve Woolgar. 2014. "Introduction: Representation in Scientific Practice Revisited." In *Representation in Scientific Practice Revisited,* ed. Catelijne Coopmans, Janet Vertesi, Michael E. Lynch, and Steve Woolgar, 1–12. Cambridge, Mass.: MIT Press.

Correns, Carl. 1901. "Die Ergebnisse der neuesten Bastardforschungen für die Vererbungslehre." *Berichte der Deutschen Botanischen Gesellschaft* 19:71–94.

Crary, Jonathan. 1990. *Techniques of the Observer: On Vision and Modernity in the Nineteenth Century.* Cambridge, Mass.: MIT Press.

Crary, Jonathan. 1999. *Suspensions of Perception: Attention, Spectacle, and Modern Culture.* Cambridge, Mass.: MIT Press.

Crary, Jonathan. 2013. *24/7: Late Capitalism and the Ends of Sleep.* London: Verso.

Crick, Francis. 2004. *Of Molecules and Man.* New York: Prometheus Books.

Cross, Nigel. 1995. "Discovering Design Ability." In *Discovering Design: Explorations in Design Studies,* ed. Richard Buchanan and Victor Margolin, 105–20. Chicago: University of Chicago Press.

Cross, Nigel. 2011. *Design Thinking: Understanding How Designers Think and Work.* London: Bloomsbury.

Cubitt, Sean. 2004. *The Cinema Effect.* Cambridge, Mass.: MIT Press.

Cubitt, Sean. 2016. *Finite Media: Environmental Implications of Digital Technologies.* Durham, N.C.: Duke University Press.

Curry, Stephen. 2012. "You Can't See DNA Unless You Look Properly." *Guardian.* December 1, 2012. https://www.theguardian.com/.

Curtis, Scott. 2015. *The Shape of Spectatorship: Art, Science, and Early Cinema in Germany.* New York: Columbia University Press.

Damasio, Antonio. 2003. *Looking for Spinoza: Joy, Sorrow, and the Feeling Brain.* New York: Harcourt Inc.

Dancoff, Sydney M., and Henry Quastler. 1953. "The Information Content and Error Rate of Living Things." In *Essays on the Information Theory in Biology,* ed. Henry Quastler, 263–73. Urbana: University of Illinois Press.

Dardot, Pierre, and Christian Laval. 2017. *The New Way of the World: On Neoliberal Society.* Trans. Gregory Elliot. London: Verso Books.

Daston, Lorraine. 1991. "Marvelous Facts and Miraculous Evidence in Early Modern Europe." *Critical Inquiry* 18, no. 1: 93–124.

Daston, Lorraine, and Peter Galison. 2007. *Objectivity.* New York: Zone Books.

Davidson, Eric H. 2006. *The Regulatory Genome: Gene Regulatory Networks in Development and Evolution.* Amsterdam: Elsevier.

Day, Mellani, Mary C. Boardman, and Norris F. Krueger, eds. 2017. *Handbook of Research Methodologies and Design in Neuroentrepreneurship.* Cheltenham, UK: Edward Elgar.

Debaise, Didier. 2017. *Nature as Event: The Lure of the Possible.* Trans. Michael Halewood. Durham, N.C.: Duke University Press.

De Boever, Arne, and Warren Neidich, eds. 2012. *The Psychopathologies of Cognitive Capitalism, Part One.* Berlin: Archive Books.

Debord, Guy. 2002. *Society of the Spectacle.* Trans. Ken Knabb. London: Rebel Press.

Deichmann, Ute. 2016. "Epigenetics: The Origins and Evolution of a Fashionable Topic." *Developmental Biology* 416, no. 1: 249–54.

De Jong, Cees W. 2017. *Dieter Rams: Ten Principles for Good Design.* Munich: Prestel.

Deleuze, Gilles. 1988. *Foucault.* Trans. Seán Hand. Minneapolis: University of Minnesota Press.

Deleuze, Gilles. 1989. *Cinema 2: The Time-Image.* Trans. Hugh Tomlinson and Greta Galeta. Minneapolis: University of Minnesota Press.

Deleuze, Gilles. 1990. *The Logic of Sense.* Trans. Constantin V. Boundas. New York: Columbia University Press.

Deleuze, Gilles. 1991. *Bergsonism.* Trans. Hugh Tomlinson and Barbara Habberjam. New York: Zone Books.

Deleuze, Gilles. 1992. "What Is a *Dispositif?*" In *Michel Foucault Philosopher,* ed. T. J. Armstrong, 159–68. New York: Harvester Wheatsheaf.

Deleuze, Gilles. 1993. *The Fold: Leibniz and the Baroque.* Trans. Tom Conley. Minneapolis: University of Minnesota Press.

Deleuze, Gilles. 1994. *Difference and Repetition.* Trans. Paul Patton. London: Continuum.

Deleuze, Gilles. 1995a. "Postscript on Control Societies." In *Negotiations.* Trans. Martin Joughin. New York: Columbia University Press.

Deleuze, Gilles. 1995b. "Control and Becoming." In *Negotiations.* Trans. Martin Joughin. New York: Columbia University Press.

Deleuze, Gilles, and Félix Guattari. 1987. *A Thousand Plateaus: Capitalism and Schizophrenia.* Trans. Brian Massumi. Minneapolis: University of Minnesota Press.

Deleuze, Gilles, and Félix Guattari. 1994. *What Is Philosophy?* Trans. Hugh Tomlinson and Graham Burchell. New York: Columbia University Press.

Dilts, Andrew. 2011. "From 'Entrepreneur of the Self' to 'Care of the Self': Neoliberal Governmentality and Foucault's Ethics." *Foucault Studies* 12:130–46.

Di Paolo, Ezequiel A. 2005. "Autopoiesis, Adaptivity, Teleology, Agency." *Phenomenology and the Cognitive Sciences* 4, no. 4: 429–52.

Doane, Mary Ann. 2002. *The Emergence of Cinematic Time: Modernity, Contingency, the Archive.* Cambridge, Mass.: Harvard University Press.

Doyle, Richard. 1997. *On Beyond Living: Rhetorical Transformations in the Life Sciences.* Stanford, Calif.: Stanford University Press.

Dreyfus, Hubert M., and Paul Rabinow. 1982. *Michel Foucault: Beyond Structuralism and Hermeneutics.* Chicago: University of Chicago Press.

Duhem, Pierre. 1991. *The Aim and Structure of Physical Theory.* Trans. P. Wiener. Princeton, N.J.: Princeton University Press.

Dunne, Anthony, and Fiona Raby. 2013. *Speculative Everything: Design, Fiction, and Social Dreaming.* Cambridge, Mass.: MIT Press.

Dupree, P., and D. J. Sherrier. 1998. "The Plant Golgi Apparatus." *Biochimica et Biophysica Acta* 1404, nos. 1–2: 259–70.

Dvorsky, George. 2012. "Scientists Snap a Picture of DNA's Double Helix for the Very First Time." Io9: We Come from the Future. November 29, 2012. https://io9.gizmodo.com/.

Dyer-Witheford, Nick. 1999. *Cyber-Marx: Cycles and Circuits of Struggle in High Technology Capitalism.* Urbana: University of Illinois Press.

Eagleton, Terry. 1983. *Literary Theory: An Introduction.* Oxford: Blackwell.

Easterling, Keller. 2014. *Extrastatecraft: The Power of Infrastructure Space.* London: Verso.

Easterling, Keller. 2018. *Medium Design.* Moscow: Strelka Press.

Elsaesser, Thomas, ed. 1990. *Early Cinema: Space, Frame, Narrative.* London: British Film Institute.

Elsaesser, Thomas. 1998. "Digital Cinema: Delivery, Event, Time." In *Cinema Futures: Cain, Abel or Cable?* ed. Thomas Elsaesser and Kay Hoffman, 201–22. Amsterdam: Amsterdam University Press.

Escobar, Arturo. 2018. *Designs for the Pluriverse: Radical Independence, Autonomy, and the Making of Worlds.* Durham, N.C.: Duke University Press.

Esposito, Roberto. 2004. *Bios: Biopolitics and Philosophy.* Trans. Timothy Campbell. Minneapolis: University of Minnesota Press.

Farías, Ignacio, and Alex Wilkie, eds. 2015. *Studio Studies: Operations, Topologies and Displacements.* London: Routledge.

Findeli, Alain. 1990. "Moholy-Nagy's Design Pedagogy in Chicago (1937–46)." *Design Issues* 7, no. 1: 4–19.

Flaxman, Gregory. 2012. "Out of Field: The Future of Film Studies." *Angelaki: Journal of the Theoretical Humanities* 17, no. 4: 119–37.

Fleck, Ludwik. 1979. *Genesis and Development of a Scientific Fact.* Chicago: University of Chicago Press.

Flusser, Vilém. 1999. *The Shape of Things: A Philosophy of Design.* Trans. Anthony Mathews. London: Reaktion Books.

Flusser, Vilém. 2000. *Towards a Philosophy of Photography.* Trans. Anthony Mathews. London: Reaktion Books.

Foucault, Michel. 1977. *Discipline and Punish: The Birth of the Prison.* Trans. Alan Sheridan. New York: Vintage Books.

Foucault, Michel. 1978. *The History of Sexuality, Vol.1: An Introduction.* Trans. Robert Hurley. New York: Vintage Books.

Foucault, Michel. 1980. "The Confession of the Flesh." In *Power/Knowledge: Selected Interviews and Other Writings, 1972–1977,* ed. C. Gordon, 194–228. New York: Pantheon Books.

Foucault, Michel. 1982. "The Subject and Power." In *Michel Foucault: Beyond Structuralism and Hermeneutics,* ed. Hubert M. Dreyfus and Paul Rabinow, 208–28. Chicago: University of Chicago Press.

Foucault, Michel. 1988. *The History of Sexuality, Vol. 3: The Care of the Self.* Trans. Robert Hurley. New York: Vintage Books.

Foucault, Michel. 1991. *Remarks on Marx: Conversations with Duccio Trombadori.* Trans. James Goldstein and James Cascaito. New York: Semiotext(e).

Foucault, Michel. 1994. *The Order of Things: An Archeology of the Human Sciences.* Trans. Anonymous. New York: Vintage Books.

Foucault, Michel. 2003. *Society Must Be Defended: Lectures at the Collège de France, 1975–1976.* Trans. David Macey. New York: Palgrave Macmillan.

Foucault, Michel. 2007. *Security, Territory, Population: Lectures at the Collège de France, 1977–1978.* Trans. Graham Burchell. New York: Palgrave Macmillan.

Foucault, Michel. 2008. *The Birth of Biopolitics: Lectures at the Collège de France, 1978–1979.* Trans. Graham Burchell. New York: Palgrave Macmillan.

Frankenberry, Nancy. 1983. "The Power of the Past." *Process Studies* 13:132–42.

Fraser, Nancy. 1981. "Foucault on Modern Power: Empirical Insights and Normative Confusions." *PRAXIS International* 3:272–87.

Freundlich, Martin. 1963. "The Origin of the Electron Microscope." *Science* 142 (3589): 185–88.

Fry, Tony. 2012. *Becoming Human by Design.* London: Berg.

Fry, Tony. 2014. "Timor-Leste: Unlearning in Order to Be." In *Design in the Borderlands,* ed. Eleni Kalantidou and Tony Fry, 76–90. London: Routledge.

Fry, Tony. 2017. "Design for/by the Global South." *Design Philosophy Papers* 15, no. 1: 3–37.

Fuller, Matthew. 2005. *Media Ecologies: Materialist Energies in Art and Technoculture.* Cambridge, Mass.: MIT Press.

Gabrys, Jennifer. 2011. *Digital Rubbish: A Natural History of Electronics.* Ann Arbor: University of Michigan Press.

Gabrys, Jennifer. 2016. *Program Earth: Environmental Sensing Technology and the Making of a Computational Planet.* Minneapolis: University of Minnesota Press.

Galloway, Alexander R. 2004. *Protocol: How Control Exists after Decentralization.* Cambridge, Mass.: MIT Press.

Galloway, Alexander R. 2011. "What Is New Media? Ten Years after the Language of New Media." *Criticism: A Quarterly for Literature and the Arts* 53, no. 3: 377–84.

Galloway, Alexander R. 2012. *The Interface Effect.* Cambridge, UK: Polity Press.

Galloway, Alexander R., Eugene Thacker, and McKenzie Wark. 2013. *Excommunication: Three Inquiries in Media and Mediation.* Chicago: University of Chicago Press.

Gaskill, Nicholas, and A. J. Nocek. 2014. "Introduction: An Adventure of Thought." In *The Lure of Whitehead,* ed. Nicholas Gaskill and A. J. Nocek, 1–42. Minneapolis: University of Minnesota Press.

Gaudillière, Jean-Paul, and Hans-Jörg Rheinberger, eds. 2015. *Classical Genetic Research and Its Legacy: The Mapping Cultures of Twentieth-Century Genetics.* London: Routledge.

Gaycken, Oliver. 2015. *Devices of Curiosity: Early Cinema and Popular Science.* Oxford: Oxford University Press.

Geiger, Jeffrey, and Karin Littau. 2013. "Introduction: Cinematicity and Comparative Media." In *Cinematicity in Media History,* ed. Jeffrey Geiger and Karin Littau, 1–18. Edinburgh: University of Edinburgh Press.

Gentile, Francesco, Manola Moretti, Tania Limongi, Andrea Falqui, Giovanni Bertoni, Alice Scarpellini, Stefania Santoriello, Luca Maragliano, Remo Proietti Zaccaria, and Enzo di Fabrizio. 2012. "Direct Imaging of DNA Fibers: The Visage of Double Helix." *Nano Letters* 12 no. 12: 6453–58.

Goffey, Andrew, and Matthew Fuller. 2012. *Evil Media.* Cambridge, Mass.: MIT Press.

Goldberg, Aaron D., C. David Allis, and Emily Bernstein. 2007. "Epigenetics: A Landscape Takes Shape." *Cell* 128, no. 4: 635–38.

Goldhaber, Michael. 2006. "The Value of Openness in an Attention Economy." *First Monday* 11, no. 6. June 6, 2006. http://firstmonday.org/.

Goodman, Corey S., and Bridget C. Coghlin. 2000. "The Evolution of Evo-Devo Biology." *Proceedings of the National Academy of Sciences* 9:4424–25.

Goodman, Steve. 2009. *Sonic Warfare: Sound, Affect, and the Ecology of Fear.* Cambridge, Mass.: MIT Press.

Gordon, Colin. 1991. "Governmental Rationality: An Introduction." In *The Foucault Effect: Studies in Governmentality,* ed. Graham Burchell, Colin Gordon, and Peter Miller, 1–51. Chicago: University of Chicago Press.

Gould, Stephen Jay. 1971. "D'Arcy Thompson and the Science of Form." *New Literary History* 2, no. 2: 229–58.

Greenfield, Adam. 2017. *Radical Technologies: The Design of Everyday Life.* London: Verso.

Grossman, Elizabeth. 2006. *High Tech Trash: Digital Devices, Hidden Toxics, and Human Health.* Washington, D.C.: Island Press.

Guattari, Félix. 1984. *Molecular Revolution: Psychiatry and Politics.* New York: Penguin.

Guattari, Félix. 1995. *Chaosmosis: An Ethico-Aesthetic Paradigm.* Trans. Paul Bains and Julian Pefanis. Indianapolis: University of Indiana Press.

Guattari, Félix. 2009. *Chaosophy: Texts and Interviews, 1972–1977.* Los Angeles: Semiotext(e).

Gunning, Tom. 2014. "Animating the Instant: The Secret Symmetry between Animation and Photography." In *Animating Film Theory,* ed. Karen Beckman, 37–53. Durham, N.C.: Duke University Press.

Haas, Peter M. 1992. "Introduction: Epistemic Communities and International Policy Coordination." *International Organization* 46, no. 1: 1–35.

Hacking, Ian. 1981. "Do We See through a Microscope?" *Pacific Philosophical Quarterly* 62, no. 4: 305–22.

Hacking, Ian. 1983. *Representing and Intervening: Introductory Topics in the Philosophy of Natural Science.* Cambridge: Cambridge University Press.

Hacking, Ian. 1999. *The Social Construction of What?* Cambridge, Mass.: Harvard University Press.

Hacking, Ian. 2007. "On Not Being a Pragmatist: Eight Reasons and a Cause." In *New Pragmatists,* ed. Cheryl Misak, 32–49. Oxford: Oxford University Press.

Halewood, Michael. 2011. *A. N. Whitehead and Social Theory: Tracing a Culture of Thought.* London: Anthem Press.

Halpern, Orit. 2014. *Beautiful Data: A History of Vision and Reason since 1945.* Durham, N.C.: Duke University Press.

Han, Beatrice. 2002. *Foucault's Critical Project: Between the Transcendental and the Historical.* Trans. Edward Pile. Stanford, Calif.: Stanford University Press.

Han, Byung-Chul. 2015. *The Transparency Society.* Stanford, Calif.: Stanford University Press.

Han, Byung-Chul. 2017a. *The Agony of Eros.* Trans. Erik Butler. Cambridge, Mass.: MIT Press.

Han, Byung-Chul. 2017b. *In the Swarm: Digital Prospects.* Trans. Erik Butler. Cambridge, Mass.: MIT Press.

Hansen, Mark B. N. 2004. *New Philosophy for New Media.* Cambridge, Mass.: MIT Press.

Hansen, Mark B. N. 2015. *Feed-Forward: On the Future of Twenty-First-Century Media.* Chicago: University of Chicago Press.

Haraway, Donna. 1976. *Crystals, Fabrics, and Fields: Metaphors That Shape Embryos.* Berkeley, Calif.: North Atlantic Books.

Haraway, Donna. 1988. "Situated Knowledges: The Science Question in Feminism and the Privilege of Partial Perspective." *Feminist Studies* 14:575–99.

Haraway, Donna. 1997. *Modest–Witness@Second–Millennium.FemaleMan–Meets–OncoMouse: Feminism and Technoscience.* New York: Routledge.

Harding, Sandra. 1991. *Whose Science? Whose Knowledge?: Thinking from Women's Lives.* Ithaca, N.Y.: Cornell University Press.

Hardt, Michael, and Antonio Negri. 2000. *Empire.* Cambridge, Mass.: Harvard University Press.

Harman, Graham. 2002. *Tool-Being: Heidegger and the Metaphysics of Objects.* Chicago: Open Court.

Hartmann, Frank. 2000. *Medienphilosophie.* Stuttgart: UTB für Wissenschaft Uni-Taschenbücher GmbH.

Hayek, Friedrich. 1948a. "Economics and Knowledge." In *Individualism and Economic Order,* 33–56. Chicago: University of Chicago Press.

Hayek, Friedrich. 1948b. "The Use of Knowledge in Society." In *Individualism and Economic Order,* 77–91. Chicago: The University of Chicago Press.

Hayles, N. Katherine. 1999. *How We Became Posthuman: Virtual Bodies in Cybernetics, Literature, and Informatics.* Chicago: University of Chicago Press.

Hempel, Carl G. 1970. "Fundamentals of Concept Formation in Empirical Science." In *Foundations of the Unity of Science.* Vol. 2. Ed. O. Neurath, R. Carnap, and C. Morris. Chicago: University of Chicago Press.

Henning, Brian G. 2005. "Saving Whitehead's Universe of Value: An 'Ecstatic' Challenge to the Classical Interpretation." *International Philosophical Quarterly* 45, no. 4: 445–65.

Henning, Brian G., and Adam Scarfe, eds. 2013. *Beyond Mechanism: Putting Life Back into Biology.* Plymouth, UK: Lexington Books.

Hentschel, Klaus. 2014. *Visual Cultures in Science and Technology: A Comparative History.* Oxford: Oxford University Press.

Hodges, Alan. 1983. *Alan Turing: The Enigma of Intelligence.* London: Burnett Books.

Holmes, Brian. n.d. "New Media from the Neolithic to Now." Continental Drift: The Other Side of Neoliberal Globalization. Accessed August 2018. http://brianholmes.wordpress.com/.

Holterhoff, Kate. 2014. "The History and Reception of Charles Darwin's Hypothesis of Pangenesis." *Journal of the History of Biology* 47:661–95.

Horkheimer, Max, and Theodor W. Adorno. 2002. *The Dialectic of Enlightenment: Philosophical Fragments.* Trans. Edmund Jephcott. Stanford, Calif.: Stanford University Press.

Hosken, David J., Scott S. Pitnick, and Tim Birkhead, eds. 2009. *Sperm Biology: An Evolutionary Perspective.* Amsterdam: Elsevier.

Huxley, Julian. 1936. "Eugenics and Society" (The Galton Lecture Given to the Eugenics Society), *Eugenics Review* 28, no. 1: 11–31.

Huxley, Julian. 1953. *Evolution in Action.* New York: Harper.

Ihde, Don. 1998. *Expanding Hermeneutics: Visualism in Science.* Evanston, Ill.: Northwestern University Press.

Inwood, Stephen. 2002. *The Man Who Knew Too Much: The Strange and Inventive Life of Robert Hooke, 1635–1703.* London: Macmillan.

Irigaray, Luce. 1985. *Speculum of the Other Woman.* Trans. Gillian C. Gill. Ithaca, N.Y.: Cornell University Press.

Ivakhiv, Adrian. 2013. *Ecologies of the Moving Image: Cinema, Affect, Nature.* Waterloo, Ont.: Wilfred Laurier University Press.

Iwasa, Janet. 2010. "Animating the Model Figure." *Trends in Cell Biology* 20, no. 12: 699–704.

Jacob, François, and Jacques Monod. 1961. "Genetic Regulatory Mechanisms in the Synthesis of Proteins." *Journal of Molecular Biology* 3:318–56.

Jacobs, Struan. 2000. "Spontaneous Order: Michael Polanyi and Friedrich Hayek." *Critical Review of International Social and Political Philosophy* 3, no. 4: 49–67.

Jameson, Fredric. 1991. *Postmodernism, or The Cultural Logic of Late Capitalism.* Durham, N.C.: Duke University Press.

Jasanoff, Shelia, ed. 2004. *States of Knowledge: The Co-Production of Science and Social Order.* London: Routledge.

Jay, Martin. 1988. "Scopic Regimes of Modernity." In *Vision and Visuality,* ed. Hal Foster, 3–28. Seattle, Wash.: Bay Press.

Jenkinson, Jodie, and Gaël McGill. 2012. "Visualizing Protein Interactions and Dynamics: Evolving a Visual Language for Molecular Animation." *Life Sciences Education* 11, no. 1: 103–10.

Johannsen, Wilhelm. 1909. *Elemente der exakten Erblichkeitslehre.* Jena, Ger.: Gustav Fischer.

Johannsen, Wilhelm. 1911. "The Genotype Conception of Heredity." *American Naturalist* 45, no. 531: 129–59.

Johnson, Graham T., Ludovic Autin, David S. Goodsell, Michel F. Sanner, and Arthur J. Olson. 2011. "*ePMV* Embeds Molecular Modeling into Professional Animation Software Environments." *Structure* 19, no. 2: 293–303.

Jones, Judith A. 1998. *Intensity: An Essay in Whiteheadian Cosmology.* Nashville, Tenn.: Vanderbilt University Press.

Judson, Horace Freeland. 2004. *The Eighth Day of Creation: The Makers of the Revolution in Biology.* Cold Spring Harbor, N.Y.: Cold Spring Harbor Laboratory Press.

Kalantidou, Eleni, and Tony Fry, eds. 2014. *Design in the Borderlands.* London: Routledge.

Kauffman, Stuart. 2016. *Humanity in a Creative Universe.* Oxford: Oxford University Press.

Kay, Lily E. 1997. "Cybernetics, Information, Life: The Emergence of Scriptural Representations of Heredity." *Configurations* 5, no. 1: 23–91.

Kay, Lily E. 2000. *Who Wrote the Book of Life? A History of the Genetic Code.* Stanford, Calif.: Stanford University Press.

Keller, Evelyn Fox. 2000. *The Century of the Gene.* Cambridge, Mass.: Harvard University Press.

Keller, Evelyn Fox. 2002. *Making Sense of Life: Explaining Biological Development with Models, Metaphors, and Machines.* Cambridge, Mass.: Harvard University Press.

Keller, Evelyn Fox, and Christine Grontkowski. 1983. "The Mind's Eye." In *Discovering Reality: Feminist Perspectives on Epistemology, Metaphysics, Methodology, and Philosophy of Science,* ed. Sandra Harding and Merrill Hintikka, 207–24. Dordrecht, Neth.: Springer.

Kelty, Christopher, and Hannah Landecker. 2006. "A Theory of Animation: Cells, L-systems, and Film." *Grey Room* 17:30–63.

Kember, Sarah, and Joanna Zylinska. 2012. *Life after New Media: Mediation as a Vital Process.* Cambridge, Mass.: MIT Press.

Kepes, György. 1956. *The New Landscape in Art and Science.* Chicago: Paul Theobald.

Kepes, György. 1969. *Language of Vision: Painting, Photography, Advertising-Design.* Chicago: Paul Theobald.

Kessler, Frank. 2006. "The Cinema of Attractions as Dispositif." In *The Cinema of Attractions Reloaded,* ed. Wanda Strauven, 57–70. Amsterdam: Amsterdam University Press.

Kirzner, Israel. 1978. *Competition and Entrepreneurship.* Chicago: University of Chicago Press.

Kitano, Hiroaki. 2002. "Systems Biology: A Brief Overview." *Science* 259, no. 5560: 1662–64.

Kittler, Friedrich. 1999. *Gramophone, Film, Typewriter.* Trans. Geoffrey Winthrop-Young and Michael Wutz. Stanford, Calif.: Stanford University Press.

Kohler, Robert E. 1994. *Lords of the Fly: Drosophila Genetics and the Experimental Life.* Chicago: University of Chicago Press.

Kohn, K. W. 1999. "Molecular Interaction Map of the Mammalian Cell Cycle Control and DNA Repair System." *Molecular Biology of the Cell* 10:2703–34.

Kohut, Tom. 2013. "Noumenal Data: Software Takes Command (Review Essay)." Red-Assiniboine Research Unit. November 16, 2013. https://red assiniboineresearchunit.wordpress.com/.

Kracauer, Siegfried. 1995. "The Mass Ornament." In *The Mass Ornament: Weimar Essays,* ed. Thomas Y. Levin, 75–88. Trans. Thomas Y. Levin. Cambridge, Mass.: Harvard University Press.

Krämer, Sybille. 2012. "The 'Eye of the Mind' and the 'Eye of the Body': Descartes and Leibniz on Truth, Mathematics, and Visuality." In *Sensory Perception: Mind and Matter,* ed. F. G. Barth, P. Giampieri-Deutsch, and H. D. Klein, 369–82. Vienna: Springer.

Kraus, Elizabeth. 1998. *The Metaphysics of Experience: A Companion to Whitehead's Process and Reality.* New York: Fordham University Press.

Kresge, Nicole, Robert D. Simoni, and Robert L. Hill. 2005. "Chargaff's Rules: The Work of Erwin Chargaff." *Journal of Biological Chemistry* 280, no. 24: 172–74.

Kritzman, Lawrence. 1988. "Power and Sex: An Interview with Michel Foucault." In *Michel Foucault: Politics, Philosophy, Culture: Interviews and Other Writings, 1977–1984,* ed. Lawrence Kritzman, 110–24. New York: Routledge.

Kuhn, Thomas. 1962. *The Structure of Scientific Revolutions*. Chicago: University of Chicago Press.

Laclau, Ernesto, and Chantal Mouffe. 1985. *Hegemony and Socialist Strategy: Towards a Radical Democratic Politics.* London: Verso.

LaMarre, Thomas. 2009. *The Anime Machine: A Media Theory of Animation.* Minneapolis: University of Minnesota Press.

LaMarre, Thomas. 2013. "Cartoon Animals and Natural Philosophy." In *Pervasive Animation,* ed. Suzanne Buchan, 117–42. London: Routledge.

LaMarre, Thomas. 2018. *The Anime Ecology: A Genealogy of Television, Animation, and Game Media.* Minneapolis: University of Minnesota Press.

Lambert, Gregg. 2015. "On *Vitalpolitik:* Between 'Raw' and 'Cooked' Capitalism." Keynote lecture delivered at the University at Albany, April 17, 2015.

Lambert, Gregg. 2016a. "What Is a Dispositif? Part I." *Religious Theory: E-Supplement to the Journal for Cultural and Religious Theory.* July 11, 2016. http://jcrt.org/.

Lambert, Gregg. 2016b. "What Is a Dispositif? Part II." *Religious Theory: E-Supplement to The Journal for Cultural and Religious Theory.* July 18, 2016. http://jcrt.org/.

Landecker, Hannah. 2005. "Cellular Features: Microcinematography and Film Theory." *Critical Inquiry* 31, no. 4: 903–37.

Landecker, Hannah. 2007. *Culturing Life: How Cells Became Technologies.* Cambridge, Mass.: Harvard University Press.

Landecker, Hannah. 2011. "Creeping, Drinking, Dying: The Cinematic Portal and the Microscopic World of the Twentieth-Century Cell." *Science in Context* 24, no. 3: 381–416.

Laruelle, François. 2013. *Philosophy and Non-philosophy.* Trans. Taylor Adkins. Minneapolis: Univocal.

Latour, Bruno. 1987. *Science in Action: Following Scientists and Engineers through Society.* Cambridge, Mass.: Harvard University Press.

Latour, Bruno. 2009. "A Cautious Prometheus? A Few Steps toward a Philosophy of Design (with Special Attention to Peter Sloterdijk)." In *Proceedings of the 2008 Annual International Conference of the Design History Society,* 2–10. Cornwall: Design History Society Falmouth.

Latour, Bruno. 2010. *On the Modern Cult of the Factish Gods.* Durham, N.C.: Duke University Press.

Lawrence, A., L. Clark, J. Labuzetta, B. Sahakian, and S. Vyakarnum. 2008. "The Innovative Brain." *Nature* 456, no. 7219: 168–69.

Lazzarato, Maurizio. 1996. "Immaterial Labor." In *Radical Thought in Italy: A Potential Politics,* ed. Paolo Virno and Michael Hardt, 133–47 Minneapolis: University of Minnesota Press.

Lazzarato, Maurizio. 2012a. "Life and the Living in the Societies of Control." In *Deleuze and the Social,* ed. Martin Fuglsang and Bent Meier Sorensen, 171–90. Edinburgh: Edinburgh University Press.

Lazzarato, Maurizio. 2012b. *The Making of the Indebted Man: An Essay on the Neoliberal Condition.* Trans. Joshua D. Jordan. Los Angeles: Semiotext(e).

Lazzarato, Maurizio. 2014. *Signs and Machines: Capitalism and the Production of Subjectivity.* Trans. Joshua David Jordan. Cambridge, Mass.: MIT Press.

Leclerc, Ivor. 1958. *Whitehead's Metaphysics: An Introductory Exposition.* Bloomington: Indiana University Press.

LeDoux, Joseph. 2003. *The Synaptic Self: How Our Brains Became Who We Are.* New York: Penguin Books.

Lefebvre, Thierry. 1993. "Contribution à l'histoire de la microcinématographie: De François-Franck à Comandon." *1895, revue d'histoire du cinema* 14:35–46.

Lemke, Thomas. 2001. "'The Birth of Bio-politics': Michel Foucault's Lecture at the Collège de France on Neo-liberal Governmentality." *Economy and Society* 30, no. 2: 190–207.

Lenoir, Timothy. 1980. "Kant, Blumenbach, and Vital Materialism in German Biology." *ISIS* 71, no. 1: 77–108.

Levitt, Deborah. 2017. *The Animatic Apparatus: Animation, Vitality, and the Futures of the Image.* Alresford, UK: Zero Books.

Lévy, Pierre. 1998. *Becoming Virtual: Reality in the Digital Age.* New York: Plenum Trade.

Lok, Corie. 2011. "Biomedical Illustration: From Monsters to Molecules." *Nature* 477, no. 7364: 359–61.

Longo, Giuseppe. n.d. "Letter to Alan Turing." Le Département d'Informatique de l'ENS. Accessed October 2018. https://www.di.ens.fr/.

Longo, Giuseppe, and Maël Montévil. 2014. *Perspectives on Organisms: Biological Time, Symmetries and Singularities.* Berlin: Springer.

Longo, Giuseppe, Maël Montévil, and Stuart Kauffman. 2012. "No Entailing Laws, but Enablement in the Evolution of the Biosphere." arXiv.org. January 12, 2012. https://arxiv.org/.

Losev E., C. A. Reinke, J. Jellen, D. E. Strongin, B. J. Bevis, and B. S. Glick. 2006. "Golgi Maturation Visualized in Living Yeast." *Nature* 22, no. 7096: 1002–6.

Lukens, Jonathan, and Carl DiSalvo. 2013. "Speculative Design and Technical Fluency." *International Journal of Learning and Media* 3, no. 4: 23–40.

Lynn, Greg. 1999. *Animate Form.* New York: Princeton Architectural Press.

Malabou, Catherine. 2008. *What Should We Do with Our Brain?* Trans. Sebastian Rand. New York: Fordham University Press.

Mallarino, Ricardo, and Arhat Abzhanov. 2012. "Paths Less Traveled: Evo-Devo Approaches to Investigating Animal Morphological Evolution." *Annual Review of Cell and Developmental Biology* 28:743–63.

Malpass, Matt. 2017. *Critical Design in Context: History, Theory, and Practices.* London: Bloomsbury.

Manovich, Lev. 2001. *The Language of New Media.* Cambridge, Mass.: MIT Press.

Manovich, Lev. 2013. *Software Takes Command.* New York: Bloomsbury.

Manovich, Lev. n.d. "Avant-garde as Software." Manovich. Accessed August 2018. http://manovich.net/.

Manovich, Lev, and Andreas Kratky. 2005. *Soft Cinema: Navigating the Database.* Cambridge, Mass.: MIT Press.

Manzini, Ezio. 2015. *Design, When Everybody Designs: An Introduction to Design for Social Innovation.* Cambridge, Mass.: MIT Press.

Marenko, Betti, and Jamie Brassett, eds. 2015. *Deleuze and Design.* Edinburgh: Edinburgh University Press.

Marks, John. 2006. "Information and Resistance: Deleuze, the Virtual, and Cybernetics." In *Deleuze and the Contemporary World,* ed. Adrian Parr and Ian Buchanan, 194–213. Cambridge: Cambridge University Press.

Martin, Reinhold. 2003. *The Organizational Complex: Architecture, Media, and Corporate Space.* Cambridge, Mass.: MIT Press.

Marx, Karl. 1977. *Capital: A Critique of Political Economy.* Vol 1. Trans. Ben Fowkes. New York: Vintage Books.

Marx, Karl. 1986. *Grundrisse.* In *Karl Marx and Frederick Engels Collected Works.* Vol. 28. Moscow: Progress.

Marx, Karl. 1992. *Capital: A Critique of Political Economy.* Vol 2. Trans. David Fernbach. London: Penguin Group.

Matsuura-Tokita, Kumi, Masaki Takeuchi, Akira Ichihara, Kenta Mikuriya, and Akihiko Nakano. 2006. "Live Imaging of Yeast Golgi Cisternal Maturation." *Nature* 441:1007–10.

Maturana, Humberto, and Francisco J. Varela. 1980. *Autopoiesis and Cognition: The Realization of the Living.* Dordrecht, Neth.: D. Reidel.

McGill, Gaël. 2008. "Molecular Movies . . . Coming to a Lecture near You." *Cell* 133, no. 7: 1127–32.

McMullan, D. 2006. "Scanning Electron Microscopy, 1928–1965." *Scanning* 17, no. 3: 175–85.

Medawar, Peter B. 1962. "D'Arcy Thompson and 'Growth and Form.'" *Perspectives in Biology and Medicine* 5, no. 2: 220–32.

Medawar, Peter B. 1986. *Memoir of a Thinking Radish: An Autobiography.* Oxford: Oxford University Press.

Meillassoux, Quentin. 2008. *After Finitude: An Essay on the Necessity of Contingency.* Trans. Ray Brassier. London: Continuum.

Meillassoux, Quentin. 2012. "Interview with Quentin Meillassoux." In *New Materialism: Interviews and Cartographies,* ed. Rick Dolphijn and Iris van der Tuin, 71–81. Ann Arbor, Mich.: Open Humanities Press.

Melucci, Alberto. 1989. *Nomads of the Present: Social Movements and Individual Needs in Contemporary Society.* Philadelphia: Temple University Press.

Meyer, Steven, ed. 2018. *The Cambridge Companion to Literature and Science.* Cambridge: Cambridge University Press.

Mihailova, Mihaela, and John Mackay. 2014. "Frame Shot: Vertov's Ideologies of Animation." In *Animating Film Theory,* ed. Karen Beckman, 145–66. Durham, N.C.: Duke University Press.

Min, Wei. 2015. "Seeing Molecular Vibrations: Optical Imaging of Small Molecules for Biomedicine." *Optics in the Life Sciences, OSA Technical Digest paper JW2B.3.*

Mitchell, Robert. 2011. *Bioart and the Vitality of Media.* Seattle: University of Washington Press.

Mitchell, Robert. 2013. *Experimental Life: Vitalism in Romantic Science and Literature.* Baltimore, Md.: Johns Hopkins University Press.

Mitchell, Robert. 2016. "Biopolitics and Population Aesthetics." *The South Atlantic Quarterly* 115, no. 2: 367–98.

Mitchell, Robert, and Phillip Thurtle, eds. 2003. *Data Made Flesh: Embodying Information.* New York: Routledge.

Moholy-Nagy, László. 1932. *The New Vision: From Material to Architecture.* Trans. Daphne M. Hoffmann. New York: Brewer, Warren, and Putnam.

Moholy-Nagy, László. 1947. *Vision in Motion.* Chicago: Paul Theobald.

Moholy-Nagy, László. 1969. *Painting, Photography, Film.* London: Lund Humphries. First published 1925.

Moholy-Nagy, Sibyl. 1969. *Moholy-Nagy: Experiment in Totality.* Cambridge, Mass.: MIT Press.

Moles, C. G., P. Mendes, and J. R. Banga. 2003. "Parameter Estimation in Biochemical Pathways: A Comparison of Global Optimization Methods." *Genome Research* 13:2467–74.

Moran, Stacey. 2017. *Feminist Materials: Quantum Physics and Critical Writing Practices for New Material Feminism.* PhD dissertation, University of California, Berkeley.

Morgan, Thomas Hunt. 1926. *The Theory of the Gene.* New Haven, Conn.: Yale University Press.

Morgan, Thomas Hunt. 1935. "The Relation of Genetics to Physiology and Medicine." In *Les prix Nobel en 1933,* 1–16. Stockholm: Imprimerie Royale.

Morgan, Thomas H., Alfred H. Sturtevant, Herman J. Muller, and Calvin B. Bridges. 1915. *The Mechanism of Mendelian Heredity.* New York: Henry Holt.

Morris, Phillip C. 1991. *Process Philosophy and Political Ideology: The Social and Political Thought of Alfred North Whitehead and Charles Hartshorne.* Albany: State University of New York Press.

Morton, Timothy. 2007. *Ecology without Nature: Rethinking Environmental Aesthetics.* Cambridge, Mass.: Harvard University Press.

Moulier-Boutang, Yann. 2011. *Cognitive Capitalism.* Trans. Ed Emery. Cambridge, UK: Polity Press.

Müller-Wille, Staffan, and Hans-Jörg Rheinberger. 2012. *A Cultural History of Heredity.* Chicago: University of Chicago Press.

Mulvey, Laura. 1975. "Visual Pleasure and Narrative Cinema." *Screen* 16, no. 3: 6–18.

Murphie, Andrew. 2005. "Differential Life, Perception and the Nervous Elements: Whitehead, Bergson and Virno on the Technics of Living." *Culture Machine.* http://www.culturemachine.net/.

Muzzey, D., and A. van Oudenaarden. 2009. "Quantitative Time-Lapse Fluorescence Microscopy in Single Cells." *Annual Review of Developmental Biology* 25:301–27.

Myers, Natasha. 2015. *Rendering Life Molecular: Models, Modelers, and Excitable Matter.* Durham, N.C.: Duke University Press.

Nealon, Jeffrey T. 2008. *Foucault beyond Foucault: Power and Its Intensification since 1984.* Stanford, Calif.: Stanford University Press.

Nealon, Jeffrey T. 2015. *Plant Theory: Biopower and Vegetable Life.* Stanford, Calif.: Stanford University Press.

Needham, Joseph. 1941. "A Biologist's View of Whitehead's Philosophy." In *The Philosophy of Alfred North Whitehead,* ed. Philip Arthur Schilpp, 241–71. Evanston, Ill.: Northwestern University Press.

Negri, Antonio. 2004. *Negri on Negri: In Conversation with Anne Dufourmentelle.* Trans. M. B. DeBevoise. New York: Routledge.

Neidich, Warren. 2012 ."Neuropower: Art in the Age of Cognitive Capitalism." In *The Psychopathologies of Cognitive Capitalism, Part One,* ed. Arne De Boever and Warren Neidich, 219–66. Berlin: Archive Books.

Neidich, Warren, ed. 2013. *The Psychopathologies of Cognitive Capitalism, Parts 1 and 2.* Berlin: Archive Books.

Neidich, Warren. 2014. "The Architectonics of the Mind's Eye in the Age of Cognitive Capitalism." In *Brain Theory*, ed. Charles T. Wolfe, 264–86. London: Palgrave Macmillan.

Nobo, Jorge Luis. 1986. *Whitehead's Metaphysics of Extension and Solidarity.* Albany: State University of New York Press.

Nocek, A. J. 2014a. "Biomedia and the Pragmatics of Life in Architectural Design." *Inflexions* 7:8–58.

Nocek, A. J. 2014b. "Imaginative Chemistry: Synthetic Ecologies and the Construction of Life." In *The Lure of Whitehead,* ed. Nicholas Gaskill and A. J. Nocek, 379–413. Minneapolis: University of Minnesota Press.

Nocek, A. J. 2017. "On Symbols, Propositions, and Idiocies: Toward a Slow Technoscience." In *Rethinking Whitehead's Symbolism,* ed. Roland Faber, Jeffrey A. Bell, and Joseph Petek, 186–207. Edinburgh: Edinburgh University Press.

Nocek, A. J. 2018. "On the Risk of Gaia for an Ecology of Practices." *SubStance: A Review of Theory and Literary Criticism* 47, no. 1: 96–111.

Nocek, A. J. 2019. "Designing Propositions." In *Propositions in the Making: Experiments in a Whiteheadian Laboratory,* ed. Roland Faber, Michael Halewood, and Andrew Davis, 61–84. Lanham, Md.: Rowman and Littlefield.

Noë, Alva. 2004. *Action in Perception.* Cambridge, Mass.: MIT Press.

Norman, D. A., and S. W. Draper, eds. 1986. *User Centered System Design: New Perspectives on Human-Computer Interaction.* Mahwah, NJ: Lawrence Erlbaum.

Olsen, Erik. 2010. "Where Cinema and Biology Meet." *New York Times.* November 15, 2010.

Orlando, David A., Siobhan M. Brady, Thomas M. A. Fink, Philip N. Benfey, and Sebastian E. Ahnert. 2010. "Detecting Separate Time Scales in Genetic Expression Data." *BMC Genomics* 381, no. 1: 1–11.

Ostherr, Kristin. 2013. *Medical Visions: Producing the Patient through Film, Television, and Imaging Technologies.* Oxford: Oxford University Press.

Painlevé, Jean. 2000. *Science Is Fiction: The Films of Jean Painlevé.* Ed. Andy Masaki Bellows, Marina McDougall, and Brigitte Berg. Trans. Jeanine Herman. San Francisco: Brico Press.

Parikka, Jussi. 2010. *Insect Media: An Archaeology of Animals and Technology.* Minneapolis: University of Minnesota Press.

Parikka, Jussi. 2012a. "New Materialism as Media Theory: Medianatures and Dirty Matter." *Communication and Critical/Cultural Studies* 9, no. 1: 95–100.

Parikka, Jussi. 2012b. *What Is Media Archeology?* Cambridge, UK: Polity Press.

Parikka, Jussi. 2015. *A Geology of Media.* Minneapolis: University of Minnesota Press.

Parisi, Luciana. 2013. *Contagious Architecture: Computation, Aesthetics, and Space.* Cambridge, Mass.: The MIT Press.

Pasquinelli, Matteo. 2015. "What an Apparatus Is Not: On the Archeology of the Norm in Foucault, Canguilhem, and Goldstein." *Parrhesia* 22:79–89.

Patterson, G. H. 2009. "Fluorescence Microscopy below the Diffraction Limit." *Seminars in Cell and Developmental Biology* 20, no. 8: 886–93.

Pauling, Linus, and Robert B. Corey. 1953. "A Proposed Structure for the Nucleic Acids." *Proceedings from the National Academy of Sciences.* 32, no. 2: 84–97.

Pauly, Philip. 1994. "Modernist Practice in American Biology." In *Modernist Impulses in Human Sciences,* ed. Dorothy Ross, 272–89. Baltimore, Md.: Johns Hopkins University Press.

Pease, Roland. 2012. "DNA Imaged with Electron Microscope for the First Time." *New Scientist.* November 28, 2012. https://www.newscientist.com/.

Peters, John Durham. 2015. *The Marvelous Clouds: Toward a Philosophy of Elemental Media.* Chicago: University of Chicago Press.

Peterson, Erik L. 2011. "The Excluded Philosophy of Evo-Devo? Revisiting C. H. Waddington's Failed Attempt to Embed Alfred North Whitehead's 'Organicism' in Evolutionary Biology." *History and Philosophy of the Life Sciences* 33, no. 3: 301–20.

Peterson, Erik L. 2016. *The Life of the Organic: The Theoretical Biology Club and the Roots of the Epigenetic.* Pittsburgh, Penn.: University of Pittsburgh Press.

Pickett, Brent L. 1996. "Foucault and the Politics of Resistance." *Polity* 28, no. 4: 445–66.

Pignarre, Philippe, and Isabelle Stengers. 2011. *Capitalist Sorcery: Breaking the Spell.* Trans. Andrew Goffey. New York: Palgrave Macmillan.

Pisters, Patricia. 2012. *The Neuro-Image: A Deleuzian Film-Philosophy of Digital Screen Culture.* Stanford, Calif.: Stanford University Press.

Plotnitsky, Arkady. 2013. *Niels Bohr and Complementarity: An Introduction.* New York: Springer.

Polanyi, Michael. 1951. *The Logic of Liberty: Reflections and Rejoinders.* New York: Routledge.

Pomeroy, Anne Fairchild. 2004. *Marx and Whitehead: Process, Dialectics, and the Critique of Capitalism.* Albany: State University of New York Press.

Prado, Luiza, and Pedro J. S. Vieira de Oliveira. 2016. "Breaking the Cycle of Macondo: Design and Decolonial Futures." *XRDS* 22, no. 4: 28–32.

Qiu, Weihong, Nathan D. Derr, Brian S. Goodman, Elizabeth Villa, David Wu, William Shih, and Samara L. Reck-Peterson. 2012. "Dynein Achieves Processive Motion Using Both Stochastic and Coordinated Stepping." *Nature Structural and Molecular Biology* 19:193–200.

Quastler, Henry, ed. 1953. *Essays on the Use of Information Theory in Biology.* Urbana: University of Illinois Press.

Rabinow, Paul. 2005. "Artificiality and Enlightenment: From Sociobiology to Biosociality." In *Anthropologies of Modernity: Foucault, Governmentality, and Life Politics,* ed. Jonathan Xavier Inda, 181–93. Oxford: Blackwell.

Rajan, Kaushik Sunder. 2006. *Biocapital: The Constitution of Postgenomic Life.* Durham, N.C.: Duke University Press.

Rasmussen, Nicolas. 1997. *Picture Control: The Electron Microscope and the Transformation of Biology, 1940–1960.* Stanford, Calif.: Stanford University Press.

Raunig, Gerald. 2016. *Dividuum: Machinic Capitalism and Molecular Revolution.* Vol. 1. Trans. Aileen Derieg. Los Angeles: Semiotext(e).

Rheinberger, Hans-Jörg. 2015. "Re-discovering Mendel: The Case of Carl Correns." *Science and Education* 24, nos. 1–2: 51–60.

Rheinberger, Hans-Jörg, Staffan Müller-Wille, and Robert Meunier. 2015. "Gene." *The Stanford Encyclopedia of Philosophy.* February 19, 2015. https://plato.stanford.edu.

Roach, Leigh Ann. 2010. *A Positive, Popular Art: Sources, Structure, and Impact of György Kepes's Language of Vision.* PhD dissertation, Florida State University.

Roberts, Anthony J., T. Kon, P. J. Knight, K. Sutoh, and S. A. Burgess. 2013. "Functions and Mechanics of Dynein Motor Proteins." *Nature Reviews. Molecular Cell Biology* 14, no. 11: 713–26.

Rodowick, D. N. 2007. *The Virtual Life of Film.* Cambridge, Mass.: Harvard University Press.

Rose, Nikolas. 2001. "The Politics of Life Itself." *Theory, Culture and Society* 18, no. 6: 1–30.

Rosen, Philip, ed. 1986. *Narrative, Apparatus, Ideology: A Film Theory Reader.* New York: Columbia University Press.

Ruska, Ernst. 1987. "The Development of the Electron Microscope and of Electron Microscopy (Nobel Lecture)." *Reviews of Modern Physics* 59, no. 3: 627–38.

Rüstow, Alexander. 1980. *Freedom and Domination: A Historical Critique of Civilization.* Trans. Salvator Attanasio. Princeton, N.J.: Princeton University Press.

Ryan, Morgan, Gaël McGill, and Edward O. Wilson. 2014. E. O. Wilson's Life on Earth (e-book series). The E. O. Wilson Biodiversity Foundation. https://eowilsonfoundation.org/.

Sankey, Howard. 2012. "Reference, Success, and Entity Realism." *Kairos* 5: 31–42.

Schindler, David. 1983. "Whitehead's Inability to Affirm a Universe of Value." *Process Studies* 13, no. 2: 112–31.

Schön, Donald. 1983. *The Reflective Practitioner.* New York: Basic Books.

Schwartz, James. 2008. *In Pursuit of the Gene: From Darwin to DNA.* Cambridge, Mass.: Harvard University Press.

Sha, Xin Wei. 2013. *Poiesis and Enchantment in Topological Matter.* Cambridge, Mass.: MIT Press.

Shannon, Claude E. 1948. "A Mathematical Theory of Communication." Bell Labs Innovations. Reprinted with corrections from *Bell System Technical Journal* 27:379–423, 623–56.

Sharpe, Jason, Nicholas Woolridge, and Charles Lumsden. 2008. *In Silico: 3D Animation and Simulation of Cell Biology with Maya and MEL.* Amsterdam: Elsevier.

Shaviro, Steven. 2009. *Without Criteria: Kant, Whitehead, Deleuze, and Aesthetics.* Cambridge, Mass.: MIT Press.

Shaviro, Steven. 2010. *Post-cinematic Affect.* Winchester, UK: Zero Books.

Shaviro, Steven. 2014. *The Universe of Things: On Speculative Realism.* Minneapolis: University of Minnesota Press.

Sherburne, Donald W. 1967. "Whitehead without God." *Christian Scholar* 50, no. 3: 251–72.

Shoham, Y. 1990. "Agent-Oriented Programming." *Technical Report STAN-CS-90–1335.* Stanford University, Calif.: Computer Science Department.

Shukin, Nicole. 2009. *Animal Capital: Rendering Life in Biopolitical Times.* Minneapolis: University of Minnesota Press.

Simondon, Gilbert. 1989. *L'Individuation psychique et collective.* Paris: Aubier.

Simondon, Gilbert. 2016. *On the Mode of Existence of Technical Objects.* Trans. Cecile Malaspina and John Rogove. Minneapolis: Univocal.

Smith, Daniel W. 2012. "On the Nature of Concepts." *Parallax* 18, no. 1: 62–73.

Sneath, P. H. A. 1967. "Trend-Surface Analysis of Transformation Grids." *Journal of Zoology* 151, no. 1: 65–122.

Snow, C. P. 2012. *The Two Cultures.* Cambridge: Cambridge University Press.

Spiller, Neil. 2002. *Cyber Reader: Critical Writings for the Digital Era.* London: Phaidon Press.

Steinberg, Marc. 2012. *Anime's Media Mix Franchising Toys and Characters in Japan.* Minneapolis: University of Minnesota Press.

Stengers, Isabelle. 2005. "The Cosmopolitical Proposal." In *Making Things Public: Atmospheres of Democracy.* Cambridge, Mass.: MIT Press.

Stengers, Isabelle. 2010. *Cosmopolitics I.* Trans. Robert Bononno. Minneapolis: University of Minnesota Press.

Stengers, Isabelle. 2011a. "Another Science Is Possible! A Plea for Slow Science." Faculté de Philosophie et Lettres, ULB. December 13, 2011. Inauguration lecture Chair Willy Calewaert 2011–2012 (VUB).

Stengers, Isabelle. 2011b. *Thinking with Whitehead: A Free and Wild Creation of Concepts.* Trans. Michael Chase. Cambridge, Mass.: Harvard University Press.

Stengers, Isabelle. 2014. "A Constructivist Reading of *Process and Reality.*" In *The Lure of Whitehead,* ed. Nicholas Gaskill and A. J. Nocek, 43–64. Minneapolis: University of Minnesota Press.

Stengers, Isabelle. 2015. *In Catastrophic Times: Resisting the Coming Barbarism.* Trans. Andrew Goffey. Ann Arbor, Mich.: Open Humanities Press.

Stephens, David J., and Victoria J. Allen. 2003. "Light Microscopy Techniques for Live Cell Imaging." *Science* 300, no. 5616: 82–86.

Sterling, Bruce. 2005. *Shaping Things.* Cambridge, Mass.: MIT Press.

Stevens, Charles F. 2004. "Systems Biology *versus* Molecular Biology." *Current Biology* 14, no. 2. https://www.cell.com.

Steyerl, Hito. 2012. *The Wretched of the Screen.* Berlin: Sternberg Press.

Suppe, Fredrick, ed. 1977. *The Structure of Scientific Theories.* Urbana: University of Illinois Press.

Tanay, Amos, Israel Steinfeld, Martin Kupiec, and Ron Shamir. 2005. "Integrative Analysis of Genome-Wide Experiments in the Context of a Large High-Throughput Data Compendium." *Molecular Systems Biology* 1, no. 1. http://msb.embopress.org/.

Tasker, Roy, and Rebecca Dalton. 2008. "Visualizing the Molecular World—Design, Evaluation, and Use of Animations." In *Visualization: Theory and Practice in Science Education,* ed. John K. Gilbert, Miriam Reiner, and Mary Nakhleh, 103–31. New York: Springer.

Terranova, Charissa. 2016. *Art as Organism: Biology and the Evolution of the Digital Image.* London: I. B. Tauris.

Terranova, Tiziana. 2003. "Free Labor: Producing Culture for the Digital Economy." *Electronic Book Review.* June 20, 2003. https://electronicbookreview.com/essay/free-labor-producing-culture-for-the-digital-economy/.

Terranova, Tiziana. 2004. *Network Culture: Politics for the Information Age.* London: Pluto Press.

Terranova, Tiziana. 2012. "Attention, Economy and the Brain." *Culture Machine* 13:1–19.

Thacker, Eugene. 2004. *Biomedia: The Merging of Computer Science and Molecular Biology, Genetic Codes and Computer Codes.* Minneapolis: University of Minnesota Press.

Thacker, Eugene. 2005. *The Global Genome: Biotechnology, Politics, Culture.* Cambridge, Mass.: MIT Press.

Tharp, Bruce M., and Stephanie M. Tharp. 2019. *Discursive Design: Critical, Speculative, and Alternative Things*. Cambridge, Mass.: MIT Press.

Thompson, D'Arcy Wentworth. 1942. *On Growth and Form*. 2nd ed. Cambridge: Cambridge University Press.

Thompson, Evan. 2007. *Mind in Life: Biology, Phenomenology, and the Sciences of Mind*. Cambridge, Mass.: Harvard University Press.

Thompson, Kristin. 1980. "Implications of the Cel Animation Technique." In *The Cinematic Apparatus*, ed. T. de Lauretis and S. Heath, 106–20. London: Palgrave Macmillan.

Thurtle, Phillip. 2007. *The Emergence of Genetic Rationality: Space, Time, and Information in America, 1870–1920*. Seattle: University of Washington Press.

Thurtle, Phillip. 2014. "Animation and Vitality." *Inflexions* 7 (Animating Biophilosophy): 98–117.

Thurtle, Phillip. 2018. *Biology in the Grid: Graphic Design and the Envisioning of Life*. Minneapolis: University of Minnesota Press.

Tonkinwise, Cameron. N.d. "Speculative Practice Needs Diverse Cultures." Interview for *Speculative*. Accessed October 2018. http://speculative.hr/.

Toscano, Alberto. 2012. "The Disparate: Ontology and Politics in Simondon." *Pli: The Warwick Journal of Philosophy* 20, no. 8: 107–17.

Touraine, Alain. 1971. *The Post-industrial Society: Tomorrow's Social History; Classes, Conflicts and Culture in the Programmed Society*. New York: Random House.

Trewavas, Anthony. 2006. "A Brief History of Systems Biology: 'Every Object That Biology Studies Is a System of Systems.' François Jacob (1974)." *Plant Cell* 18, no. 10: 2420–30.

Tronti, Mario. 1973. "Social Capital." *Telos* 17:98–121.

Turing, Alan M. 1952. "The Chemical Basis of Morphogenesis." *Philosophical Transactions of the Royal Society, London B* 237:37–72.

Twemlow, Alice. 2017. *Sifting the Trash: A History of Design Criticism*. Cambridge, Mass.: MIT Press.

Väliaho, Pasi. 2014. *Biopolitical Screens: Image, Power, and the Neoliberal Brain*. Cambridge, Mass.: MIT Press.

van Dijk, José. 2005 *The Transparent Body: A Cultural Analysis of Medical Imaging*. Seattle: University of Washington Press.

Van Regenmortel, Marc H. V. 2004. "Reductionism and Complexity in Molecular Biology." *EMBO Reports* 11, no. 5: 1016–20.

Varela, Francisco J. 1992. "Autopoiesis and the Biology of Intentionality." In *Proceedings of the Workshop "Autopoiesis and Perception,"* ed. Barry McMullin, 4–13. Dublin: Dublin City University. http://cepa.info/.

Varela, Francisco J., Evan Thompson, and Eleanor Rosch. 1991. *The Embodied Mind: Cognitive Science and Human Experience.* Cambridge, Mass.: MIT Press.

Vidal, Fernando. 2009. "Brainhood, Anthropological Figure of Modernity." *History of the Human Sciences* 22, no. 1: 5–36.

Virilio, Paul. 1989. *War and Cinema: The Logistics of Perception.* London: Verso.

Virilio, Paul. 1997. *Pure War.* 2nd ed. New York: Semiotext(e).

Virno, Paolo. 2001. *A Grammar of the Multitude: For an Analysis of Contemporary Forms of Life.* Trans. Isabella Bertoletti. Cambridge, Mass.: MIT Press.

von Andrian, Ulrich H., and Charles Mackay. 2000. "T-cell Function and Migration: Two Sides of the Same Coin." *New England Journal of Medicine.* 343, no. 14: 1020–34.

von Bertalanffy, Karl Ludwig. 1968. *General System Theory: Foundations, Development, Application.* New York: George Braziller.

von Dassow, George, Eli Meir, Edwin H. Munro, and Garrett M. Odell. 2000. "The Segment Polarity Network Is a Robust Developmental Module." *Nature* 406, no. 6792: 188–92.

von Mises, Ludwig. 1996. *Human Action: A Treatise on Economics.* 4th ed. San Francisco: Fox and Wilkes.

Waddington, C. H. 1953. "The Origin of Biological Pattern." *New Biology* 15:117–25.

Waldby, Catherine, and Robert Mitchell. 2006. *Tissue Economies: Blood, Organs, and Cell Lines in Late Capitalism.* Durham, N.C.: Duke University Press.

Watson, James, and Francis Crick. 1953a. "A Structure for Deoxyribonucleic Acid." *Nature* 171: 737–38.

Watson, James, and Francis Crick. 1953b. "Genetical Implications of the Structure of Deoxyribonucleic Acid." *Nature* 171: 964–67.

Wellmann, Janina. 2017a. "Animating Embryos: The In Toto Representation of Life." *British Society for the History of Science* 50, no. 3: 521–35.

Wellmann, Janina. 2017b. *The Form of Becoming: Embryology and the Epistemology of Rhythm, 1760–1830.* Cambridge, Mass.: MIT Press.

Westerhoff, Hans, and Bernard O. Palsson. 2004. "The Evolution of Molecular Biology into Systems Biology." *Nature Biotechnology* 22, no. 10: 1249–52.

Whitehead, Alfred North. 1927. *Symbolism: Its Meaning and Effect.* New York: Macmillan.

Whitehead, Alfred North. 1964. *The Concept of Nature.* Cambridge: Cambridge University Press. First published 1920.

Whitehead, Alfred North. 1967a. *Adventures of Ideas.* New York: Free Press. First published 1933.

Whitehead, Alfred North. 1967b. *Science and the Modern World.* New York: Free Press. First published 1925.

Whitehead, Alfred North. 1968. *Modes of Thought.* New York: Free Press. First published 1938.

Whitehead, Alfred North. 1978. *Process and Reality: An Essay in Cosmology.* Corrected edition. Ed. David Ray Griffin and Donald W. Sherburne. New York: Free Press. First published 1929.

Whitehead, Alfred North. 1984. "Autobiographical Notes." *Science and Philosophy.* New York: Philosophical Library.

Whitehead, Alfred North. 1996. *Religion in the Making.* New York: Fordham University Press. First published 1926.

Whitehead, Alfred North, and Bertrand Russell. 1997. *Principia Mathematica.* Vols. 1–3. Cambridge: Cambridge University Press.

Wiener, Norbert. 1954. *The Human Use of Human Beings: Cybernetics and Society.* 2nd ed. Garden City, N.Y.: Doubleday.

Wiener, Norbert. 1956. "Pure Patterns in a Natural World." *The New Landscape in Art and Science.* Ed. György Kepes. Chicago: Paul Theobald.

Wiener, Norbert. 1965. *Cybernetics: Or Control and Communication in the Animal and the Machine.* Cambridge, Mass.: MIT Press.

Wild, Alex. 2012. "What DNA Actually Looks Like." *Scientific American.* November 30, 2012. https://blogs.scientificamerican.com/.

Wilkie, Alex, Martin Savransky, and Marsha Rosengarten, eds. 2017. *Speculative Research: The Lure of Possible Futures.* London: Routledge.

Willis, Anne-Marie. 2006. "Ontological Designing." *Design Philosophy Papers* 4, no. 2: 69–92.

Winograd, Terry, and Fernando Flores. 1986. *Understanding Computers and Cognition: A New Foundation for Design.* Norwood, N.J.: Ablex.

Wolfe, Cary. 2010. *What Is Posthumanism?* Minneapolis: University of Minnesota Press.

Wolfe, Cary. 2012. *Before the Law: Humans and Other Animals in a Biopolitical Frame.* Chicago: University of Chicago Press.

Yuh, Chiou-Hwa, H. Bolouri, and E. H. Davidson. 1998. "Genomic Cis-Regulatory Logic: Experimental and Computational Analysis of a Sea Urchin Gene." *Science* 279, no. 5358: 1896–1902.

Zald, D. H., R. L. Cowan, P. Riccardi, R. M. Baldwin, M. S. Ansari, R. Li, and R. M. Kessler. 2008. "Midbrain Dopamine Receptor Availability Is Inversely Associated with Novelty-Seeking Traits in Humans." *Journal of Neuroscience* 28, no. 53: 14372–378.

Zielinski, Siegfried. 2006. *Deep Time of the Media.* Cambridge, Mass.: MIT Press.

INDEX

Page numbers in italics refer to illustrations.

(*continued from page ii*)

50 *Anthropocene Poetics: Deep Time, Sacrifice Zones, and Extinction*
David Farrier

49 *Metaphysical Experiments: Physics and the Invention of the Universe*
Bjørn Ekeberg

48 *Dialogues on the Human Ape*
Laurent Dubreuil and Sue Savage-Rumbaugh

47 *Elements of a Philosophy of Technology: On the Evolutionary History of Culture*
Ernst Kapp

46 *Biology in the Grid: Graphic Design and the Envisioning of Life*
Phillip Thurtle

45 *Neurotechnology and the End of Finitude*
Michael Haworth

44 *Life: A Modern Invention*
Davide Tarizzo

43 *Bioaesthetics: Making Sense of Life in Science and the Arts*
Carsten Strathausen

42 *Creaturely Love: How Desire Makes Us More and Less Than Human*
Dominic Pettman

41 *Matters of Care: Speculative Ethics in More Than Human Worlds*
Maria Puig de la Bellacasa

40 *Of Sheep, Oranges, and Yeast: A Multispecies Impression*
Julian Yates

39 *Fuel: A Speculative Dictionary*
Karen Pinkus

38 *What Would Animals Say If We Asked the Right Questions?*
Vinciane Despret

37 *Manifestly Haraway*
Donna J. Haraway

36 *Neofinalism*
Raymond Ruyer

35 *Inanimation: Theories of Inorganic Life*
David Wills

34 *All Thoughts Are Equal: Laruelle and Nonhuman Philosophy*
John Ó Maoilearca

33 *Necromedia*
Marcel O'Gorman

32 *The Intellective Space: Thinking beyond Cognition*
Laurent Dubreuil

31 *Laruelle: Against the Digital*
Alexander R. Galloway

30 *The Universe of Things: On Speculative Realism*
Steven Shaviro

29 *Neocybernetics and Narrative*
Bruce Clarke

28 *Cinders*
Jacques Derrida

27 *Hyperobjects: Philosophy and Ecology after the End of the World*
Timothy Morton

26 *Humanesis: Sound and Technological Posthumanism*
David Cecchetto

25 *Artist Animal*
Steve Baker

24 *Without Offending Humans: A Critique of Animal Rights*
Élisabeth de Fontenay

23 *Vampyroteuthis Infernalis: A Treatise, with a Report by the Institut Scientifique de Recherche Paranaturaliste*
Vilém Flusser and Louis Bec

22 *Body Drift: Butler, Hayles, Haraway*
Arthur Kroker

21 *HumAnimal: Race, Law, Language*
Kalpana Rahita Seshadri

20 *Alien Phenomenology, or What It's Like to Be a Thing*
Ian Bogost

19 *CIFERAE: A Bestiary in Five Fingers*
Tom Tyler

18 *Improper Life: Technology and Biopolitics from Heidegger to Agamben*
Timothy C. Campbell

17 *Surface Encounters: Thinking with Animals and Art*
Ron Broglio

16 *Against Ecological Sovereignty: Ethics, Biopolitics, and Saving the Natural World*
Mick Smith

15 *Animal Stories: Narrating across Species Lines*
Susan McHugh

14 *Human Error: Species-Being and Media Machines*
Dominic Pettman

13 *Junkware*
Thierry Bardini

12 *A Foray into the Worlds of Animals and Humans,* with *A Theory of Meaning*
Jakob von Uexküll

11 *Insect Media: An Archaeology of Animals and Technology*
Jussi Parikka

10 *Cosmopolitics II*
Isabelle Stengers

9 *Cosmopolitics I*
Isabelle Stengers

8 *What Is Posthumanism?*
Cary Wolfe

7 *Political Affect: Connecting the Social and the Somatic*
John Protevi

6 *Animal Capital: Rendering Life in Biopolitical Times*
Nicole Shukin

5 *Dorsality: Thinking Back through Technology and Politics*
David Wills

4 *Bíos: Biopolitics and Philosophy*
Roberto Esposito

3 *When Species Meet*
Donna J. Haraway

2 *The Poetics of DNA*
Judith Roof

1 *The Parasite*
Michel Serres

Adam Nocek is assistant professor in the philosophy of technology and science and technology studies in the School of Arts, Media, and Engineering at Arizona State University, where he is also the founding director of the Center for Philosophical Technologies.